# Advances in Experimental Medicine and Biology

Volume 1009

More information about this series at http://www.springer.com/series/5584

Barnali Chaudhuri • Inés G. Muñoz
Shuo Qian • Volker S. Urban
Editors

# Biological Small Angle Scattering: Techniques, Strategies and Tips

Springer

*Editors*
Barnali Chaudhuri
G N Ramachandran Protein Center
CSIR Institute of Microbial Technology
Chandigarh, India

Shuo Qian
Center for Structural
Molecular Biology and Neutron
Scattering Division
Oak Ridge National Laboratory
Oak Ridge, TN, USA

Inés G. Muñoz
Structural Biology Programme
Spanish National Cancer Research
Centre (CNIO)
Madrid, Spain

Volker S. Urban
Center for Structural
Molecular Biology and Neutron
Scattering Division
Oak Ridge National Laboratory
Oak Ridge, TN, USA

ISSN 0065-2598 ISSN 2214-8019 (electronic)
Advances in Experimental Medicine and Biology
ISBN 978-981-10-6037-3 ISBN 978-981-10-6038-0 (eBook)
DOI 10.1007/978-981-10-6038-0

Library of Congress Control Number: 2017956347



Printed on acid-free paper

This Springer imprint is published by Springer Nature
The registered company is Springer Nature Singapore Pte Ltd.
The registered company address is: 152 Beach Road, #21-01/04 Gateway East, Singapore 189721, Singapore

# Preface

The last decade has seen remarkable growth in small-angle scattering (SAS) applications in structural biology. Perhaps the most important driver of this growth is the desire to characterize and understand biomolecular complexes and assemblies that require the use of multiple techniques in order to overcome the challenges of size, complexity and dynamics. Aiding the growth has been the proliferation of available instrumentation with multiple small-angle X-ray scattering (SAXS) beam-lines at synchrotrons, more powerful small-angle neutron scattering (SANS) beam lines at reactor and spallation neutron sources, and diverse offerings of commercial laboratory-based SAXS instruments. Developments in in-line sample purification has dramatically expanded the success rate for obtaining high quality SAS data from targets that were previously intractable due to polydispersity or limits in solubility and stability. Finally, there have been substantial developments in computational tools for SAS data analysis and 3D structural modelling.

This volume focuses on the applications of SAS to biomolecules in solution. The introductory chapter provides an excellent historical context for the development of SAS and is followed by a chapter focused on planning, preparing and performing a basic solution SAXS measurement. The emphasis on sample preparation and the challenges of obtaining an accurate SAXS profile from precisely matched solvent and solvent plus biomolecule of interest is well placed. This focus continues in Chap. 3, but in the context of a combined SEC-SAXS experiment where samples eluted from the SEC are directly injected into the SAXS measurement chamber as a final step of purification and potentially separation of species. The SEC-SAXS topic is taken up again in Chap. 11 with a focus on quality and examples that show one can go beyond separating species to learning more about the system of interest.

The analysis and accurate structural interpretation of solution SAS data is non-trivial, and Chaps. 4 and 6 combined provide important insights and guidance on data analysis, how to minimize the influence of experimental artefacts, and demonstrate the data supports the structural interpretation presented. As the solution SAS experiment provides structural information that is reduced to one-dimension due to the random orientation of the biomolecule, approaches to three-dimensional modelling require the application of restraints if conformational space is to be adequately sampled. The question of uniqueness of the solution must also be addressed. In the case of atomistic models, complementary structural data are used in hybrid modelling schemes. Hybrid structural modelling is maturing as a field and Chap. 13 considers the critical issues of the completeness of conformational space sampling, model validation, and data compatibility and provides elegant examples of hybrid modelling using SAXS and SANS combined with NMR and crystallography. Hybrid modelling combining SAXS data with hydroxyl radical footprinting and computational docking simulations is the subject of Chap. 14.

While the great majority of SAS experiments utilize X-rays, the unique properties of neutrons that facilitate contrast variation studies provide a powerful probe of the structures and arrangements of components within biomolecular complexes or assemblies. How to design and execute SANS studies of soluble complexes is the focus of Chap. 5, while Chap. 12 describes the application of SANS on membrane protein structural analysis.

Biomolecular functions are dynamic processes that depend upon conformational flexibility and there is increasing recognition that many proteins contain regions of intrinsic disorder that are important for function. Solution SAS can probe flexible and dynamic systems over a very broad range of sizes. The characterization of highly flexible proteins is described in Chap. 7, while Chap. 9 describes how to optimize experiments aimed at studying the fibrillation process, providing the basic principles behind the analysis of SAS data. The breadth in the range of distances that can be probed with SAS (essentially atomic to micrometer) is why the technique is well-suited for studies of intrinsically disordered proteins, conformational flexibility in individual folded biomolecules and complexes, and assemblies such as fibrils – the power of this capability to inform our understanding of biology is the subject of Chap. 10. Extending to wider angles, Chap. 8 shows how wide-angle X-ray scattering (WAXS) can provide insights into the secondary, tertiary and quaternary structural organization of macromolecules.

The volume finishes with an overview of current applications of SAS in drug research (Chap. 15), how it can be used in optimizing pharmaceutical efficacy at its most fundamental level. Specific examples of pharmaceutically relevant research on novel systems and the role SAS are described, with some practical advice for selecting scattering techniques for this important area of biomedical research.

I expect that this volume, with its breadth of topics penned by leaders in the field, will be a valuable resource for both expert and aspiring small-angle scatterers as biomolecular solution SAS continues to develop and grow.

University of Sydney, Australia Jill Trewhella
June, 2017

# Note from the Editors

It is customary in the SAS literature to express its central physical quantity, the scattering vector or momentum transfer, by one of a few different symbols: Q, q, S, s, or k. Different authors of the chapters in this book likewise have adopted different versions of this convention, as defined in each of the chapters. Another frequently used quantity, the pair distance distribution function, is symbolized by either P(r) or alternatively p(r).

# Contents

# 1 Small Angle Scattering: Historical Perspective and Future Outlook

Thomas M. Weiss

**Abstract**

Small angle scattering (SAS) is a powerful and versatile tool to elucidate the structure of matter at the nanometer scale. Recently, the technique has seen a tremendous growth of applications in the field of structural molecular biology. Its origins however date back to almost a century ago and even though the methods potential for studying biological macromolecules was realized already early on, it was only during the last two decades that SAS gradually became a major experimental technique for the structural biologist. This rise in popularity and application was driven by the concurrence of different key factors such as the increased accessibility to high quality SAS instruments enabled by the growing number of synchrotron facilities and neutron sources established around the world, the emerging need of the structural biology community to study large multi-domain complexes and flexible systems that are hard to crystalize, and in particular the development and availability of data analysis software together with the overall access to computational resources powerful enough to run them. Today, SAS is an established and widely used tool for structural studies on bio-macromolecules. Given the potential offered by the next generation X-ray and neutron sources as well as the development of new, innovative approaches to collect and analyze solution scattering data, the application of SAS in the field of structural molecular biology will certainly continue to thrive in the years to come.

**Keywords**

SAXS • SANS • Solution SAS • Biological SAS • History of SAS

T.M. Weiss (✉)
Stanford Synchrotron Radiation Lightsource, Stanford University, Menlo Park, CA, USA
e-mail: weiss@slac.stanford.edu





B. Chaudhuri et al. (eds.), *Biological Small Angle Scattering: Techniques, Strategies and Tips*, Advances in Experimental Medicine and Biology 1009, DOI 10.1007/978-981-10-6038-0_1

## 1.1 Introduction

Small angle scattering (SAS) is a powerful and versatile analytical technique that can provide detailed structural information from non-crystalline samples at moderate resolution on the length scale of typically a few nanometers. The laws of the underlying scattering process are such that the angular dependence of the scattering intensity is directly connected to the spatial distribution of scattering centers in the sample by a Fourier transformation. Reversing this transformation and translating the information from the reciprocal space in which the scattering intensities are measured into the real space of the sample is the goal in any structural investigation using scattering techniques. However, as we are only able to measure intensities (that is the square of the wave amplitude), the phases necessary for a straightforward reversal of the Fourier transformation are lost in this measurement process. To complicate things further, in non-crystalline, disordered samples typically used for small angle scattering the orientational averaging of the signal together with the limited range in which data is measurable in the reciprocal space due to weak signal-to-background intensity on the high Q side and interference with the primary beam on the low Q side leads to a further loss of structural information. Nevertheless, due to the conceptual simplicity of the experiment and its general applicability, the ease of sample preparation and speed of measurement SAS remains a highly useful and valuable tool for the structural biologist.

Typically small angle scattering experiments are performed using X-ray or neutron beams with wavelengths ranging from approximately one up to several Ångstroms. SAS can be applied to virtually any kind of material as long as it exhibits structural inhomogeneities at length scales ranging from one up to hundreds of nanometers. Due to the inverse relationship between the angular dependence of the scattering signal and the real space distances between scattering centers in the sample (originating from the Fourier relationship between real and reciprocal space) it follows that for wavelengths on the order of Ångstroms significant scattering intensity will be observable only at low scattering angles close to the incoming beam direction.

In its application to problems in structural biology the term small angle scattering is most often used as a synonym for solution scattering where the sample consists of dilute solutions of biological macromolecules or macromolecular complexes. In this case the dissolved macromolecules themselves are at the origin of the scattering signal and the observable intensity as a function of momentum transfer contains the information on the size, shape and structure of these molecules or molecular assemblies.

Although the underlying nature of the interaction leading to the scattering is different for neutrons and X-rays (as X-rays are scattered predominantly by the electrons in the sample while neutrons are interacting with the nuclei and spins) the mathematical formalism describing the scattering process is fundamentally the same. In fact Debye (Debye 1915) has shown that for scattering objects for which the coordinates of the atomic positions are known the measureable scattering intensity can be written down as:

$$I(Q) = \sum_i \sum_j f_i f_j \frac{\sin QR_{ij}}{QR_{ij}}$$

where the summation goes over the constituent atoms of the scattering object in question and $R_{ij}$ is the distance between the atomic pair $i$ and $j$, $Q$ is the momentum transfer defined as $Q = 4\pi sin\theta/\lambda$ (with $2\theta$ being the scattering angle and $\lambda$ the wavelength of the radiation). The factor $f_i$ is the so-called scattering length of the atom $i$ and describes the amount of radiation scattered by the respective atom. In the case of X-rays this factor is a function of the total number of electrons in the atom and therefore of its position in the periodic table of elements. For neutrons on the other hand the elemental scattering length exhibits a rather complex behavior and seems to randomly change from atom to atom. Most notably the neutron scattering length is generally different for different isotopes of the same chemical element and so neutron scattering is in principle sensitive to the isotope composition of the scattering object. This unique property of neutrons is exploited in the so-called contrast variation and contrast matching experiments in which the systematic variation of the isotopic composition of either the macromolecule or of the buffer medium is used to single out specific parts within the molecule (Stuhrmann and Kirste 1965; Ibel and Stuhrmann 1975). Most importantly for structural biology, there is a large difference between the scattering power of hydrogen and deuterium. With hydrogen being the most abundant atom in any biological macromolecule, the technique can virtually be used for all macromolecules or macromolecular complexes of biological interest.

As an experimental technique solution SAS is conceptually a rather simple experiment and it requires only a minimum amount of instrumentation, namely a collimated source, a sample container and a detector. However, the contrast of biological macromolecules in salt buffers is rather small and the concentration of the macromolecules needs to be kept low enough in order to avoid inter-particle interference and aggregation issues in the sample. Thus the useful signal from such dilute samples is rather weak and on top of a high background level. A SAS instrument must therefore be carefully optimized to minimize the contribution of parasitic instrumental background scatter.

Moreover, the Debye equation above illustrates that in the course of the orientational averaging of the measurement process caused by the unrestricted rotational motion of the macromolecule in solution the resulting scattering intensity becomes a function of the magnitude of the momentum transfer only, causing the signal to be circularly symmetric around the beam direction, and the structural information contained in the scattering signal is reduced to the value of the interatomic distances while any directional information between the different atoms is lost.

The structural parameters as determined by SAS for the dissolved macromolecules are ensemble values and are averaged over all the macromolecules present in the illuminated volume. Therefore the most accurate and precise structural information that can be extracted from a SAS measurement stems from samples that contain identical particles only. On the other hand, due to the ensemble average in the measurement process SAS data can also inform about the ensemble nature of the scattering objects in the sample such as their size distribution or possible conformational heterogeneities and flexibility.

In the more recent past SAS has seen an impressive growth of application from within the structural molecular biology community (see Fig. 1.1) and today it is one of the major experimental tools in structural biology.

## 1.2 The Beginning of SAS

Small angle scattering is not a new technique but has a long history going back to the first half of the last century. Following the discovery of the X-rays by William Conrad Roentgen in 1895, it took not quite two decades of scientific experiments and discussions about the nature of the X-radiation until W. Friedrich, P. Knipping and M. Laue observed the first diffraction pattern of X-rays from a crystal (Friedrich et al. 1912) establishing that the X-rays described by Roentgen are a specific kind of electromagnetic wave with very short wavelengths. Shortly after this W. L. Bragg and W. H. Bragg (father and son) formulated what is today known as Bragg's law of diffraction which describes the diffraction condition of X-rays by a crystalline lattice (Bragg and Bragg 1913). They showed that the scattering of such waves from the crystal planes found in a well-ordered crystal are responsible for the pronounced diffraction patterns observed and with this laid the foundation to the field of X-ray crystallography.

From then on it took another 16 years until the first experimental observation of small angle scattering was reported by P. Krishnamurti and C.V. Raman in 1929 (Raman and Krishnamurti 1929). In their X-ray experiments they were investigating graphite powders using the X-ray

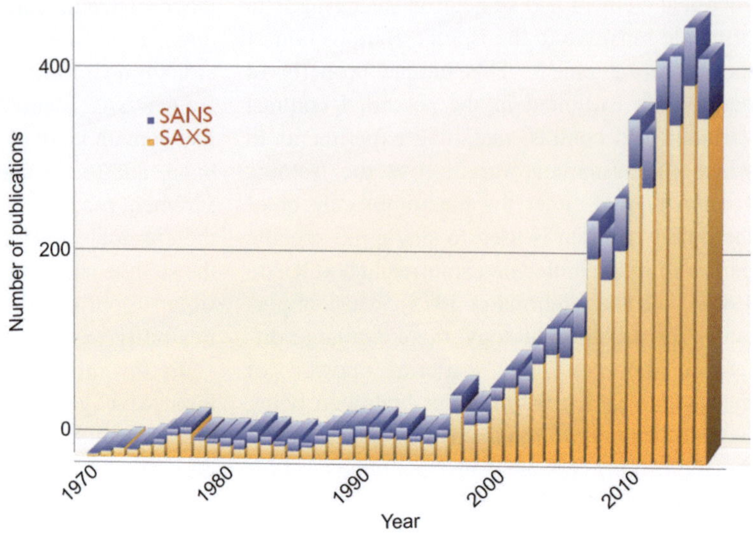


**Fig. 1.1** Number of publications in the PubMed database related to SANS and SAXS applications in structural molecular biology from 1970 to 2016

diffraction and observed strong scattering intensities close to the primary beam that became more pronounced with decreasing particle sizes of the powder. Similar scattering signal close to the primary beam was observed from samples of amorphous carbon of various origin (Krishnamurti 1930). A few years later B. E. Warren reported the same phenomenon as he observed it in the X-ray scattering intensities measured from carbon black samples (Warren 1934). At this time both Warren and Krishnamurti were already aware of and recognized the fact that the measured diffuse small angle scattering intensities near the origin were related to the size of small particulates present in the sample.

A few years later it was A. Guinier during his doctoral thesis, who started to investigate the scattering of X-rays at small angles more systematically (Guinier 1939). He worked out a quantitative theoretical framework for the interpretation and understanding of the observed diffuse X-ray scattering signals found close to the primary beam. Guinier realized that for densely packed particles the inter-particle interference effects will dominate the particulate scattering and concluded that if details of the particles themselves were to be investigated the sample needs to be measured under dilute conditions. He also recognized that the particles in the specimen are at the origin of the measured small angle scattering signal and that independent of the particle shape the scattering close to the origin can be approximated by an exponential function giving rise to his well-known theorem relating the scattering intensity near the origin to the radius of gyration of the particles in solution, which remains to be at the start of every SAS data analysis, even today. Furthermore he realized that proteins, protein complexes and other biological macromolecules in solution would constitute ideal systems to be studied by the then new method. On the one hand because their typical size ranges coincided with the sizes accessible by SAS, on the other hand because they can be prepared at a very high degree of purity which is necessary for a stringent analysis of the experimental SAS data. In the following years the theoretical framework for the interpretation of small angle scattering continued to be refined with important additional contributions from G. Fournet, O. Kratky, G. Porod, and many others. Together with Fournet, Guinier published the first monograph on SAS (Guinier and Fournet 1955). All of these theoretical developments were based on X-ray scattering experiments, while the discovery of the neutron by Chadwick (Chadwick 1932) and the first demonstrations of its wave properties by diffraction (von Halban and Preiswerk 1936; Mitchell and Powers 1936) occurred four decades after the discovery of X-rays. Neutron scattering or diffraction experiments were conducted by several scientists using beam ports of the first nuclear reactors. The exact timeline of discoveries during this period is somewhat obscured, because the work was classified under the Manhattan project. At a 1946 conference of the American Physical Society (APS) much of the previously classified results were revealed. In the following years, Wollan and Shull systematically developed the neutron diffraction technique (Shull and Wollan 1948) and laid the groundwork for small angle neutron scattering as a valuable and powerful tool for investigations on biological macromolecular solutions. The low flux and poor energy resolution of the available neutron sources and their high instrumental background made the application to biological solution samples quite a bit more challenging than in the case of X-rays and it was not until the late 1960s that the first neutron small angle scattering measurement on proteins in solution was reported by J. Schelten and coworkers (Schelten et al. 1969).

Although SAS in the early days was generally recognized by members of the biophysical community as a useful tool for investigating proteins in solution, technical limitations made its experimental applications to biological macromolecular solutions in the 1950s and 1960s rather difficult, with the measured data exhibiting poor signal-to-noise and the results extracted showing only limited accuracy, going beyond the radius of gyration.

## 1.3 The 1970s and 1980s: Slowly Gearing Up

The situation started to change gradually in the 1970s and early 1980s with the advent of high flux neutron reactors and spallation sources and the application of synchrotron radiation as a powerful and high brightness X-ray source. These new radiation sources were bright enough to produce well-collimated beams with sufficient flux, which is particularly important for SAS instruments as the collimation makes it possible to measure intensities at low momentum transfer. For X-rays in particular the several order of magnitude increase in flux due to the use of synchrotron radiation was flanked by the inherently low divergence of the beam making it an ideal source for SAXS experiments. Further technical advances in the detection of X-rays led to the development of one- and two-dimensional position sensitive X-ray detectors (Gabriel and Dupont 1972), which significantly boosted the data collection efficiency and signal-to-noise in the experimental data. In both cases the new sources provided a considerably higher flux at the sample and allowed to build instruments with lower instrument background leading to a much improved signal-to-noise in the data.

The new sources most often provided dedicated and optimized instruments for SAS and were operated as user facilities, where scientists from outside institutions could do their experiments at the available instruments. Access to these instruments was typically granted by a proposal system based on scientific merit and thus open for the general research community making these instruments, and with them also the SAS technique, more widely available. In addition to the instrumental accessibility, highly skilled scientific staff at the different experimental stations provided support for the outside scientists helping out with the experimental planning, setup and execution.

In parallel to these technological advances and the progress made on the instrumental side, significant steps forward were also made in developing new approaches for SAS data analysis. One of them was the introduction of spherical harmonics in the treatment of SAS data (Harrison 1969; Stuhrmann 1970a, b) and its application for the direct and model independent shape reconstruction of the underlying particles from the measured scattering intensity (Stuhrmann 1970c). Another important theoretical development during that time was the introduction of the indirect Fourier transform method (Glatter 1977). This enabled the reliable and accurate determination of the pair-distance distribution function of the isolated scattering object, giving a real space representation of the particle similar to the Patterson function used in crystallography. A further key development especially important for SANS was the continued refinement of the method of contrast variation (Stuhrmann and Kirste 1965; Ibel and Stuhrmann 1975) and its application particularly in biology. While the method was first developed and systematically applied using SAXS it turned out to be much more powerful and applicable in neutron scattering experiments on biological material due the large scattering length difference between hydrogen and deuterium, given the widespread presence of hydrogen atoms in biological macromolecules as well as in the buffer solution. Choosing for example different $H_2O/D_2O$ ratios for the buffer the investigation of bimolecular complexes of protein and DNA or RNA can be studied in detail by matching out the scattering of the protein or the DNA/RNA. In this way it was shown for example that the DNA is wrapped around the nucleosome (Pardon et al. 1975). The contrast variation method was also employed in the determination of the overall shape of the 50S ribosome (Stuhrmann et al. 1977), and was key to the systematic triangulation of protein and RNA subunits within the ribosome leading to a comprehensive map of the 30S ribosomal subunit (Capel et al. 1987) and later also a partial map of the 50S subunit (May et al. 1992). All of these results settled key questions in structural biology long before high-resolution crystal structures of these complexes became available.

However, the advent of synchrotron radiation as powerful X-ray source for structural biology research not only boosted the application of SAXS in the field, but also greatly benefitted macromolecular crystallography further enforcing it as the gold standard method for structural studies on bio-macromolecules. So regardless of the significant advances in instrumentation and progress in data analysis methods as well as the impressive successes in the determination of large multicomponent complexes such as the ribosome subunits, biological SAS remained mostly a niche technique without widespread applications and acceptance. Used mostly by scattering experts with scientific interest in biology and a few daring structural biologists who were exploring the potential of small angle scattering to gain insight into their scientific problem.

## 1.4 The 1990s: Taking Off

This situation started changing around the mid to end 1990s. At that point the application of SAS as a method for structural biology research increased dramatically and its acceptance within the community as a powerful complementary tool for structural studies began to broaden. This change was ultimately driven by a concurrence of several different key factors:

On the one hand there was the increased accessibility to high quality SAS instruments at the growing number of synchrotron facilities and neutron sources around the world, many offering dedicated instrumentation for solution small angle scattering.

Secondly, there was the emerging need of the structural biology community to study large multi-domain complexes of biomolecules and systems with flexible parts or unstructured termini with functional relevance. While crystallographic approaches for such systems were challenging and sometimes just not feasible at all due to the lack of crystals, SAXS could provide accurate structural information for these macromolecules in solution albeit at reduced resolution.

The third and arguably the most influential factor for this steep increase in applications of small angle scattering in structural biology was the development and availability of data analysis software with user friendly interfaces together with the easy access to computational resources powerful enough to run these tools. This enabled the general user to apply these new analysis methods to their SAS data without becoming an expert in small angle scattering analysis methods.

A particularly important step in the development of SAS data analysis methods was the ability to reconstruct the particle shape directly from the scattering curve without referring to a specific geometrical model. This *ab-initio* approach allowed the reconstruction of three-dimensional models of the macromolecular structure at nanometer resolution. Although the resolution obtained by this method was quite limited it nevertheless allowed the visualization of the general particle shape and enabled a direct comparison with higher resolution structural data if available (e.g. crystallographic or NMR structures) by overlaying the two and ultimately provided a better understanding of functional aspects of these macromolecules and complexes.

The first such *ab initio* method for structure reconstruction from SAS data used a spherical harmonics decomposition of the particle shape. Although the method had already been developed in the early 1970s (Stuhrmann 1970a, b), it was refined, implemented and made available to the biology community in the 1990s (Svergun and Stuhrmann 1991; Svergun 1997). The method is in principle applicable to arbitrary particle shapes but it exhibits difficulties with topologically more complicated geometries such as e.g. particles with internal cavities. A more widely applicable method based on bead modeling was subsequently proposed (Chacon et al. 1998). In this approach the particle is modeled as a collection of a large number of beads describing the particle shape. The arrangement of the beads in space is then iteratively modified to find the structure with the best fit to the experimental scattering data. Different versions of this general algorithm were implemented by a variety of authors (Svergun 1999; Chacon et al. 2000; Bada et al. 2000; Walther et al. 2000; Franke and Svergun 2009).

In most cases these analysis programs can be applied to both X-ray and neutron scattering data. Arguably the most comprehensive and popular set of programs for the analysis of SAS data is the ATSAS program suite developed by Svergun and coworkers at the EMBL in Hamburg (Petoukhov et al. 2012) but other programs are available from different sources as well.

Driven by the growing demand for beam time substantial efforts were undertaken by the facilities to push the efficiency at the beam line and speed up the measurement process to enable higher sample throughput. In particular at synchrotrons where the exposure times were becoming significantly shorter than the time necessary for exchanging the sample, automated sample handling was implemented. Today most biological SAXS beam lines at synchrotron facilities have some kind of automated sample loading robotics (Round et al. 2015; Martel et al. 2012; Hura et al. 2009; Nielsen et al. 2012). These robotic systems not only reduce the overall time spent on changing the sample but also eliminate human errors in the loading process and thus have made the data collection more robust and the data more consistent. It also allowed continuous data collection for extended periods of time opening the door for large-scale high-throughput studies. In most cases these robotic systems are coupled to an automated software pipeline for data reduction and analysis providing real-time data analysis during the measurement.

Although sample preparation for solution SAS samples is minimal when compared to crystallography or electron microscopy, the purity and monodispersity of the sample is extremely important. Proteins in solution are prone to aggregate and even small amounts of aggregates can render the data unsuitable for further analysis. Unfortunately, for some systems even the brief period between purification in the wet lab and measurement at the beam line can be sufficient to compromise the sample. In order to circumvent this and provide the highest sample purity possible for the SAS experiment one can combine the sample purification by size exclusion chromatography with the SAS measurement (SEC-SAS). Due to the high flux necessary to enable short enough exposure times providing sufficient chromatographic resolution this approach is currently limited to synchrotron sources (Mathew et al. 2004; Watanabe et al. 2009; David and Perez 2009; Perez and Nishino 2012). However, it has been shown that with careful choice of experimental conditions the technique can be ported to currently available high flux SANS beam lines as well (Jordan et al. 2016).

SEC-SAS is currently available at most biological solution scattering beam lines at synchrotrons around the world and special software tools are becoming available to specifically analyze SEC-SAS data taking full advantage of the additional chromatographic information obtainable from such experiments (Brookes et al. 2016). At most facilities additional characterization tools such as multi angle light scattering (MALS) or refractive index (RI) can be added into the flow line forming an extensive online biophysical characterization pipeline, providing a wealth of complementary information beyond the SAS data. The idea of directly coupling a size exclusion column to the SAS instrument has recently also been extended to other types of chromatography (Hutin et al. 2016). The potential of these types of combined online methods is nicely illustrated in a recent application of SEC-SAXS to the membrane protein aquaporin-0 (Berthaud et al. 2012).

The rising interest in SAS and the increasing number of publications presenting structural models based on SAS data have raised concerns within the community about model validation, data quality assessment and archiving of such models. In SAS this is particularly problematic due to the lack of objective statistical measures within the data analysis (such as e.g. the R factors in crystallography) that would allow an assessment of the quality of the analysis and the resulting structural model by comparing a set of numbers. To address this important point efforts have recently been started to define standards for publication and reporting and pave the road to a worldwide repository similar to (or potentially as part of) the world wide Protein Data Bank (wwPDB) already existing for crystallographic data (Jacques and Trewhella 2010; Trewhella et al. 2013).

## 1.5 Current State and Future Outlook

Today SAS is an established method in the experimental toolbox of the structural biologist. It's versatility, fast speed, moderate requirements on sample amount and relative ease of use makes it an attractive technique for routine structural characterization of biomolecules. Given the current level of automation in the sample handling as well as data reduction and analysis, SAS measurements can easily be performed by non-experts having only limited knowledge of the method. The complementarity of SAS to other structural methods has led to many successful studies using a hybrid approach for structural investigations of large biomolecular assemblies and complexes combining other techniques such as crystallography, NMR and electron microscopy with SAS while computational methods and simulations provide the glue between the different experimental techniques (Alber et al. 2008; Russel et al. 2012).

With the increasing brightness of current and future synchrotron radiation sources and the evolving use of microfluidic sample handling the necessary volumes and concentration requirements are expected to reduce further to the point where micrograms of sample will provide enough material for large-scale studies and allow its application to systems previously not amenable to the technique due to sample amount requirements. For the use at free electron lasers experimental protocols and analysis routines for SAS will have to be adjusted to take into account the new capabilities these sources offer. Methods based on fluctuation scattering (Kam 1977; Kam et al. 1981) are currently being explored as an extension to regular SAS data collection and analysis (Kirian et al. 2011; Malmerberg et al. 2015) at these new generation sources. The high intensity of these sources will also allow a further substantial reduction of the sample volume and more importantly enhance the capabilities to study the dynamics of macromolecules and the kinetic of macromolecular reactions in time-resolved solution scattering experiments.

New spallation neutron sources with increased flux will improve signal-to-noise and also help to reduce sample size for SANS experiments making the technique more widely applicable. Emerging techniques such as online chromatography coupled SANS will become more generally available at these sources. Other contrast methods (Stuhrmann 2012) might also provide new opportunities for SANS in structural biology.

To analyze the increasingly complex data from the new experimental applications of SAS, advanced computational methods will undoubtedly play an important role. Large-scale molecular dynamics simulations will allow better refinement of high-resolution structures against their solution scattering data with the inclusion of higher angle data. Such simulations will also help to interpret time-resolved data and model macromolecular kinetics and dynamics.

Robotics and smart data acquisition systems will be developed and implemented. These systems will be able to suggest changes in the data collection strategy on the fly based on the analyzed data and implement them as the data is being collected to optimize the data quality, making the SAS measurement even more automated.

Overall, the technological advances in the years to come will offer exciting new possibilities for SAS in its application to problems in structural biology. The technique will further grow and as a complementary method that is capable of measuring the bio-macromolecules in close to their physiological environment, SAS will play an increasingly important role in refining the solution state of bio-macromolecules and macromolecular complexes.

## References


Alber F, Forster F, Korkin D, Topf M, Sali A (2008) Annu Rev Biochem 77:443

Bada M, Walther D, Arcangioli B, Doniach S, Delarue M (2000) J Mol Biol 300:563

Bernado P, Mylonas E, Petoukhov MV, Blackledge M, Svergun DI (2007) J Am Chem Soc 129(17):5656

Berthaud A, Manzi J, Perez J, Mangenot S (2012) J Am Chem Soc 134:10080

Bragg WH, Bragg WL (1913) Proc Roy Soc Lon A 88:428

Brookes E, Vachette P, Rocco M, Pérez J (2016) J Appl Crystallogr 49:1827

Capel MS, Engelman DM, Freeborn BR, Kjeldgaard M, Langer JA, Ramakrishnan V, Schindler DG, Schneider DK, Schoenborn BP, Sillers IY, Moore PB (1987) Science 238:1403

Chacon P, Moran F, Diaz JF, Pantos E, Andreu JM (1998) Biophys J 74:2760

Chacon P, Diaz JF, Moran F, Andreu JM (2000) J Mol Biol 299:1289

Chadwick J (1932) Nature 129:312

David G, Perez J (2009) J Appl Crystallogr 42:892

Debye P (1915) Ann Phys 28:809

Franke D, Svergun DI (2009) J Appl Crystallogr 42:342

Friedrich, W., Knipping, P., Laue, M. (1912), *Sitzungsberichte der K.B. Akademie der Wissenschaften,* 302

Gabriel A, Dupont Y (1972) Rev Sci Instrum 43:1600

Glatter O (1977) J Appl Crystallogr 10:415

Guinier A (1939) Ann Phys 12:161

Guinier A, Fournet G (1955) Small angle scattering of x-rays. Wiley, New York

von Halban H, Preiswerk P (1936) C R Acad Sci 203:73–75

Harrison SC (1969) J Mol Biol 42:457

Hura GL, Menon AL, Hammel M, Rambo RP, Poole FL II, Tsutakawa SE, Jenney FE Jr, Classen S, Frankel KA, Hopkins RC, Yang S, Scott JW, Dillard BD, Adams MWW, Tainer JA (2009) Nat Methods 6 (8):606

Hutin S, Brennich M, Maillot B, Round A (2016) Acta Cryst D 72:1090

Ibel K, Stuhrmann HB (1975) J Mol Biol 93:255

Jacques DA, Trewhella J (2010) Protein Sci 19:642

Jordan A, Jacques M, Merrick C, Devos J, Forsyth VT, Porcar L, Martel A (2016) J Appl Crystallogr 49:2015

Kam Z (1977) Macromolecules 10(5):927–934

Kam Z, Koch MH, Bordas J (1981) Proc Natl Acad Sci 78 (6):3559

Kirian RA, Schmidt KE, Wang X, Doak RB, Spence JCH (2011) Phys Rev E 84:011921

Krishnamurti P (1930) Ind J Phys 5:473

Malmerberg E, Kerfeld CA, Zwart PH (2015) IUCrJ 2 (3):309

Martel A, Liu P, Weiss TM, Niebuhr M, Tsuruta H (2012) J Synchrotron Radiat 19:431

Mathew E, Mirza A, Menhart N (2004) J Synchrotron Radiat 11:314

May RP, Nowotny V, Nowotny P, Voss H, Nierhaus KH (1992) EMBO J 11:373

Mitchell DP, Powers PN (1936) Phys Rev 50:486

Nielsen SS, Moller M, Gillilan RE (2012) J Appl Crystallogr 45(2):213

Pardon JF, Worcester DL, Wooley JC, Tatchell K, Vanderholde KE, Richards BM (1975) Nucleic Acids Res 2:2163

Perez J, Nishino Y (2012) Curr Opin Struct Biol 22:670
Petoukhov MV, Svergun DI (2005) Biophys J 89:1237
Petoukhov MV, Franke D, Shkumatov AV, Tria G, Kikhney AG, Gajda M, Gorba C, Mertens HDT, Konarev PV, Svergun DI (2012) J Appl Crystallogr 45:342
Pynn, Fender (1985) Physics today
Raman CV, Krishnamurti P (1929) Nature 124:53
Round A, Felisaz F, Gobbo A, Huet J, Fodinger L, Villard C, Pernot P, Blanchet C, Roessle M, Svergun D, Cipriani F (2015) Acta Cryst D71:67
Russel D, Lasker K, Webb B, Velázquez-Muriel J, Tjioe E, Schneidman-Duhovny D, Sali A (2012) PLoS Biol 10(1):e1001244
Schelten J, Mayer A, Schmatz W, Hossfeld F (1969) Z Physio Chem 350:851
Shull CG, Wollan EO (1948) Science 108:69–75
Stuhrmann HB (1970a) Z Phys Chem (neue folge) 72:177
Stuhrmann HB (1970b) Acta Cryst A26:227
Stuhrmann HB (1970c) Z Phys Chem (neue folge) 72:185
Stuhrmann HB (2012) J Phys 351:012002
Stuhrmann HB, Kirste RG (1965) Z Phys Chem (neue folge) 46:247
Stuhrmann HB, Koch MHJ, Parfait R, Haas J, Ibel K, Crichton RR (1977) Proc Natl Acad Sci U S A 74:2316
Svergun DI (1997) J Appl Crystallogr 30:792
Svergun D (1999) Biophys J 76:2879
Svergun DI, Stuhrmann HB (1991) Acta Cryst A47:736
Trewhella J, Hendrickson WA, Kleywegt GJ, Sali A, Sato M, Schwede T, Svergun DI, Tainer JA, Westbrook J, Berman HM (2013) Structure 21:875
Tria G, Mertens HDT, Kachala M, Svergun DI (2015) IUCrJ 2:207
Walther D, Cohen FE, Doniach S (2000) J Appl Crystallogr 33:350
Warren BE (1934) J Chem Phys 2:551
Watanabe Y, Inoko Y (2009) J Chromatogr A 1216:7461

# Sample and Buffer Preparation for SAXS 2

Melissa A. Graewert and Cy M. Jeffries

**Abstract**

In this book chapter, a practical approach for conducting small angle X-ray scattering (SAXS) experiments is given. Our aim is to guide SAXS users through a three-step process of planning, preparing and performing a basic SAXS measurement. The minimal requirements necessary to prepare samples are described specifically for protein and other macromolecular samples in solution. We address the very important aspects in terms of sample characterization using additional techniques as well as the essential role of accurately subtracting background scattering contributions. At the end of the chapter some advice is given for trouble-shooting problems that may occur during the course of the SAXS measurements. Automated pipelines for data processing are described which are useful in allowing users to evaluate the quality of the data 'on the spot' and consequently react to events such as radiation damage, the presence of unwanted sample aggregates or miss-matched buffers.

**Keywords**

Sample preparation • Quality control • Solvent matching

## 2.1 Planning a SAXS Experiment

Many researchers are becoming increasingly inspired by the growing number of success stories that use SAXS to analyze the structures of macromolecules in solution (see reviews such as Vestergaard 2016; Trewhella 2016; Graewert and Svergun 2013). They ask themselves: SAXS is clearly beneficial, but what are the minimal requirements needed to prepare samples for solution scattering experiments? Figure 2.1 lists the fundamental points to consider when planning SAXS measurements that encompasses:

- Sample purity and polydispersity.
- Sample quantities.
- Sample stability.

M.A. Graewert (✉) • C.M. Jeffries
European Molecular Biology Laboratory (EMBL)
Hamburg Outstation, DESY, Hamburg, Germany
e-mail: m.graewert@embl-hamburg.de;
cy.jeffries@embl-hamburg.de





B. Chaudhuri et al. (eds.), *Biological Small Angle Scattering: Techniques, Strategies and Tips*, Advances in Experimental Medicine and Biology 1009, DOI 10.1007/978-981-10-6038-0_2

**Assessment of sample suitability for SAXS**

- Sample is of a sufficient quality (purity > 95%)?
- Sample is sufficiently monodisperse (polydispersity < 0.1%)?
- Sample is available in sufficient quantity (100-200 μl at 0.5-8 mg/ml)?
- Sample is stable at a required concentration, i.e., does not aggregate?
- Sample is stable over time, that includes storage, shipment, and transport?

**Sample AND buffer preparation for SAXS experiments**

~ approximately 1 week before the experiment until the day of the experiment.

- Purify/obtain fresh samples in sufficient amounts (calculate the material required for all experiments, e.g., one sample plus four dilutions).
- Check for sample purity/polydispersity.
- Use dialysis or size exclusion chromatography (SEC) to match the sample buffer with the sample.
- Set aside sufficient, exactly-matched buffer for the SAXS measurements (e.g., 5-10 ml, for sample dilutions)
- When posting packages to a large-scale facility allow enough time for delivery.

**Sample handling during the SAXS experiment**

~ on the day of the experiment

- If applicable, remove the samples from dialysis. Remember to keep the post-dialysis buffer for background SAXS measurements and sample dilutions.
- If applicable, add ligands or other small compounds to the sample; add the exact quantity of these compounds to the matched buffer.
- Prepare a dilution series and measure the sample at different concentrations.
- Quickly evaluate the SAXS data immediately after measurement to assess concentration dependencies, miss-matched buffer or radiation damage. Refer to Table 3 'First-aid at the beam line.'

**Fig. 2.1** Time-line for planning/preparation/performing successful SAXS measurements for protein and other biological macromolecules

- Sample handling and transport.
- Obtaining an exactly-matched buffer for accurate background subtraction.
- Measuring dilution series.
- Trouble shooting at the beam line, e.g., overcoming radiation damage.

### 2.1.1 Sample Purity and Polydispersity

A biological solution SAXS experiment is very straightforward. A sample containing a macromolecule of interest is held directly in an X-ray beam with a defined energy (i.e., at a specific wavelength, λ) and the intensities of the scattered X-rays, $I$, are recorded as a function of the angle, $q$, to ultimately produce a plot of $I(q)$ vs $q$; where $q = 4\pi\sin\theta/\lambda$ ($2\theta$ is the scattering angle). In a second step, the scattering of an identical solution that does not contain the sample, i.e., the supporting solvent, is collected and the scattering intensities are subtracted from the sample scattering to yield the scattering contributions from the macromolecule of interest. However, as easy as these two measurements may sound, it is just as easy to collect meaningless scattering data as any material placed into the beam will scatter X-rays (Jacques and Trewhella 2010). Therefore, it is important that the composition of both the sample and the matched buffer are known and well characterized.

With respect to sample quality, the degree to which known and unknown contamination affects the outcome of the experiment depends on a number of factors. Of particular importance is the size, or more specifically the volume of contaminating species. For example, a protein has to be purified to at least 95% as assessed using sodium dodecyl sulfate polyacrylamide gel electrophoresis (SDS PAGE; Jacques et al. 2012a). In some cases a higher degree of purity is required depending on the size of the contaminating proteins. As the total scattering is proportional to the square of the particle volume ($V^2$) even trace amounts of species present in a sample that are larger than the macromolecule of interest can 'swamp' the scattering signal rendering the data (often) uninterpretable. Producing samples that are free of higher molecular weight species and free of aggregates represents the biggest challenge when preparing SAXS samples. Figure 2.2 summarizes a number of techniques that are especially suited for detecting and quantifying higher oligomeric species (described in more detail below).

Sample polydispersity can also significantly impact the interpretation of SAXS data and, if possible, should be minimized as data interpretation and modeling is greatly simplified for monodisperse samples. However, for many biological systems polydispersity is often an intrinsic property of the sample such as monomer-oligomer equilibrium or the formation of complexes with low affinity constants. Such samples will generate scattering profiles representing the summed, volume-fraction weighted contribution of each species in the mixture. Recording SAXS data from a dilution series and evaluating changes in the scattering profiles (that includes determining the molecular weight, MW) is one way to evaluate whether a sample forms a concentration dependent mixture. In addition, the application of size exclusion chromatography (SEC) immediately prior to SAXS allows for the separation and sequential measurement of the separated mixture components. Furthermore, many recent methodological advances have made the analysis of mixtures relatively straightforward (Petoukhov et al. 2013). These include the analysis of intrinsically-disordered or flexible macromolecular systems that are by definition, polydisperse

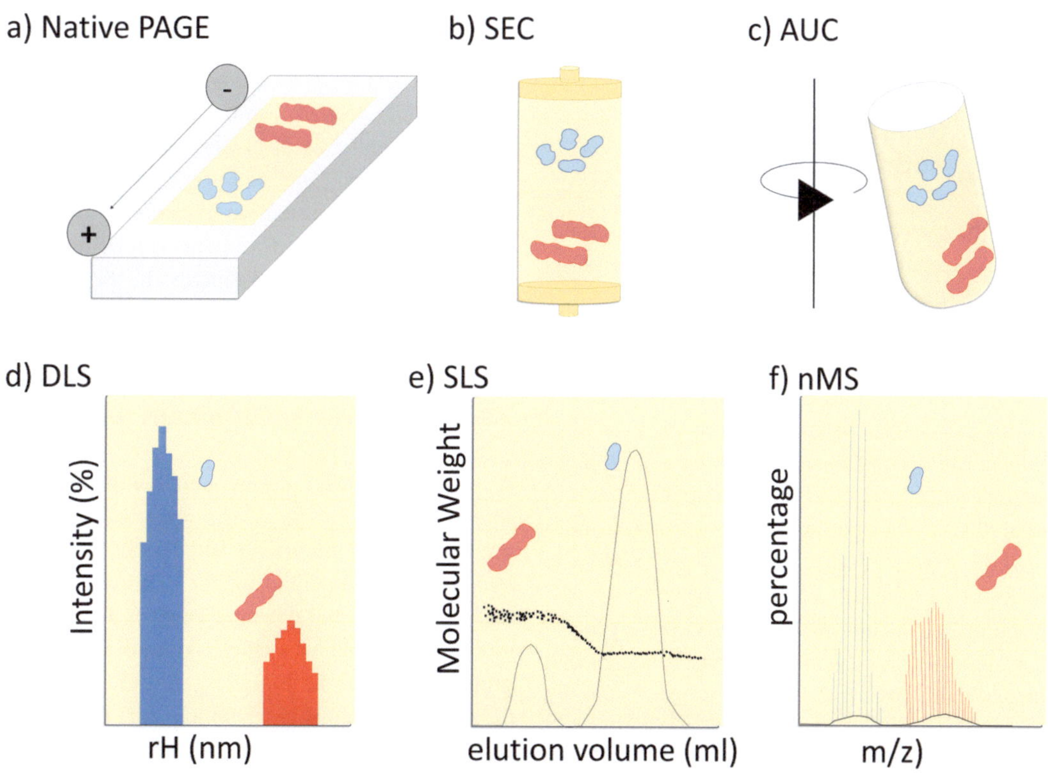


**Fig. 2.2** Analytical methods for studying sample polydispersity

**Table 2.1** Rough estimation of sample requirements and amounts

| Experiment | Sample amounts/ volumes[a] | Duration per measurement[b] | Comments |
|---|---|---|---|
| Lab source | 20–50 μl; >2 mg/ml | 15–90 min; depending on type of lab source | Less sensitive to radiation damage, but check in advance that samples will be stable over the time of the experiment |
| Synchrotron (proteins) | | | |
| No flow | 5–20 μl; >0.5 mg/ml | Approx. 1–5 min[c] | Radiation sensitive |
| Flow | 20–50 μl; >0.5 mg/ml | Approx. 1–5 min[c] | Less sensitive to radiation damage at the cost of more sample |
| SEC-SAXS | 50–100 μl; >5 mg/ml | 10–90 min; depending on column | Strong dilution of the sample |
| | | | Capillary fouling can occur during the elution/X-ray exposure process |
| Synchrotron (nucleic acids) | 5–50 μl; >0.25 mg/ ml | Approx. 5 min[c] | Take difference of electron density compared to protein into account for MW calculations |
| | | | If measuring RNA, ensure Rnase–free environment |

[a]For complete data set concentration series (at least four different concentrations) should be measured
[b]Includes measurement of sample and buffer
[c]Includes automated washing of the measurement cell

(Kikhney and Svergun 2015). Yet, for both monodisperse and polydisperse samples the general over-arching rule of sample preparation remains the same: it is crucial to prepare samples that are as pure as possible and free of unwanted contaminants.

### 2.1.2 Sample Quantities and Concentrations

Once the question of purity has been addressed, the next question becomes: how much sample is needed? Approximate guidelines on the quantity of material required for SAXS experiments are listed in Table 2.1. New developments in sample-delivery robotics and their installation at many biological SAXS beam lines not only prevent human errors but also ensure precise and efficient sample loading (David and Perez 2009; Round et al. 2015; Blanchet et al. 2015). Thus, usually 5–25 μl of protein sample are sufficient to fill the measuring cell at concentrations between 0.5 and 8 mg/ml. As a rule-of-thumb, a suitable protein concentration for synchrotron-based SAXS can be described by:

$$\mathrm{Concentration(mg/ml)} \approx 100/\mathrm{MW(kDa)}.$$

For example, a protein concentration of 2 mg/ml is most likely adequate for studying a monodisperse 50 kDa protein. However, for polydisperse samples that undergo concentration dependent oligomerization, it may be necessary to increase or decrease the sample concentration to either assemble or disassemble the oligomers, respectively. For more advanced measurements such as SEC-SAXS, more sample is likely required. Finally, it is always prudent to have some additional material on-hand in cases where a sample is sensitive to radiation damage so ‘at the beam line’ adjustments can be made to reduce the effects of this damage to the sample (see below).

### 2.1.3 Sample Environment (Buffer)

A solution SAXS measurement comprises two essential steps: (i) the measurement of the scattering data from the sample and; (ii) the measurement of scattering data from an identical, exactly matched buffer that does not contain the macromolecule of interest which is used for background subtraction. Imprecise buffer matching is a frequent stumbling block for first time SAXS users. Only after data collection and processing does it become apparent just how sensitive the method is to small discrepancies between correct and incorrectly-matched buffers. While planning a SAXS measurement the practical question that has to be addressed is: How is a suitable buffer obtained? For example, if the final purification step for a protein is ion-exchange chromatography, then it becomes difficult to evaluate the exact salt concentration at which the sample elutes from the column and then to prepare an exact replica of this buffer for SAXS. Or, in other words, as X-rays scatter from electrons, it is difficult to manually prepare a buffer with matched X-ray scattering and absorption properties as the supporting solvent of the sample. In such circumstances, a dialysis step is strongly recommended to obtain a good matching buffer, or to use buffer-exchange using SEC.

The atomic composition, or more precisely, the electron density of a chosen buffer is indeed a crucial aspect to consider when preparing samples for SAXS. As it happens, the average electron density of water (0.33 electrons/Å$^3$) is not that much lower than the average electron density of protein (~0.43 electrons/Å$^3$) and it is this very small difference that, after background subtraction, gives rise to a coherent SAXS pattern at low angle that can be used to extract structural information. However, the small difference in electron density decreases even further with the addition of components to the buffer, for example high-salt, glycerol, sucrose, etc. If the concentration of these buffer components becomes too great the X-ray contrast will limit to zero and the net scattering from the macromolecule will be effectively negated. For example the addition of either ~35% v/v glycerol, ~3 M NaCl, or ~1.2 M sucrose to a buffer will result in an approximate 50% reduction in the net scattering intensity measured from proteins in solution (see Box 2.1).

**Box 2.1: Calculating the Contrast, Δρ, of a Sample**

One of the advantages of solution SAXS is that a diverse range of solution conditions can be screened to assess the effects of changing sample environment on the structures of macromolecules. However, the addition of high concentrations of small molecules (e.g., 2 M NaCl) or the addition of electron-dense molecules to the supporting solvent will reduce the difference in electron density between the solvent and the macromolecules of a sample. As a result, the net scattering intensities derived from the macromolecules of a sample, i.e., the scattering contributions after the buffer scattering has been subtracted, will decrease. This could become important to consider, for example, when adding small molecules to a sample that limit the effects of radiation damage (e.g., electron-dense polyols). Adding too much will eventually result in the 'matching out' of the scattering signal.

The effect of changing the electron density of a buffer on the overall magnitude of the scattering intensities can be assessed in advance by calculating the contrast of a sample (Δρ), for example using the program MULCh (modules for the analysis of small-angle neutron contrast variation data from biomolecular assemblies (Whitten et al. 2008)). The Δρ is the difference between the average scattering length density of a macromolecule and the average scattering length density of the buffer which relates to the difference in electron density between a macromolecule and the buffer. The magnitude of the net small-angle scattering intensities from the macromolecules of the sample will be proportionate to $\Delta\rho^2$. The CONTRAST module of MULCh is specifically tailored for calculating X-ray (and neutron) scattering contrasts of a macromolecular system. For this calculation, the scattering data is not required. CONTRAST simply uses protein, RNA or DNA sequences in combination with the atomic formulae and concentrations of small molecules in the solvent. Using this information, CONTRAST calculates the X-ray and neutron-scattering-length densities of

(continued)

**Box 2.1** (continued)
the macromolecule and solvent (ρ) and subtracts these values to obtain Δρ of the sample. The entire MULCh package, which includes CONTRAST, can be downloaded as an off-line tool (with instructions) or used interactively online *via* http://smb-research.smb.usyd.edu.au/NCVWeb/. All you need as input is: (i) the list of solvent/buffer components (atomic formulae) and their molar concentrations: (ii) the one-letter amino-acid code or one-letter DNA/RNA code of the macromolecules; (iii) the atomic formulae of any small molecules bound to the macromolecule of interest— e.g., metal ions, cofactors.

### 2.1.4 Sample Stability

Further considerations have to include the assessment of sample stability during the measurement. SAXS experiments performed 'in house' using a laboratory X-ray source may require higher sample concentrations combined with longer exposure times. Thus, sample conditions may have to be found where the sample is both concentration and time-stable, specifically in regard to the formation of aggregates, during potentially prolonged measurements (e.g., up to 1 h). The brilliance afforded by synchrotron based SAXS means that samples can be measured using very short exposure times (in the order of milliseconds to seconds) and at low concentration. However, synchrotron SAXS poses a different set of challenges. Although radiation damage is a universal problem for both lab-based and synchrotron SAXS, the rate of damage using a synchrotron X-ray source may be more apparent even during very short exposure periods. Predicting whether a sample might be prone to radiation damage prior to a SAXS experiment is difficult and has to be treated on a case-by-case basis. For example, some proteins such as biomolecules with metal centers, may be particularly sensitive to radiation damage (e.g., cytochrome C, that binds Fe-heme), then again others are not (e.g., glucose isomerase, that binds $Mg^{2+}$ or $Mn^{2+}$). In Box 2.2 and 2.3 the means of dealing with radiation damage at the beam line are listed.

**Box 2.2: Addition of Small Molecules to Limit X-Ray Radiation Damage**
There a few 'tricks' that can be used to limit the effects of X-ray radiation damage by adding small-molecule free radical scavengers or polyols to a sample. Unfortunately, there is no single 'tried-and-true' method that can be applied and, somewhat annoyingly, it is impossible to predict before a SAXS experiment whether such measures will be effective. There a few considerations that can be helpful to determine which (if any) scavenger might be compatible to the system being studied. As a reminder, care has to be taken when adding accurate and equal measures of additive to both the sample and to the corresponding solvent blank. Ideally, a dialysis of the sample should be performed against the buffer with the added scavenger. However, dialysis might not be feasible as beam-time and sample quantities might be limited. In such cases, well calibrated pipettes or a microbalance should be used to add an equal volume or mass of concentrated additive stock solutions. Extreme care has to be applied, especially when adding viscous polyol solutions such as glycerol or sucrose. The main disadvantage of the solution additive approach is the increased risk of altering the chemical or physical properties of a macromolecule.

<u>DTT, @1–5 mM:</u>
Dithiotheritol has been often described as a useful scavenger, as it is not overly expensive and available in most molecular biology/structural biology laboratories. However, one must keep in mind that DTT is a reducing agent and is therefore not suitable for systems in which disulfide bonds play an essential role. Reduction of disulfide bonds can resulting in undesirable changes in

(continued)

**Box 2.2** (continued)

structure. One must also remember that DTT has a short shelf life (just up to a few hours) and should first be added directly before the measurement. In this sense, it is not suitable for SEC-SAXS. In addition, DTT undergoes oxidation and changes its ultraviolet (280 nm) absorption properties that may affect protein concentration estimates.

Ascorbic acid, @ 1–2 mM

Ascorbic acid is a 'classic' free radical scavenger that can be added to a sample to limit radiation damage. As the name suggests, ascorbic acid is acidic and thus one must be acutely aware that adding 'neat' ascorbic acid to a sample can significantly lower the pH which may then induce chemical alterations to a sample. Ascorbic acid also changes the UV absorbance properties so that concentration determination using UV methods may be hindered.

Glycerol, @ 3–5% v/v

Glycerol is not scavenger *per se* but is very good at limiting X-ray induced aggregation in solution. The addition of glycerol will increase the electron-density of the solvent, i.e., reduce the contrast of a sample, which needs to be considered with respect to maintaining the SAXS signal intensities (see Box 2.1). Glycerol can also influence protein–solvent/protein–proteins interactions that may affect concentration dependent oligomerization. In addition, due to its high viscosity, glycerol is difficult to add in exactly-equivalent amounts to sample and to the corresponding solvent/buffer blank needed for the SAXS measurements. Therefore, it is often preferable to make up a 10–20% v/v glycerol dilution in the buffer of choice (checking that the pH does not change) and then add the more diluted glycerol stock to the SAXS sample and buffer using a microbalance or a pipette (with the pipette tip-end clipped off). For SEC-SAXS, glycerol is often very effective in reducing radiation damage in the often slower sample flows through the X-ray beam line, but care must be taken that the SEC columns can withstand the increased pressure caused by the addition of glycerol to the mobile phase.

**Box 2.3: Tips for Performing a SAXS Experiment at a Synchrotron Beam Line**

Tip 1: Take time to think about and plan the experiment. Importantly ask yourself the question. What question do I want to focus on when using SAXS to probe the structure(s) of my samples?

Tip 2: Contact a local beam line responsible, or someone with experience, to coordinate the experiment in regard to sample handling, shipment, any additional equipment or necessary paperwork at a facility.

Tip 3: Thoroughly characterize the sample, this includes measuring small test batch-samples using SAXS to assess the susceptibility of samples to radiation damage prior to SEC-SAXS.

Tip 4: Determine the sample concentration immediately prior to the SAXS measurements to account for any sample loss during storage/transportation. If applicable, have some back-up material at hand for unforeseen problems (e.g., to add free radical scavengers to a sample if radiation damage is observed).

Tip 5: Remind yourself that it is crucial to have sufficient matching buffer for background subtraction. Make sure to

(continued)

**Box 2.3** (continued)

set aside a large volume of exactly-matched buffer for each sample for all of the SAXS measurements (e.g., dilution series, SEC-SAXS, etc.).

Tip 6: Prioritize your experiments. It is often preferable to measure a smaller number of characterized samples well during a beam-line shift compared to collecting data from as many poorly-characterized samples as possible ('garbage-in-garbage-out').

Another challenge faced by synchrotron SAXS users concerns the storage and shipping of samples to large-scale facilities. The general shelf-life of a sample and the tendency to form aggregates over time needs to be assessed. Effects of long-term storage at 4 °C or freeze-thawing under different conditions should be inspected. The practical aspects of transporting or shipping the sample to the synchrotron facility must also be considered. The stability of a sample can easily be tested in advance. Analytical size exclusion chromatography (SEC) and dynamic light scattering (DLS; both described in more detail below) are convenient methods for such an assessment, especially for detecting the presence of time-dependent aggregates. Small aliquots of a test sample can be screened using different handling protocols, such as fast/slow freezing, fast/slow thawing, plus or minus salt, glycerol, etc. and then be compared with each other. The objective is to identify the condition that prevents the sample from aggregating.

### 2.1.5 Support SAXS with Sample Characterization Data During Sample Preparation

During the planning stage of a SAXS experiment that more-often-than-not involves optimizing sample conditions, it is recommended to gather as much information as possible to support the conclusions from the SAXS investigation. This includes but is not limited to: what is the estimated MW of my sample (e.g., determined using light scattering techniques, for example multi-angle, right-angle laser light or static light laser scattering, MALLS, RALLS, SLS); what is the oligomeric state and does it change with different pH values and/or salt concentration (e.g., using SEC)? How is the system influenced by small-molecule ligands, temperature, etc. (e.g., using DLS or thermofluor assays to assess stability/aggregation (Boivin et al. 2013))? How flexible/folded is the system (e.g., using circular dichronism CD)?

In summary, a basic SAXS experiment is conceptually simple, but can be demanding in terms of preparing quality samples and matched buffers; not necessarily in regard to obtaining the required amount of sample, but regarding the quality and stability of the sample. In the next section, protocols for preparing the sample and the buffer for a solution SAXS experiment are described in more detail.

## 2.2 Preparation for SAXS Measurement

Here, we discuss general options and techniques that can be performed in the laboratory to achieve the goal of producing SAXS-quality samples and matched sample buffers under the general concepts of:

- Sample characterization using gel electrophoresis, SEC, light scattering techniques, analytical ultra-centrifugation and mass spectrometry.
- Assessing sample concentration using spectrophotometry or refractive index.
- Preparing the matched sample buffer.
- Organizing SAXS experiments

### 2.2.1 Sample Characterization: Assessing Polydispersity

It is very important to set aside sufficient time to thoroughly characterize those samples that will be used for the SAXS measurements (Jacques et al. 2012b). Due to the often irrecoverable and deleterious influence on the scattering data by aggregates or other large MW species it is very important to determine the association processes of a sample in solution. The ultimate aim should be to present the results obtained from a SAXS investigation in such a manner that the quality of the samples used to obtain the data can be assessed. For this, the sample purification procedure must be documented and reported, along with an estimate of the final purity of the sample. Available methods for such assessments are summarized in Table 2.2 as well as Fig. 2.2 and are shortly outlined here.

**Gel electrophoresis** provides an invaluable tool to assess the purity of the native proteins and complexes. Denaturing SDS-PAGE (both reducing and non-reducing) is excellent to evaluate whether a sample contains additional higher MW contaminants or if the target protein is affected by non-specific disulphide cross links. Native PAGE (run without SDS and non-reducing/non-degrading conditions) is useful to assess whether higher oligomeric species and to some extent self-associated aggregates are present in the sample. A big advantage of PAGE is that only small volumes of sample are required and a number of samples can be analyzed in parallel. A drawback of native-PAGE is that the separation is dependent on the size as well as the overall net charge of the molecule. Thus, the success of the separation is dependent on the isoelectric point (pI) of the protein as well as the behavior of the protein in the somewhat limited choice of native PAGE buffer systems (that may be very different to the final buffer selected for SAXS). However, in general, both SDS- and native-PAGE are exceptionally useful for routinely checking the purity and the stability of a sample over time.

In **Size Exclusion Chromatography** (SEC) a solvent carrying the sample, or mobile phase, passes through porous particles (the matrix, or stationary phase)–typically supported in a column–in which smaller particles are trapped for a short time resulting in a shift in their retention time. Consequently, larger particles (e.g., particles with a larger molecular weight or

**Table 2.2** Comparison of bioanalytical techniques used to study polydispersity

| Bioanalytical technique | Sample requirement | Experiment duration | Separation resolution |
|---|---|---|---|
| NAGE: native gel electrophoresis | 5–20 µl, >20 µg<br>buffer requires electrolytes | 60–300 min; or overnight at 4 °C | Separation not only by size but also surface charge |
| SEC: size exclusion chromatography | 50–100 µl<br>No severe aggregates (column clogging) wide range of buffers (might require addition of salt, e.g., 200 mM NaCl) | 15–90 min | Resolution depends on column length, buffer composition, flow rate, sample load volume and concentration |
| AUC: analytical ultra-centrifugation | ~50 to 400 µl<br>0.1–2 mg/ml (absorbance)<br>0.05–30 mg/ml (interference) wide range of buffers | Sedimentation velocity: 3–6 h<br>Sedimentation equilibrium: 2–5 days | Potentially high-resolution. Experiment has to be designed well to obtain the resolution required for the specific system e.g. to study monomer/dimer equilibrium |
| DLS: dynamic light scattering | 5–20 µl<br>>0.5 mg/ml wide range of buffers | 1–30 min | Low resolution technique; monomer-dimer not distinguishable |
| SEC-SLS: static light scattering | 50–100 µl<br>>2 mg/ml<br>coupled to SEC for fraction separation | As for SEC | As for SEC; combine with RI or UV to obtain MW estimates |
| MS: native mass spectrometry | ~50 µl<br>10–50 µM, in terms of monomer aqueous solution containing a volatile salt (e.g., ammonium acetate) | <1 min (longer preparation time) | High resolution, even small changes in size such as ligand binding can be detected |

**Table 2.3** Trouble shooting at the beamline

| Problem | Detection | Possible measures | Word of caution |
|---|---|---|---|
| Radiation damage | Discrepancy when comparing individual frames | Alter data collection strategy (e.g., attenuate the X-ray beam) | Reduction in signal-to noise ratio |
| | | Addition of radical scavengers or polyols | Potentially deleterious chemical alterations; problems with solvent matching if scavengers/polyols not added precisely |
| Aggregation | Non-linearity in the Guinier region | Dilution series | Reduction in signal-to noise ratio at low concentration (can be overcome with merging low- and high-concentration datasets) |
| | | (Ultra-) centrifugation | Not always sufficient |
| | | Filter sample | Possible binding of sample to filter membrane |
| | | SEC-SAXS | Time-consuming, dilution of sample; local/expert assistance often required (booking in advance) |
| Miss-matched buffer | Deviations (e.g., negative intensities) in the higher $q$-range; questionable Kratky plots; difficulties in obtaining $p(r)$ functions with $p(0) = 0$ | Dialysis | Time consuming and high buffer consumption |
| | | Diafiltration | Possible binding of sample to filter membrane; contamination with membrane preservatives |
| | | Strong dilution | Decreased signal to noise ratio |
| Interparticle repulsion | Decrease of $R_g$ vs increase in concentration | Repeat-measure low concentration samples | Decrease in signal to noise |
| | | Add salt to shield surface charge | Higher background scattering |

hydrodynamic radius) migrate through the separation matrix faster and are separated from the smaller species, assuming that there are no significant interactions with the column matrix. The column resolution (separation of two individual peaks) depends on a number of controllable parameters such as the choice of column (material, pore size, length) and running conditions (flow rate, mobile phase, loading volume). A major advantage of SEC is that it can be performed in a number of buffer conditions that can be screened and optimized for maintaining a sample in a desired state. There are, however, some limitations. Unavoidable interactions between the macromolecules of the sample and the stationary phase can result in adsorption, shifts in retention time, elution peak tailing/asymmetry, or even to changes in the three dimensional conformation (Hong et al. 2012). These undesirable interactions can often be prevented through the addition of salts to the running buffer. In addition, the type of separation matrix can be chosen; silica-based SEC columns represent a good choice for samples that may interact with dextran-based matrices used in the most common sepharose columns.

Another useful advantage of SEC is that it can be used in combination with UV-spectroscopy to qualitatively evaluate the oligomerization or aggregation state(s) of a protein sample. Although quantitative MW estimates based on the retention volume are not reliable, UV-SEC provides a means through which to visualize and detect the presence of the aggregates and higher oligomeric species and how these species may change in the sample over time or in different buffer conditions. In some circumstances, the equilibrium driven self-association of the individual components within a sample, even after purification, may be unavoidable. If this is observed, one can profit from SEC-SAXS set-ups which are now offered at almost all biological SAXS beam lines whereby the SEC column outlet is directly connected in-line to the SAXS capillary so that scattering data can be collected from the freshly-separated components as they elute from the column (Mathew et al. 2004; David and Perez 2009; Round et al. 2013; Graewert et al. 2015). As a SEC run is often accompanied with a solvent exchange, collecting SAXS data from the buffer that has run through the column acts as a convenient means to obtain the scattering required for background subtraction. However, and once again, care maybe required when selecting the correct buffer for SEC-SAXS background subtraction. Unknowable buffer-matrix interactions may cause extremely subtle alterations to the electron density of the exchanged solvent flowing through the column that may have a slightly different X-ray scattering and absorption properties compared to the buffer of the eluted macromolecules.

Sample component separation using SEC is normally monitored with UV detectors. There are, however, advantages of adding other detectors such as **static light scattering** (SLS) to the system. Using RALLS or MALLS in combination with the concentration estimates derived from UV or refractive index (RI) measurements, the molecular mass of the samples can be determined, independent from the elution volume and can be used to validate the MW of the SAXS samples (Graewert et al. 2015). This approach is ideal to determine to exact oligomeric state(s) of the SEC-separated components of a protein sample. In the simplest case, light scattering is detected for just one angle (90° in right angle laser light scattering, RALLS, or <7° for low angle scattering, LALLS). However, using multiple of detectors (up to 18) placed at different angles (multi angle laser light scattering, MALLS) significantly increases the sensitivity for the high MW aggregates as well as accuracy of the MW estimations, especially for larger complexes (Ahrer et al. 2003).

In a different type of laser light scattering experiment, **Dynamic Light scattering (DLS)**, the correlations between the fluctuations in the intensity of scattered light from macromolecules relating to their movement (Brownian motion) in solution are analyzed. As this is an exceptionally sensitive technique it can be employed to detect aggregates as well as to estimate polydispersity. Large globular particles not only scatter more strongly compared to smaller counterparts, but their Brownian motion is decreased due to their increased mass (as they dwell longer within the illuminated area). Consequently, relative to a starting time, $t0$, the fluctuations in the intensities for larger globular particles will be correlated relative to $t0$ for longer time periods compared to smaller particles, before exponentially decaying to zero, i.e., will eventually become uncorrelated relative to the initial time point. As the polydispersity and/or aggregation of a sample increases, the time taken for the auto-correlation to decay increases and it no longer smoothly decays towards zero. Accordingly, information can be obtained from the auto-correlation function by fitting the data (for example using a sphere model) from which hydrodynamic radius distributions of the particles and sample polydispersity can be estimated. An examination of the populations present within a sample using DLS, especially the presence of aggregates, is a good way to evaluate sample quality for SAXS.

There is no limitation for the size or type of particles that can be studied with DLS (peptides, proteins, polymers, micelles, carbohydrates, nanoparticles, etc.), however the resolution of the technique is quite low, i.e., it is not possible to distinguish between a sample consisting of monomers from a sample in monomer-dimer equilibrium. Its power as a characterization technique for SAXS samples is that DLS is exceptional for detecting trace amounts of aggregates (if aggregates are detected in DLS then they will also interfere with the SAXS) as well as for stability testing and condition screening. As DLS is a non-destructive method it is very suitable to examine the samples directly before they are loaded into the capillary of the beam line for example with a 96-well plate reader.

**Analytical Ultracentrifugation** is based on the sedimentation of macromolecules in their native state, often under extreme *g*-forces, is followed using an optical device (UV light absorption, fluorescent system or Rayleigh interferometer). The separation of the sample components within a mixture is dependent on the hydrodynamic and thermodynamic properties/association states of the macromolecules and enables the analysis of the shape, size distribution and molar masses of the sample components. Using a different approach, called a sedimentation equilibrium experiment, the final steady-state of the components are analyzed, where their sedimentation is balanced by diffusion opposing the concentration gradients. From this, one can retrieve information directly on the MW of the macromolecules as it is insensitive to shape. The amount of required sample is low (<0.5 mg). If the experiment is designed well molecules between 100 Da and 10 GDa and the size resolution can be chosen to detect even small mass changes. However, sophisticated equipment and technical know-how is required so that it is not typically accessible to the inexperienced user (Lebowitz et al. 2002). The experiment itself can be lengthy, depending on type of run (3–6 h for sedimentation velocity analysis, and even days for a sedimentation equilibrium experiment). However, ultracentrifugation can be exceptionally informative and a valuable asset in interpreting and supporting the conclusions reached from a SAXS experiment.

Structural biology studies can also profit from **native mass spectrometry (nMS)**. Here, electrospray ionization techniques are commonly employed so that the tertiary and quaternary protein structures are preserved. Very accurate mass estimates of proteins and the stoichiometry of subsequent assemblies can be determined. In turn, information can be gained regarding quaternary structure stability, dynamical behavior, conformation(s), subunit interaction sites, glycosylation state(s) and the topological arrangement of the individual proteins within a complex (Sharon and Horovitz 2015). However, as with ultracentrifugation, sophisticated equipment and technical know-how are absolutely required.

### 2.2.2 Sample Characterization: Assessing Concentration

Along with assessing the polydispersity of a sample, it is also very important to experimentally determine the concentration of a sample used for a SAXS experiment. The method for concentration determination should be chosen such that the accuracy of the method is sufficient to derive the MW of the sample from the SAXS measurements, specifically from the forward scattering at zero angle, $I(0)$. In this respect, the recent wwwPDB SAS task force has emphasized the importance on reporting which concentration technique was employed for a specific experiment (Jacques et al. 2012b). The concentration determination of proteins based on UV absorption at 280 nm is the most frequently used method as it is often the most understood and the fastest technique that has low sample consumption. However, and especially for proteins that completely lack or only have a few aromatic amino acid side chains (e.g., tryptophan) that consequently do not absorb strongly at 280 nm or for proteins that bind to non-protein ligands that absorb strongly in the UV region (e.g., heme) the concentration estimates using UV at 280 nm tend to become inaccurate. An alternative is differential refractometry (RI) which is far less dependent on amino acid sequence composition. Importantly, RI can be employed to measure the concentration of hetero complexes, or intrinsically disordered proteins (that often lack or are aromatic amino acid poor). With refractometry, the degree to which light bends as it passes through the interface between two substances is measured. The physical characteristic is dependent on the protein concentration and the protein's refractive increment (dn/dc). However, for most proteins in standard aqueous solutions dn/dc is 0.187 mL/g, allowing one to determine the concentration of the sample.

In some cases, using colorimetric assays may produce the most reliable results. In these instances, it is always beneficial to correlate and/or standardize the results obtained from colorimetric assays to UV or RI measurements. Correlating the results to UV also means that a fast assessment of the sample concentration, for example after transport, can easily be performed directly at a SAXS beam line, as most facilities offer access to an UV spectrometer. For the determination of nucleotide concentration commonly three different methods can be used: (i) UV absorbance at 260 nm (specific absorption peak of purine and pyrimidine rings), (ii) fluorescence (the amount of binding of fluorescent dye to double stranded DNA is determined with a fluorimeter and compared to reference measurements) and (iii) lesser available method: diphenylamine reaction (Li et al. 2014). For the latter, DNA is heated under acidic conditions to obtain 2-deoxyribose, which after dehydration can react with diphenylamine to produce a blue substance with an absorption maximum at 595 nm.

### 2.2.3 Buffer Preparation

As mentioned above, buffer matching is crucial for any SAXS measurement. SAXS is sensitive to even small changes in the composition of a buffer, in particular the electron density and X-ray absorption properties so that even small discrepancies between a sample and its corresponding buffer can lead to an erroneous background subtraction. In our experience the best way to obtain the optimal buffer is to perform dialysis (Fig. 2.3a) and this should always be the first method of choice. There are, however, cases in which dialysis is not feasible (for example a sample is prone to time-induced self-association/aggregation) and alternative approaches are available (Fig. 2.3b). In the following section we discuss:

- Dialysis.
- How to accurately add small molecules or expensive ligands to a sample (sample 'spiking').
- Diafiltration.
- Desalting columns.

In **dialysis** small molecules are exchanged into or out of a macromolecular sample by size-restricted diffusion through a porous membrane. The sample is placed on one side of the membrane and a buffer solution, the so-called dialysate, on the other side. Usually the dialysate is 200–500 times the volume of the sample. Sample molecules that are larger than the molecular-weight cutoff (MWCO) of the semi-permeable membrane are retained on the sample side of the membrane while smaller components (specifically buffer components) freely diffuse through the membrane and approach an equilibrium concentration with the dialysate. The process of buffer exchange may require 4–24 h to complete depending on the dialysate viscosity and typically produces an optimal matched buffer for SAXS background subtraction. Aside from using regular dialysis tubing, a number of devices are now commercially available which are essentially ready to use and resist sample leakage (Fig. 2.3a). The dialysis of small sample volumes (~200 μl) can benefit from both commercial or home-made "cup devices" which can be placed inside a medium size reagent tube containing the dialysate (for instance a 50 ml Falcon® tube) which enables easy transport. To ensure complete buffer exchange, different aspects such as ratio of sample volume to dialysate volume as well as the surface area of the membrane and factors including temperature, viscosity, mixing, etc., have to be taken into account. When choosing the membrane for the dialysis one has to be aware that the MWCO is not a sharply defined value; in general, the MWCO should be 5× smaller than the expected MW of the macromolecule.

Preparing sufficient dialysate volume (200–500 times larger than the sample) is not always feasible especially when expensive ligands are required. In such cases, **spiking** the sample and buffer will be necessary. If the sample is stable in the absence of the ligand, then the apo, or unligated, variant of the sample should be

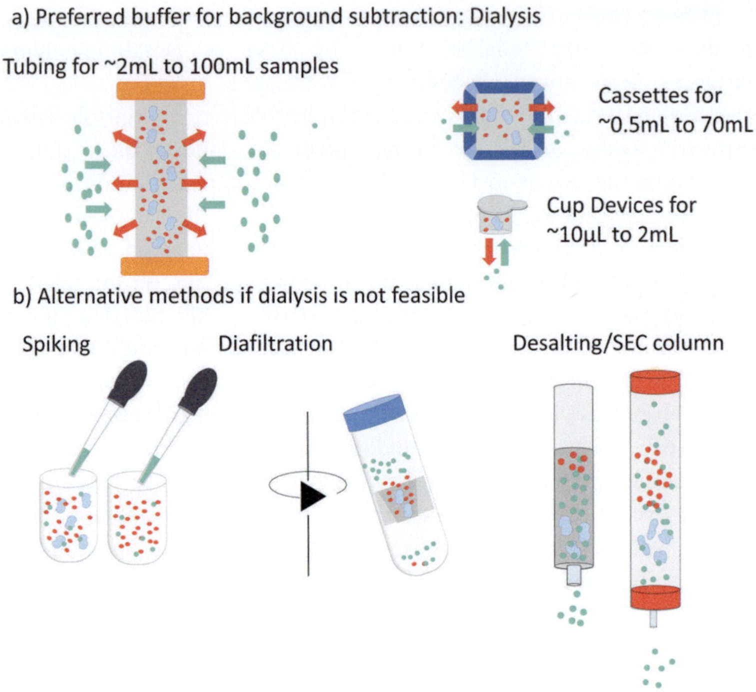


**Fig. 2.3** Preparation of matching buffer for background subtraction

prepared and dialysed against the ligand-free buffer. Afterwards the ligand can be added in small amounts from a stock solution to the sample as well as the buffer (dialysate) using accurately calibrated pipettes or an electronic microbalance. If the ligand has to be added throughout the preparation or the sample is generally prone to time-dependent aggregation, then **diafiltration** might be an alternative option for obtaining a matched buffer. As in dialysis, a semi-permeable membrane is used to separate macromolecules from low molecular-weight compounds. However, instead of relying on passive diffusion, the solutions are forced through the membrane by pressure or centrifugation. As the water along with the low molecular-weight solutes collect on one side of the membrane, the macromolecules are concentrated on the opposite side. Thus, diafiltration devices are often employed as concentrators, but can–if used with caution–also be used for buffer exchange. For this, the sample undergoes successive rounds of dilution in a buffer of choice, concentration, followed by subsequent rounds of dilution and concentration. If this method is used to adjust sample solvent to the desired buffer for background subtraction it is very important to thoroughly rinse the membrane before use to remove any preservatives on that membrane that are included in the manufacturing process (e.g., glycerol, azide, etc.). This is done, by passing larger volumes of buffer through the membrane before adding the sample. In addition, it is preferable to perform a series of short centrifugation/concentration steps, with careful mixing of the sample using a pipette, as opposed to one long centrifugation step that may result in an unwanted concentration gradient at the sample-membrane interface which may cause the sample to aggregate. Therefore, it may be necessary to monitor the sample e.g., using DLS, to ensure that aggregates do not form during the concentration steps.

Buffer exchange/adjustment can also be performed with a **desalting** column that is a similar process that occurs during SEC. By choosing the correct column length and sample load, the macromolecules of a sample will be too large to enter the pores of the desalting resin and will quickly pass through the column. Buffer salts and other small molecules will, on the other hand, enter the pores of the resin. After equilibrating the desalting column in a desired buffer for the SAXS measurements and passing the sample over it, the original buffer components will remain trapped by the resin, while the macromolecule of interest will flow through and be recovered in the SAXS buffer. The disadvantage of using a desalting column is that the sample undergoes significant dilution. In addition, our experience shows that with this method as well as during SEC, the buffer does not always undergo complete exchange.

### 2.2.4 Organization of the Experiment

Once the sample is obtained in sufficient quantities, final arrangements for shipping/transporting the samples can be made. Figure 2.4 demonstrates some scenarios on how to organize the experiment depending on the nature of the sample.

Practical aspects for shipping samples can include taking care of required papers/documentations for customs and/or the transport of dried ice, etc. For synchrotron SAXS, it is always advisable to contact the local beam line responsible in advance and ask for the specific shipping address (to avoid delays in delivery). Furthermore, it is always important to include clear instructions on the storage conditions for the person receiving the sample package. If further sample preparation is required, e.g., overnight dialysis, access to the samples and laboratories for after working hours should be

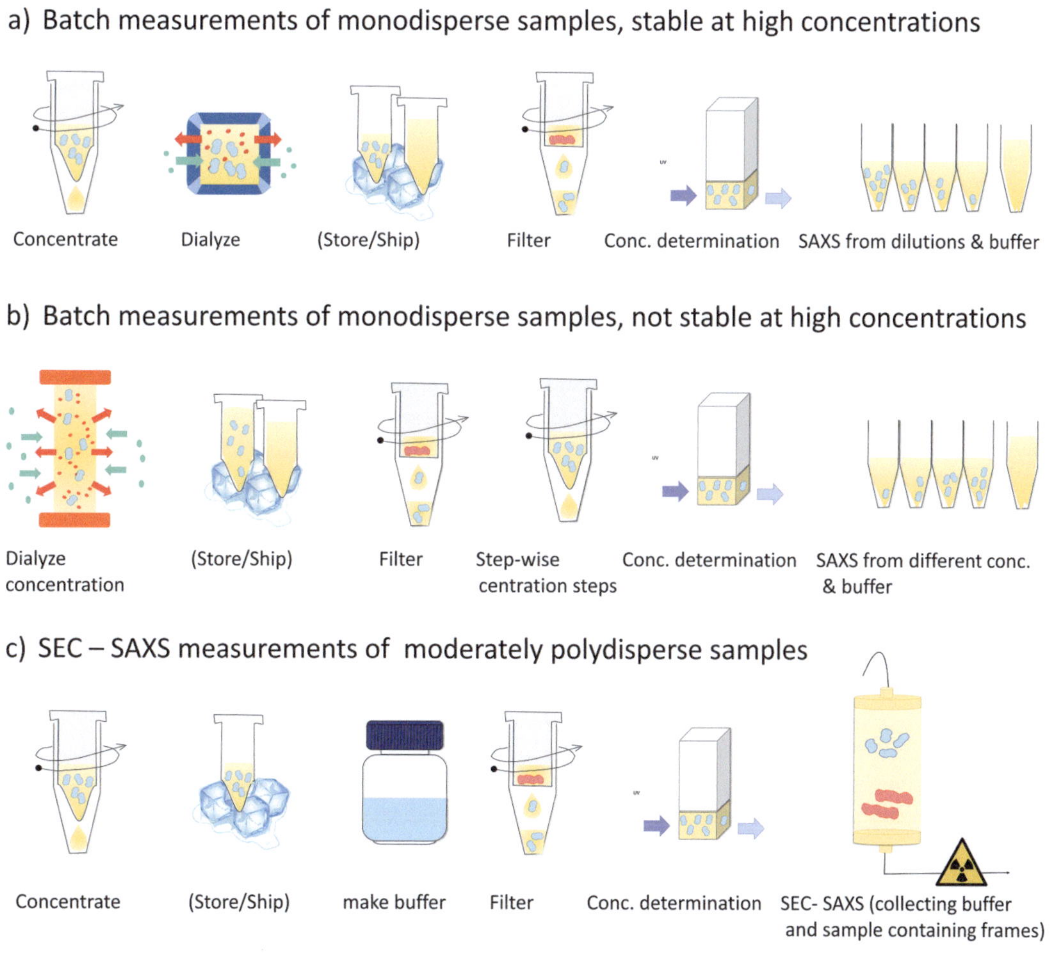


**Fig. 2.4** Sample preparation strategies

arranged in advance. In this respect, consumables (dialysis accessories, syringes and needles, concentrators, etc.) should be shipped with the samples. For some samples it might become necessary to perform the final purification steps on-site. Many beam lines offer access to laboratory facilities, which can be booked prior to arrival (for an example see Boivin et al. 2016). In such instances, it is best to discuss the experimental procedures in detail with the local responsible to ensure that all equipment and consumables that are needed, are indeed available.

Finally, and very importantly, it can be very easy under the stresses of sample preparation to focus entirely on the sample for a SAXS experiment and forget about the buffer. Remember to include sufficient buffer for your standard SAXS measurements (for example 50 ml). For SEC-SAXS up to 1 liter may be required, or alternatively, pack the dried ingredients to reconstitute the desired buffer on-site, immediately prior to the experiment.

## 2.3 Trouble-Shooting at the Beam Line

Once at the beam line make sure all your equipment has arrived and take time to orient yourself around the facility and laboratories. If applicable, thaw sample and buffer keeping in mind that this can take some time. Before continuing, it is advised to give the samples a quick spin in a centrifuge (e.g., 10,000×*g* for 10 min) or pass the samples through a 0.22 μm filter to remove any large aggregates or insoluble particulate matter (this is particularly important for SEC-SAXS samples). UV spectroscopy is a good method to determine if any sample was lost during transport and storage. Synchrotron SAXS experiments can be hectic affairs as there is often pressure to measure as many samples as possible within an allocated beam time period. Consequently, there may be little time to conduct a thorough analysis of the data. However, thanks to automated pipelines at most SAXS beam lines, data is analyzed on-the-fly in near real-time and as a consequence if problems arise regarding the sample (e.g., the presence of aggregates) decisions can be made on-the-spot to ameliorate the sample conditions to improve data quality (Blanchet et al. 2015). Basic trouble-shooting approaches include

- Detecting and handling radiation damage.
- Dealing with aggregation.
- Dealing with miss-matched buffer
- Dealing with concentration dependent effects.

### 2.3.1 Dealing with Radiation Damage

Unavoidably, some samples will be susceptible to radiation damage that manifest as a continual increase in scattering intensity during the measurement. To detect such an effect, a number of frames are typically collected during one exposure, e.g., instead of recording a single continuous one second exposure, twenty 50 ms frames are recorded over one second. The individual frames are then compared to each other and those frames showing traces of radiation damage are discarded (Franke et al. 2015). In some cases, the onset of radiation damage may occur late during the measurement so that a sufficient number of frames can be averaged to produce a SAXS profile with an acceptable signal-to-noise ratio. However, if a sample is overly susceptible to radiation damage, it maybe that only one frame can be collected, that may or may not, be influenced by damaged species. In this scenario, there are several adaptions that can be made to reduce the effects of radiation damage that include altering the data collection strategy or changing the sample environment (Jeffries et al. 2015). If the amount of sample is sufficient, the flow-rate of the sample through the X-ray beam can be increased; the beam can be attenuated to reduce the X-ray flux, or the beam can be defocused. Alternatively, the chemical composition of the sample can be changed *via* the addition of equal quantities of small-molecules to the sample and matched buffer. These small molecules are referred to as radical scavengers and may include–but are not limited to–dithiothreitol (DTT) and ascorbic acid. Although not scavengers *per se*, polyols such as glycerol, ethylene glycol as well as sucrose influence long-range protein-protein interactions and are effective in reducing radiation-induced aggregation. In Fig. 2.5a the effect of adding DTT and glycerol to the sample are shown.

All of these options to curb the effects of radiation damage come with an associated cost. Changes in data collection strategy may require the repeated measurement of more sample; addition of polyols result in a reduced contrast; the addition of classical scavengers can potentially cause deleterious chemical alterations (e.g. DTT may reduce disulfide bonds).

### 2.3.2 Dealing with Aggregation

Another common observation during the measurements is the presence of aggregates. Aggregates are often detected by non-linear or 'upturn' features in the Guinier plot of the SAXS data ($\ln(I(q))$ vs $q^2$) at very low-angles ($qR_g < 1.3$; where $R_g$ is radius of gyration) or an greater than expected MW estimate of the sample determined from the SAXS data (Guinier 1939). Figure 2.5b shows an example in which the presence of aggregates is detectable in one of two storage conditions. In these cases, a dilution series of the sample becomes important; it may be that aggregates are concentration-dependent and simple dilution can remove their influence. However, if the aggregates remain persistent, a few options can be explored to remove them from solution. Giving the sample a strong spin with a (ultra-) centrifuge can sometimes be useful (e.g., $20.000 \times g$ for 15 min); the insoluble aggregates collect at the bottom of the tube and the soluble protein remains in the supernatant. A faster option is to try filtering the sample; either using a syringe filter tip or small centrifugal device (keeping in mind that some proteins may interact with membranes and be adsorbed, altering their structure). Another option is the use of online SEC-SAXS. Here, aggregates are separated from the sample during the size-exclusion step. This option of course requires more time, sample volume and concentration and often has to be organized in advance, but can be well worth the investment (especially for samples not prone to radiation damage).

### 2.3.3 Dealing with Miss-Matched Buffer

As mentioned above, incorrectly-matched sample buffers required for the accurate subtraction of background scattering contributions is often reason why SAXS experiments are unsuccessful. The detection of miss-matched buffer is unfortunately not always that straight-forward and not always unambiguous. There are, however, a few indications when looking at the distinct scattering plots (often automatically generated by integrated pipelines) especially for over-subtracted buffer contributions that produce negative scattering intensities at high-angle (i.e., at high-$q$) in the subtracted SAXS profiles. Figure 2.5c, shows the effect of under- as well as over-subtracting the background (blue and red curves, respectively). Once buffer miss-match is detected and time permits then a new dialysis can be set-up. Alternatively, a diafiltration device as described above can be used if time is scarce.

### 2.3.4 Dealing with Concentration Dependent Effects

As mentioned above it may be important to measure the sample at different concentrations in order to detect and consequently ameliorate concentration dependent effects in the sample that impact the scattering data. For example, as depicted in Fig. 2.5d, a significant decrease in $R_g$ as the sample concentration increases may indicate that repulsive interactions are present between the macromolecules of the sample (i.e., coulombic repulsion). Conversely, if the $R_g$ significantly increases as the sample concentration goes up, attractive interactions between the sample macromolecules are likely present that may result in sample oligomerization, sample

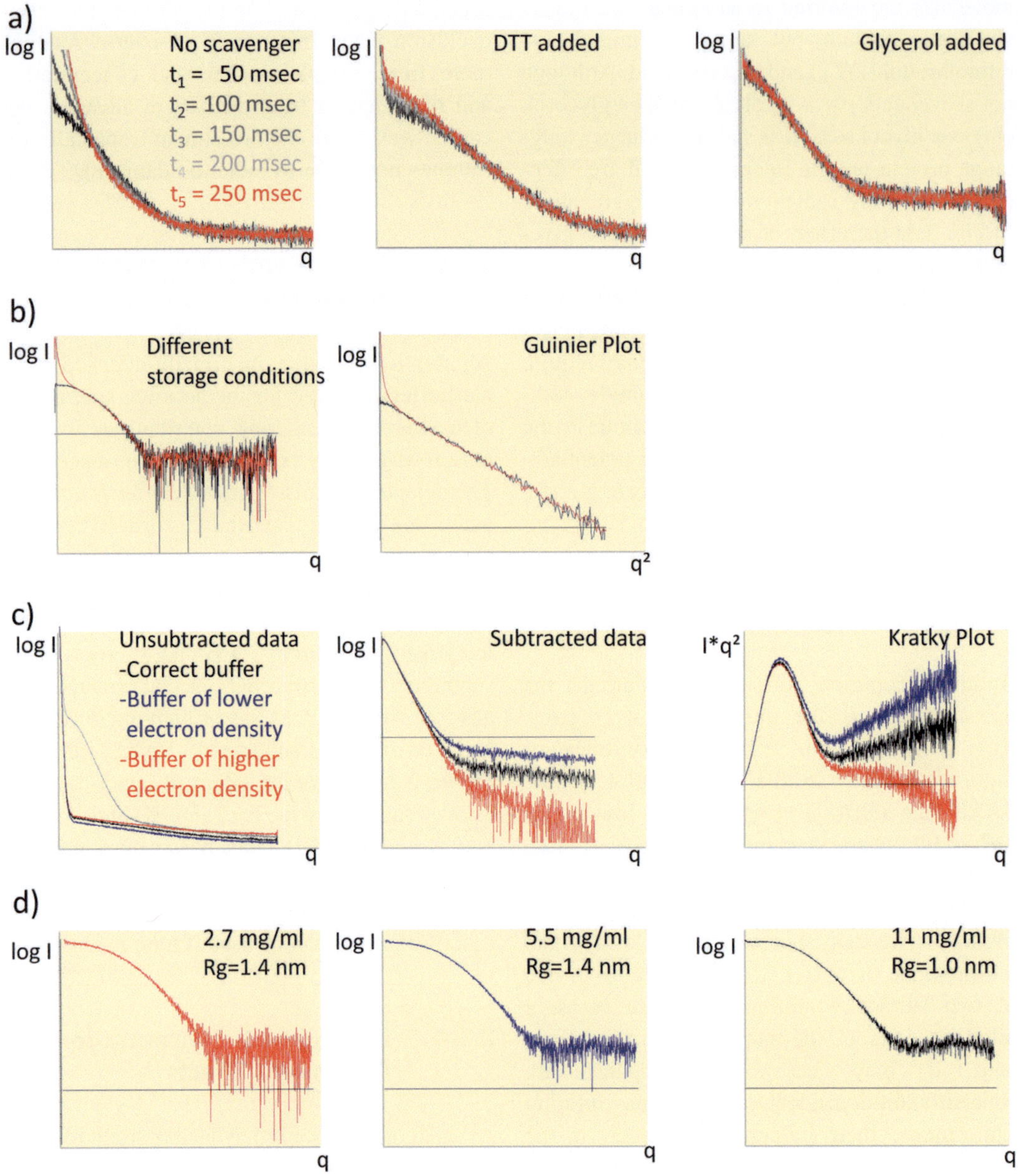


**Fig. 2.5** Preliminary sample/data evaluation

polydispersity, or in the worst case scenario, sample aggregation. As a result, samples may require additional dilutions to nullify concentration effects or, more radically, alterations to the supporting solvent such as the addition of salts and changes in pH to alter protein surface potentials responsible for concentration-dependent effects.

### 2.3.5 Other Considerations

Finally, it is advised to always check the structural parameters derived from experimental scattering data with the results obtained from complementary methods. This is especially important for validating the molecular weight of the sample. For one thing, errors such as accidental swapping of tubes, loading of air bubbles, questionable dilution series etc. can be best identified at the beam line allowing for 'on-the-spot' corrections.

In summary, SAXS experiments are straightforward and the demands for sample preparation are feasible for most systems. There are, however, a few essential steps to keep in mind when planning, preparing and performing the SAXS experiments. In understanding how these sample preparative steps directly influence the scattering process, enables the SAXS user to successfully produce high quality samples and ultimately high quality SAXS data.

**Acknowledgments** This work was supported by the Bundesministerium fur Bildung und Forschung (BMBF) project BIOSCAT, grant 05K12YE1 by iNEXT, grant number 653706, funded by the Horizon 2020 programme of the European Union and by a Human Frontier Science Program research grant, HFSP Ref: RGP0017/2012. M.A.G. was supported by an EMBL Interdisciplinary Postdoc Programme (EIPOD) and Marie Curie COFUND actions.

## References

Ahrer K, Buchacher A, Iberer G, Josic D, Jungbauer A (2003) Analysis of aggregates of human immunoglobulin G using size-exclusion chromatography, static and dynamic light scattering. J Chromatogr A 1009 (1–2):89–96

Blanchet CE, Spilotros A, Schwemmer F, Graewert MA, Kikhney A, Jeffries CM, Franke D, Mark D, Zengerle R, Cipriani F, Fiedler S, Roessle M, Svergun DI (2015) Versatile sample environments and automation for biological solution X-ray scattering experiments at the P12 beamline (PETRA III, DESY). J Appl Crystallogr 48:431–443

Boivin S, Kozak S, Meijers R (2013) Optimization of protein purification and characterization using thermofluor screens. Protein Expr Purif 91 (2):192–206

Boivin S, Kozak S, Rasmussen G, Nemtanu IM, Vieira V, Meijers R (2016) An integrated pipeline for sample preparation and characterization at the EMBL@PETRA3 synchrotron facilities. Methods 95:70–77

David G, Perez J (2009) Combined sampler robot and high-performance liquid chromatography: a fully automated system for biological small-angle X-ray scattering experiments at the synchrotron SOLEIL SWING beamline. J Appl Crystallogr 42:892–900

Franke D, Jeffries CM, Svergun DI (2015) Correlation map, a goodness-of-fit test for one-dimensional X-ray scattering spectra. Nat Methods 12(5):419–422

Graewert MA, Svergun DI (2013) Impact and progress in small and wide angle X-ray scattering (SAXS and WAXS). Curr Opin Struct Biol 23:748–754

Graewert MA, Franke D, Jeffries CM, Blanchet CE, Ruskule D, Kuhle K, Flieger A, Schäfer B, Tartsch B, Meijers R, Svergun DI (2015) Automated pipeline for purification, biophysical and x-ray analysis of biomacromolecular solutions. Sci Rep 5:10734

Guinier A (1939) La diffraction des rayons X aux tr'es petits angles: application 'a l'étude de phénomenes ultramicroscopiques ed. Univ. de Paris, Paris, pp 3, p 1, 80, 2, p 1 1

Hong P, Koza S, Bouvier ES (2012) Size-exclusion chromatography for the analysis of protein biotherapeutics and their aggregates. J Liq Chromatogr Relat Technol 35(20):2923–2950

Jacques DA, Trewhella J (2010) Small-angle scattering for structural biology–expanding the frontier while avoiding the pitfalls. Protein Sci 19:642–657

Jacques DA, Guss JM, Svergun DI, Trewhella J (2012a) Publication guidelines for structural modelling of small-angle scattering data from biomolecules in solution. Acta Crystallogr Sect D Biol Crystallogr 68:620–626

Jacques DA, Guss JM, Trewhella J (2012b) Reliable structural interpretation of small-angle scattering data from bio-molecules in solution—the importance of quality control and a standard reporting framework. BMC Struct Biol 12:9

Jeffries CM, Graewert MA, Svergun DI, Blanchet CE (2015) Limiting radiation damage for high-brilliance biological solution scattering: practical experience at the EMBL P12 beamline PETRAIII. J Synchrotron Radiat 22:273–279

Kikhney AG, Svergun DI (2015) A practical guide to small angle X-ray scattering (SAXS) of flexible and intrinsically disordered proteins. FEBS Lett 589(19 Pt A):2570–2577. doi:10.1016/j.febslet.2015.08.027. Epub 2015 Aug 29

Lebowitz J, Lewis MS, Schuck P (2002) Modern analytical ultracentrifugation in protein science: a tutorial review. Protein Sci., 2002 11(9):2067–2079

Li X, Wu Y, Zhang L, Cao Y, Li Y, Li J, Zhu L, Wu G (2014) Comparison of three common DNA concentration measurement methods. Anal Biochem 451:18–24

Mathew E, Mirza A, Menhart N (2004) Liquid-chromatography-coupled SAXS for accurate sizing of aggregating proteins. J Synchrotron Radiat 11:314–318

Petoukhov MV, Billas IM, Takacs M, Graewert MA, Moras D, Svergun DI (2013) Reconstruction of quaternary structure from X-ray scattering by equilibrium mixtures of biological macromolecules. Biochemistry 52(39):6844–6855

Round A, Brown E, Marcellin R, Kapp U, Westfall CS, Jez JM, Zubieta C (2013) Determination of the GH3.12 protein conformation through HPLC-integrated SAXS measurements combined with X-ray crystallography. Acta Crystallogr D 69:2072–2080

Round A, Felisaz F, Fodinger L, Gobbo A, Huet J, Villard C, Blanchet CE, Pernot P, McSweeney S, Roessle M, Svergun DI, Cipriani F (2015) BioSAXS sample changer: a robotic sample changer for rapid and reliable high-throughput X-ray solution scattering experiments. Acta Crystallogr Sect D Biol Crystallogr 71:67–75

Sharon M, Horovitz A (2015) Probing allosteric mechanisms using native mass spectrometry. Curr Opin Struct Biol 34:7–16

Trewhella J (2016) Small-angle scattering and 3D structure interpretation. Curr Opin Struct Biol 40:1–7

Vestergaard B (2016) Analysis of biostructural changes, dynamics, and interactions – small -angle X-ray scattering to the rescue. Arch Biochem Biophys 602:69–79

Whitten AE, Cai S, Trewhella J (2008) MULCh: modules for the analysis of small-angle neutron contrast variation data from biomolecular assemblies. J Appl Crystallogr 41:222–226

# Considerations for Sample Preparation Using Size-Exclusion Chromatography for Home and Synchrotron Sources

# 3

Robert P. Rambo

**Abstract**

The success of a SAXS experiment for structural investigations depends on two precise measurements, the sample and the buffer background. Buffer matching between the sample and background can be achieved using dialysis methods but in biological SAXS of monodisperse systems, sample preparation is routinely being performed with size exclusion chromatography (SEC). SEC is the most reliable method for SAXS sample preparation as the method not only purifies the sample for SAXS but also almost guarantees ideal buffer matching. Here, I will highlight the use of SEC for SAXS sample preparation and demonstrate using example proteins that SEC purification does not always provide for ideal samples. Scrutiny of the SEC elution peak using quasi-elastic and multi-angle light scattering techniques can reveal hidden features (heterogeneity) of the sample that should be considered during SAXS data analysis. In some cases, sample heterogeneity can be controlled using a small molecule additive and I outline a simple additive screening method for sample preparation.

**Keywords**

SEC • MALS • QELS • Total scattered intensity • Additives • Stability • Aggregation • Monodispersity • Sucrose • Phosphate

## 3.1 Introduction

Structural investigations of biological systems in the solution-state are investigations made from an ensemble of macromolecular particles. In biological, solution-state SAXS, the ensemble is composed of thousands of billions of macromolecules in various interchangeable, conformational states (Rambo and Tainer 2010a, b). Since domain motions range from micro- to milli-seconds (Henzler-Wildman et al. 2007), a solution-state SAXS measurement is often an

R.P. Rambo (✉)
Diamond Light Source Ltd., Harwell Science & Innovation Campus, Didcot OX11 0DE, UK
e-mail: robert.rambo@diamond.ac.uk





B. Chaudhuri et al. (eds.), *Biological Small Angle Scattering: Techniques, Strategies and Tips*, Advances in Experimental Medicine and Biology 1009, DOI 10.1007/978-981-10-6038-0_3

observation of the thermodynamic state due to the X-ray exposure times being much greater than the internal motions of the particle. The distribution of macromolecules across this conformational landscape is determined by the buffer composition and temperature that defines the sample environment.

The SAXS signal is a direct observation of this conformational landscape. If the landscape is broad and diverse, interpreting the SAXS signal using a single atomistic model will be difficult and likewise, the information quality of any *ab initio* model will be low. This type of conformational heterogeneity is difficult to assess but can be influenced by changing the composition of the buffer (Rambo and Tainer 2010a, b). Similarly, mass heterogeneity due to multimerization, aggregation or low purity will reduce the information quality of the SAXS signal and confound the structural interpretation (Jacques and Trewhella 2010). These issues highlight a fundamental property of solution-state SAXS and that is *everything scatters* in the sample. Unlike NMR and X-ray crystallography where heterogeneity will broaden and weaken the desired structural signal, heterogeneity contributes directly to the SAXS signal whose contribution is proportional to mass and concentration. Therefore, any structural modelling using SAXS data must be made from data collected from samples that are well-characterized and optimized for monodispersity and homogeneity (Rambo and Tainer 2013).

Unfortunately, the quality of the measured SAXS signal is not fully determined by sample heterogeneity. The actual SAXS signal of the ensemble is taken as the difference (Fig. 3.1) between the measured SAXS curve of the sample (i.e., particle and buffer) and the background (buffer only). Matching the buffer between the sample and background is critical to the accuracy of the recovered SAXS curve. Particularly at high scattering vectors (q), poor buffer matching often leads to under- or over-subtraction and errors in subtraction will limit the resolution of the SAXS experiment. If not properly identified

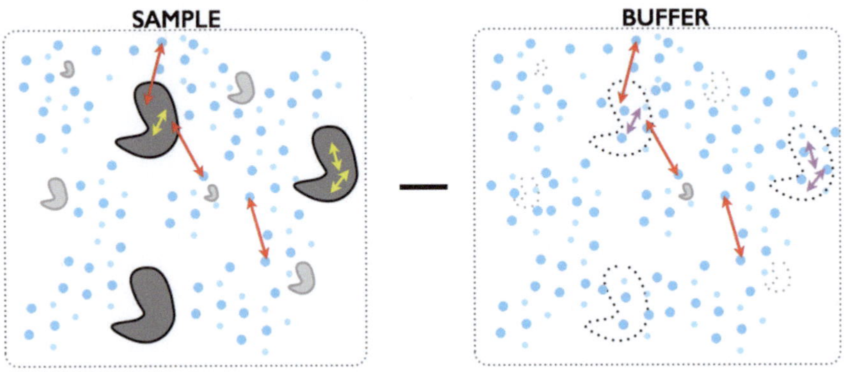


$$I_{particle}(q) = I_{sample}(q) - I_{buffer}(q)$$

**Fig. 3.1** SAXS as a difference measurement. SAXS requires two precise measurements *1* sample (particles + buffer) and *2* buffer background. For the sample, the scattering is the result of the dissolved particles, solute that participates in the hydration of the particle, and bulk solvent. For the buffer, the scattering will be the result of the bulk solvent, solute that will participate in the particle hydration, and the excluded volume. The excluded volume is the imprint of the particle in the buffer whose electron density is bulk solvent. Under dilute, monodispersed particle conditions, the observed SAXS intensity will be approximated as the scattering from a single particle scaled by concentration with corrections due to the excluded volume and hydration. At low resolution, the d-spacing vectors (2π/q) (*red double arrows*) are large and can only exists across the particle-solvent boundary whereas at higher resolutions, the d-spacing vector is smaller (*yellow and magenta double arrows*) and can exists within a single particle. To accurately minimize contributions from the bulk solvent, the composition of the bulk solvent in both the sample and background must be identical and removed from the recovered SAXS curve during post-processing, the systematic contributions from the mismatch can increase the false discovery rates in modelling and introduce artefacts in the P(r)-distribution.

An efficient and readily available technique that can assess sample quality and provide a reliable method for buffer matching is size exclusion chromatography (SEC). SEC chromatographic separation is based principally on the ability of the macromolecules to move through the pores of the chromatographic resin. If the macromolecule is physically larger than the pores, it will be excluded by the resin and elute relatively early from the SEC column whereas a macromolecule that is smaller than the pores will reside within the column longer and elute later. The standard method for monitoring an SEC chromatographic separation is UV absorption which exploits the absorption properties of the peptide backbone and aromatic rings of common amino and nucleic acids. Absorption based methods only inform on particle concentration. Regardless of size, an elution profile should be nearly symmetric for a sample consisting of homogenous and monodisperse particles. Any peak asymmetry should not go unnoticed and can be indicative of particle-column interaction, multimerization or heterogeneity. Furthermore, the use of native gels to assert a sample is free of aggregation should be avoided as the method of detection is unreliable for obvious reasons.

SEC is suitable for a wide range of macromolecular masses (10–10,000 kDa) and shapes. For globular proteins, there is a linear relationship between mass and the physical dimensions of the particle. Using a set of standards, i.e., proteins with known mass and dimensions, an SEC column can be calibrated such that the elution time corresponds to the mass of the particle. This technique assumes the unknown particle can be approximated by a simple sphere whose radius (Stoke's radius) scales linearly with mass. For asymmetric or elongated particles, a calibrated SEC column will give erroneous mass estimates, since it is essentially the largest dimension of the particle that determines how the particle will travel through the column (Fig. 3.2).

Resolving this ambiguity between mass and particle dimensions can only be made using a scattering technique such as multi-angle light scattering (MALS) or SAXS. Here, scattering measurements are made at time points along the

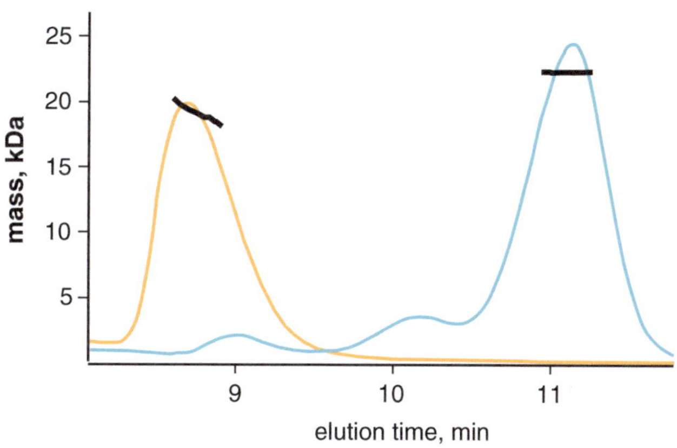


**Fig. 3.2** Anomalously eluting proteins by size exclusion chromatography. The Stoke's radius is a spherical approximation of a particle that scales linearly with globular proteins. Using proteins of known mass, an SEC column can be calibrated where elution volume (time) correlates with protein mass. This method makes a critical assumption regarding the globularity of the particle and is often erroneous with asymmetric particles. Here, the globular particle xylanase (21 kDa) elutes (*cyan*) with a stable MALS mass (y-axis, kDa) across the main peak. In comparison, is a novel protein that has a smaller MALS mass than xylanase but elutes earlier. The protein was determined to be highly asymmetric with high mass heterogeneity. In the absence of MALS, the peak mass would have been over-estimated

SEC run either by fractionation of the elution or by coupling the MALS (SEC-MALS) or SAXS (SEC-SAXS) instrument directly inline with the SEC (Perez and Nishino 2012; Gillis et al. 2014; Jeffries et al. 2016; Meisburger et al. 2016). Direct coupling of the SEC to the scattering instrument has proven to be optimal for accurate scattering measurements of the background and sample. In MALS, it is the intensity of the scattered light by the particle that is used to determine molecular weight. This intensity must be properly normalized by the particle's concentration whereas in SAXS, it is the angular dependence on the scattered intensity that is used to describe the shape and mass of the particle.

SEC-MALS is the most reliable method for assessing mass homogeneity of an SEC elution peak. The MALS instrument contains several detectors arranged in a circle around a flow cell that measures the intensity of scattered light from a laser source. In some instruments, at least one of the detectors can perform time-resolved measurements allowing for quasi-elastic light scattering (QELS) observations. QELS monitors how the solution sparkles with time and it is the decay rate of the sparkling that is proportional to particle dimension of radius-of-hydration ($r_H$). QELS will be sensitive to particle conformation and QELS measurements across an elution peak can provide additional information on conformational heterogeneity (Minton 2016).

## 3.2 Refractive Index, Ultra-Violet and Light Scattering

Standard UV absorption ($A_{260\ nm,\ 280\ nm}$) detectors found on most SEC instruments are not sufficient to monitor all classes of biomolecules or biomolecules with exceedingly low extinction coefficients. Alternatively, a refractive index (rI) detector can be used to reliably monitor particle concentration in a wide range of buffer conditions. The rI detector measures the bending of light between a reference cell (buffer only) and flow cell (SEC eluent). The differential refractive index detector will register a signal as the composition of the flow cell changes relative to the reference cell (Fig. 3.3). The detector can demonstrate small variations in refractive index due to concentration differences in dissolved gases, salts and particle concentration. For proteins and nucleic acids, the refractive index is nearly constant, irrespective of the primary, secondary or tertiary structure of the biopolymer. In contrast, UV absorption detection requires the appropriate chromophore to be present in the biopolymer whose extinction coefficient will vary with the hydrophobic environment of the chromophore. Accurate particle concentration is critical to SEC-MALS and refractive index detectors are the preferred method for concentration determination (Wyatt 1993; Tarazona and Saiz 2003).

## 3.3 Hidden Features (Known Unknowns)

The SEC elution profile of a sample can provide information regarding mass and conformational heterogeneity. Both mass and conformational heterogeneity, as well as particle-resin interaction, will cause an asymmetry in the SEC elution peak. However, the ability to discern these types of heterogeneities rely on how the chromatographic separation is being monitored (Brookes et al. 2013; Meisburger et al. 2016).

Since UV absorption ($A_{280}$) detectors mainly inform on particle concentration and light scattering based methods inform on mass, shape and concentration, an integrated approach that combines both types of measurements can provide a thorough characterization of the sample (Fig. 3.3a). The SEC elution begins with the column's void volume. The void volume is the volume of elution since sample injection that contains particles too large to be partitioned by the SEC resin. Large aggregates in a sample may go un-detected by UV methods as the concentrations may be too low for detection. However, scattering intensity is proportional to the squared mass suggesting MALS or SAXS will be most sensitive to the presence of large aggregates.

SEC-MALS analysis of dissolved glucose isomerase (GI) crystals (Fig. 3.3a) shows these

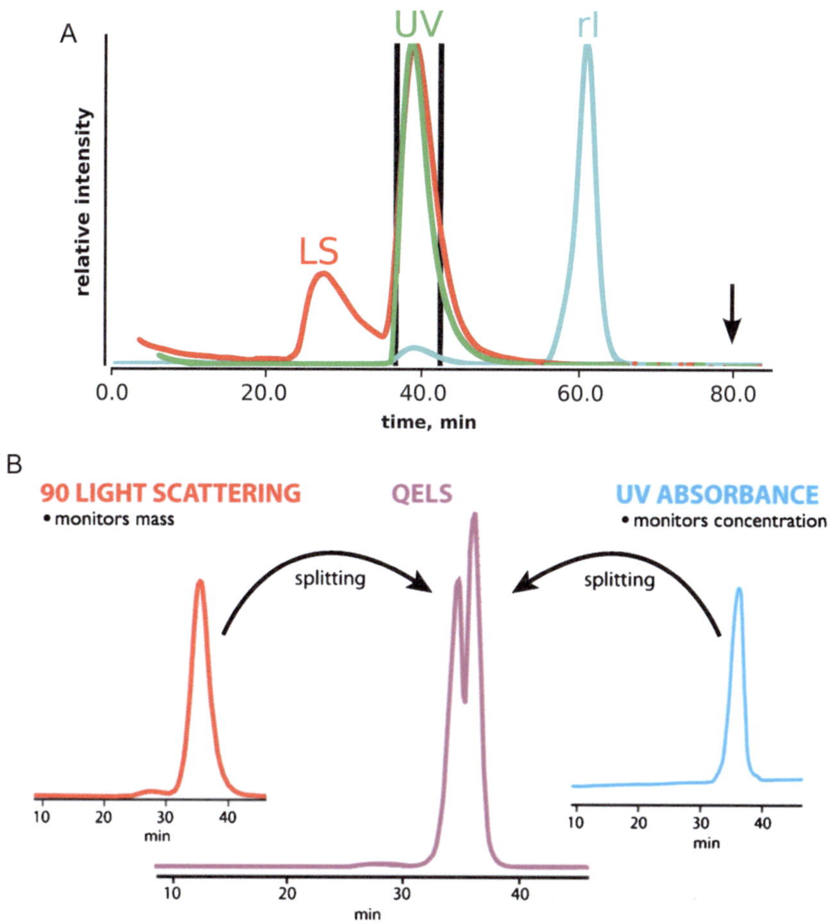


**Fig. 3.3** Light scattering, UV and refractive index detection. (**a**) Comparing all three signals can reveal hidden features of a sample. Light scattering (LS) intensity is directly proportional to the squared mass of a particle. The presence of a LS signal in the void (25 min) suggests the presence of large aggregates. In the SEC separation of dissolved glucose isomerase crystals, the samples contained significant amounts of aggregation that were not detectable by UV (flat *green* curve at 25 min). The dissolved crystals contained high concentrations of ammonium sulfate leading to a strong peak in the refractive index (rI) detector near the end of the column run (60 min) as the small molecules eluted off the column. It is recommended that buffers collected for SAXS be obtained at 1.5 column volumes (*black arrow*) where the eluent has stabilized. (**b**) Neither LS nor UV will be sensitive to particle conformation, subtle differences in a sample can be glimpsed by examining time-resolved LS measurements (QELS) across the elution peak. Here, QELS will be proportional to particle conformation and for the PYR1 protein, initial analysis of the protein always revealed a split QELS peak. After 2 weeks, the QELS peaks would resolve to a single peak and it was surmised the splitting was due to an isomerization of a proline residue

contrasting features between UV and light scattering. The injected sample demonstrates a large scattering peak at the void volume with essentially no UV signal. The lack of a notable UV signal near the void volume would give the false confidence that the sample was free of aggregation. Therefore, it is recommended that in the absence of an additional scattering detector, any observed deflection of the UV signal near the void volume should be considered significant. In these cases, if the sample is being prepared for SAXS, extensive centrifugation or the use of a spin-filter may be necessary to remove the aggregation prior to data collection (Hura et al. 2009).

Furthermore, the SEC-MALS analysis of GI utilized a refractive index detector. As discussed previously, the refractive index detector is sensitive to differences in the chemical composition of the running buffer as measured against the reference cell. Due to small differences in salt, glycerol or dissolved gases, it can be expected that the buffer composition of the injected sample will not be identical to the SEC running buffer thereby causing a notable signal in the refractive index detector near the end of the column run. Since these small molecules are invisible to UV absorption, the UV signal near the end of the column volume would appear flat giving the false impression that a background sample could be taken. While it is recommended that samples prepared for SAXS by SEC use the same running buffer as background, it is critical to the accuracy of the background subtraction that the buffer collected for the background measurement occur at least 1.5 column volumes after injection (Fig. 3.3a).

In the analysis of an SEC elution peak, the shape of the peak profile is the most informative method for indicating possible sample heterogeneity. Elution peaks that are asymmetric can suggest mass heterogeneity or particle-column interactions. However, conformational heterogeneity that is stable to partitioning can be more difficult to assess unless the structural differences are large enough to produce significant differences in the Stoke's radius of the different conformations. Here, monitoring the elution peak using QELS or by SAXS can provide additional information to determine the cause of the asymmetry in the elution peak (Fig. 3.3b). SEC-MALS with QELS studies performed on PYR1 (Nishimura et al. 2009), a 42 kDa protein, demonstrated the slightest asymmetry in the MALS and UV absorbance peaks. The MALS mass was consistent across the elution peak suggesting a homogenous sample; however, the QELS measurements showed a splitting of the peak (Fig. 3.3b) suggesting two distinct conformations were present in the single elution peak. SAXS data collected on the peak could not be fully explained by the crystal structure unless the model fitting was performed on the lagging side of the elution peak. In the absence of QELS or MALS information, a comparison of the individual SAXS frames from the leading and lagging sides of the elution profile must be inspected. At the very least, conformational heterogeneity would show the leading side to be larger than the lagging side in terms of $R_g$ and possibly $d_{max}$. Due to the thorough characterization of the sample, a multi-model fit would be necessitated to fully explain the SAXS curve.

A similar peak splitting was observed for GI (Fig. 3.4). GI was commonly used in the Tainer laboratory (Classen et al. 2013) as a mass standard for MALS calibrations. It was noticed that in moderately high pH and salt conditions, the QELS data would demonstrate a split peak. The peak splitting would disappear by lowering the salt concentration suggesting the conformational states of the protein could readily be influenced by adjusting the composition of the buffer. Similar observations were made for BSA where at pH >7.5 in PBS buffer, BSA would show a severe, asymmetric elution profile. Lowering the pH or by adding 1% sucrose to the buffer would stabilize the protein to partitioning producing the canonical monomer-dimer SEC profile of BSA.

## 3.4 Influence

The GI and PYR1 asymmetric elution peaks were due to conformational heterogeneity that could be influenced by the composition of the buffer. This type of conformational heterogeneity was stable to partitioning and produced the peak splitting in the QELS data. Nonetheless, heterogeneity can involve both conformation and mass. Mass heterogeneity is readily detected by MALS and will cause a negative slope in the mass distribution across the elution peak. The mass heterogeneity may be due to the particle in rapid equilibrium with higher order states or due to small truncations of component domains. If the heterogeneity is stable to partitioning, then the MALS data would demonstrate distinct steps in the mass distribution with distinct peaks in the

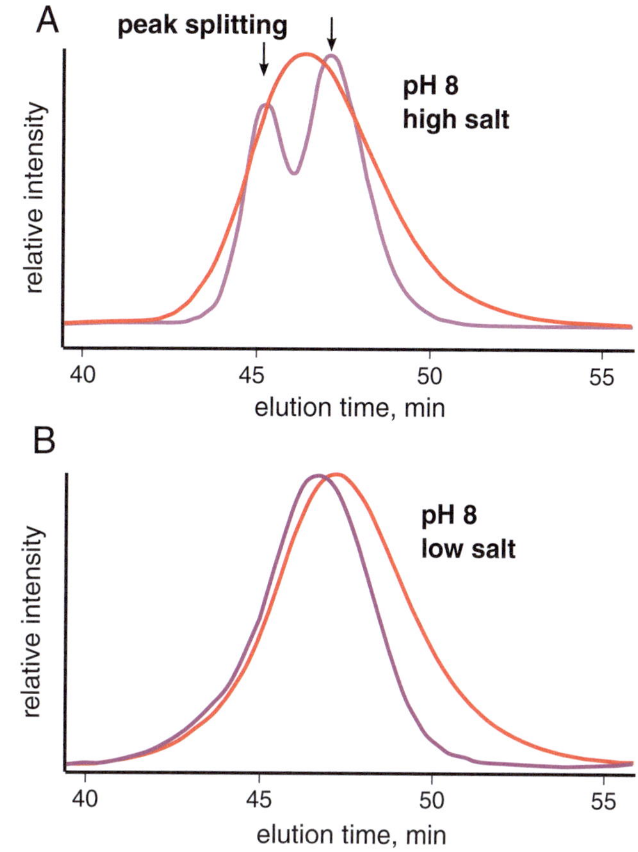


**Fig. 3.4** Conformational changes by QELS. Protein conformations are often controlled by the environment. At low pH, protonation of histidines residues change the charge distribution of the protein and can stabilize an alternate conformation. Similarly, salt can provide significant charge shielding to induce the same effects. (**a**) At pH 8 and high salt, it was observed that MALS (*red*) and QELS (*magenta*) analysis of glucose isomerase (GI) leads to a splitting of the QELS peak (*magenta*). (**b**) Resolution of split peak at lower ionic strengths. Close inspection of the MALS peak in high salt shows a slight asymmetry in the elution peak. Note that the QELS and LS signals are shifted for clarity

QELS profile (Fig. 3.5). SEC-MALS/QELS studies on a 185 kDa ATP motor protein showed such a profile. The QELS profile contained a leading, shoulder peak suggesting the elution peak contained at least two structurally distinct species. Further analysis of the MALS information showed the shoulder peak was ~22 kDa larger than the lagging side of the elution peak. We speculated the mass difference was due to limited proteolysis of the protein during purification. To test if the protein was responsive to ATP, the SEC-MALS/QELS was repeated with the protein incubated in ATP-vanadate. The vanadate locks down the protein in a phosphoryl-transfer transition state (Davies and Hol 2004) and for a motor protein, binding should demonstrate a notable conformational change. QELS results showed a decrease in the radius-of-hydration upon incubation suggesting the protein was competent to ATP binding and hydrolysis. While the sample is a mixture and remains unresolved during the SEC separation, SAXS data collected on the protein in the bound and unbound states would still be informative. In a SEC-SAXS experiment, a comparative SAXS analysis from the lagging side that uses the P(r)-distributions would characterize the conformational change in terms of compactness and dimensions. It can be expected that a measured decrease in $r_H$ would produce a notable decrease in $R_g$.

The 185 kDa protein demonstrated a compound heterogeneity involving both mass and conformation. The larger mass species was stable to partitioning by SEC thereby producing a distinct step in the MALS mass distribution across the elution peak. However, heterogeneity could be due to a rapid equilibrium between states such that the partitioning is characterized by a broad

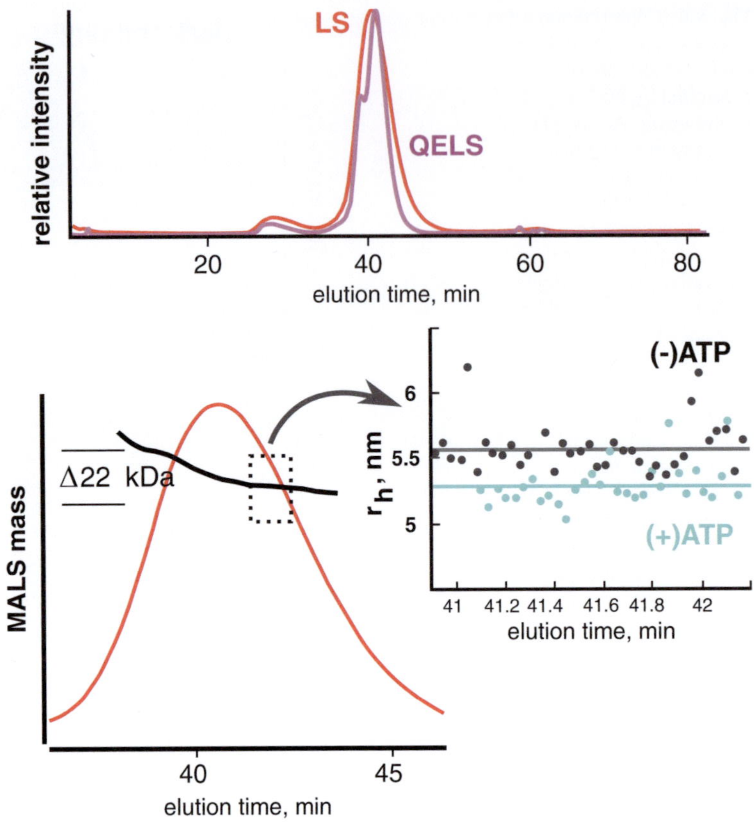


**Fig. 3.5** Stable mass heterogeneity. SEC-MALS/QELS analysis of a 185 kDa ATP binding protein reveal a leading shoulder in the QELS (*magenta*). The shoulder corresponded to a larger mass species by MALS (*black line, lower left panel*). To test if the protein was responsive to ATP, SEC-MALS/QELS was performed in the presence of 500 uM ATP-vanadate. The (+) ATP state showed a measurable and consistent decrease in radius-of-hydration ($r_H$) by QELS (*cyan*) indicating the protein undergoes as a significant conformational change. A small change in $r_H$ would translate into an observable change by SAXS

and asymmetric elution peak. MALS and QELS analysis will show a slope across the elution peak and likewise, SEC-SAXS would demonstrate a disagreement between the leading and lagging sides of the elution peak. This type of heterogeneity is particularly nefarious and suggests the biological particles are not stable to partitioning down the SEC column.

In macromolecular crystallography, conditions are sought to promote particle-particle interactions and often, the macromolecules are purified to high concentrations in a minimal buffer. These conditions may not be suitable for SAXS and can be the cause of the compound heterogeneity described above. Solution-state structural studies require buffer conditions that are stabilizing to the particle. For nucleic acid binding proteins, we have found phosphate and sucrose to be excellent additives that stabilize the protein while minimizing column-particle interactions. In most cases, nucleic acid binding proteins interact with nucleic acids through the sugar-phosphate backbone and in the absence of nucleic acids, these proteins may be charged imbalanced through the residues arginine and lysine. In the apo-state, the addition of 1% sucrose and phosphate can make a poor SAXS sample into an excellent, well-behaved SAXS sample.

The effects of additives must be evaluated using a suitable assay (Han et al. 2007; Leibly et al. 2012). If using SEC, the additives can be added to the running buffer, but this method will take hours per additive as the SEC column will have to be equilibrated for each additive. Another method for screening the effects of additives can be performed using micro-spin concentrators. The Tainer group had successfully solved the crystal structure of the exonuclease domain from the DNA repair protein WRN (Perry et al. 2006). The functional state of the domain in solution was unknown with some results suggesting the protein was a trimer

(Choi et al. 2007). The domain, as purified, would aggregate during spin-concentrating causing the flow-rate to be exceptionally slow with a significant loss of material. Concentrating the protein in high salt (1 M NaCl) failed to stabilize the protein leading to a large redistribution of the protein into the void volume (Fig. 3.6a). Therefore, we reasoned that if an additive was present during concentrating that could ameliorate the aggregation, then differences in flow-rates between additives during spin-concentrating would serve as the assay (Fig. 3.6b).

For the WRN exonuclease domain, aliquots of protein were mixed with various additives to 600 uL. The volume was transferred to a set of spin-concentrators where the weight of each tube was pre-recorded. Flow-rates were determined by weighing each tube at intervals of 5 min during the centrifugation. We found that 50 mM phosphate produced the fastest flow rate with 2% sucrose in second place. Some additives were too slow to be considered effective. These results suggested phosphate and sucrose could be stabilizing to the protein. To validate the stabilization, the protein was concentrated in a mixture of 50 mM phosphate and 1% sucrose and subject to SEC-MALS. Ideally, if a protein is concentrated tenfold, then it can be expected there would be a corresponding increase in the $A_{280}$ at the elution peak. In the absence of additives, concentrating the protein led to an $A_{280}$ peak in the void volume suggesting most of the protein was forming large aggregates. However, in the presence of phosphate and sucrose (Fig. 3.7), we see that concentrating the protein by 10× increased the $A_{280}$ nearly 10× with no increase in absorbance near the void. The stability was further demonstrated by concentrating the protein 20×. The MALS results showed the protein existed as a stable dimer and allowed for confident interpretation of the SAXS data collected from the peak.

Similar results were obtained with a small, 37.5 kDa RNA binding domain. The protein was purified for crystallization and concentrated in 5% glycerol. The glycerol was necessary to keep the protein "happy". However, SEC analysis reveal an asymmetric elution peak suggesting heterogeneity and further SAXS analysis of the SEC purified protein implied a protein with a volume of 99,687 $Å^3$, far too large for a monomer and far too small for a dimer. This ambiguity suggests the protein is a mixture of monomeric and dimeric states. We found the protein required 100 mM phosphate and 2% sucrose to be stable to SEC partitioning. QELS analysis in the phosphate-sucrose buffer showed an elution peak with a distinct leading shoulder (Fig. 3.8). SAXS of the main peak determined a particle volume of 84,000 $Å^3$ suggesting the heterogeneity of the sample could be influenced by additives. In both conditions, the SAXS sample displayed a distinct plateau in the Porod-Debye plot supporting the presence of a compact, well-folded particle (Rambo and Tainer 2011) but it was the discrepancy between the experimental volume and expected mass that confirmed the suspected heterogeneity. It can be expected that a buffer additive that alters the plateau region in the Porod-Debye plot will also effect the peak profile in a dimensionless Kratky plot.

We have found that additives such as pH, sucrose, sulfate, phosphate, ATP, GTP, proline, arginine or heparin-like sulfated carbohydrates to be effective additives to a wide range of proteins. Sulfate or sulfated-carbohydrates were effective in stabilizing extracellular matrix proteins whereas ATP/GTP-vanadate was important for specific motor proteins including dynamin and DNA repair proteins. If the protein of interest contains a Walker A/B motif, it may be important to lyse the cells in a high phosphate containing buffer as the phosphate ions may slow the release of nucleotide diphosphate (a simple application of Le Chatelier's principle).

## 3.5 Choosing the Right Column

SEC columns have a useful separation range that is described in terms of molecular weights. As mentioned previously, there can be a linear relationship between particle mass and Stoke's radius, but a proper description of separation range would be in terms of Stoke's radius. The ubiquitous Superdex 75, 200 and Superose

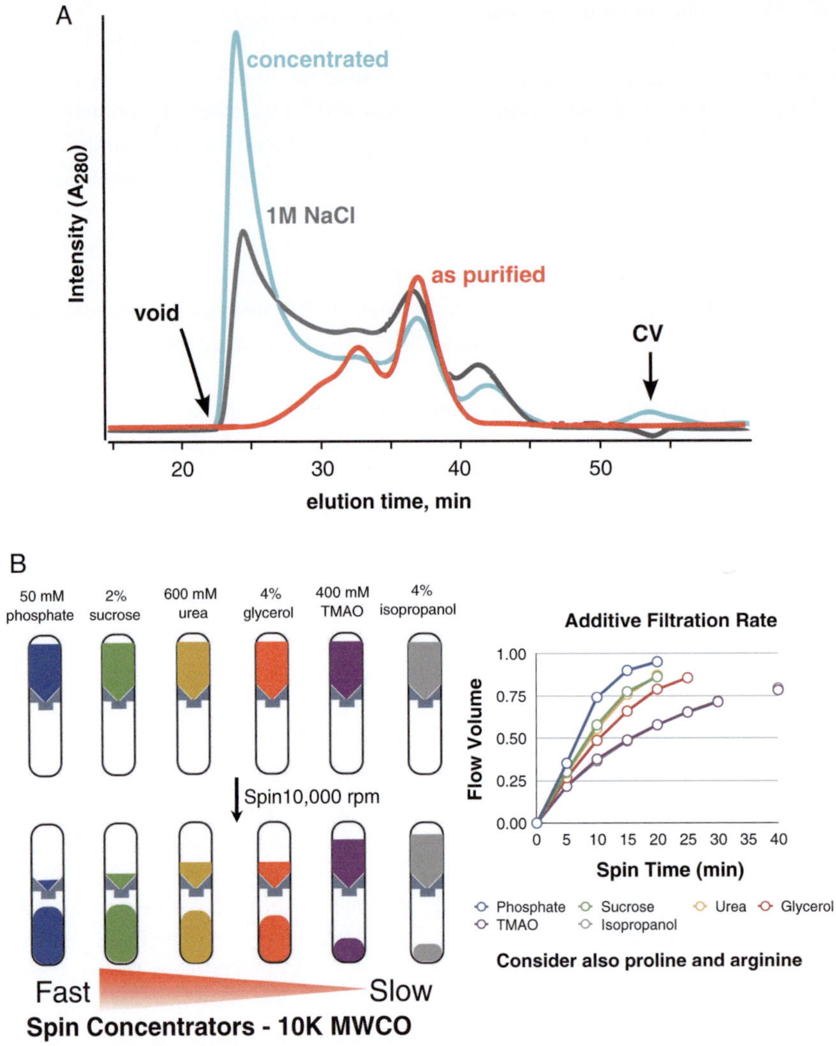


**Fig. 3.6** Sample instability during concentrating. Protein aggregation during concentrating is a routinely encountered problem and leads to exceedingly long concentrating times when using a spin-concentrator. The issue may not be relevant for most biochemistry experiments, however, it is a critical problem for SAXS. The WRN exonuclease domain was purified and concentrated in 20 mM Tris (7.6), 200 mM NaCl, and 5 mM beta-mercaptoethanol. (**a**) The protein was subjected to SEC analysis using a Superdex 200 PC 3.2 column in the same buffer (*red trace*). Upon concentrating the protein, the injected sample produced a significant UV signal in the void volume (*gray trace*). Concentrating the protein in buffer with 1 M NaCl increased the aggregation peak (*cyan trace*). (**b**) Additive screen using 10 K MWCO spin-concentrators. Various additives were added to the buffer and used to dilute the protein. For each additive, the filtration rate (volume of material that flowed through during centrifugation) was recorded at 5 min intervals. Samples contained phosphate (*blue*) and sucrose (*green*) had the fastest flow rates (Figure adapted from Kevin Dyer, Advanced Light Source, SIBYLS beamline, Berkeley, CA)

6 columns (GE Healthcare Life Sciences) have dominated SEC of biological macromolecules. These columns contain polymeric resins derived from cross-linked agarose and are chemically inert. The Superdex 75 is recommended for proteins less than 70 kDa, whereas the Superdex 200 is recommended for proteins less than 200 kDa and the Superose 6 is for complexes

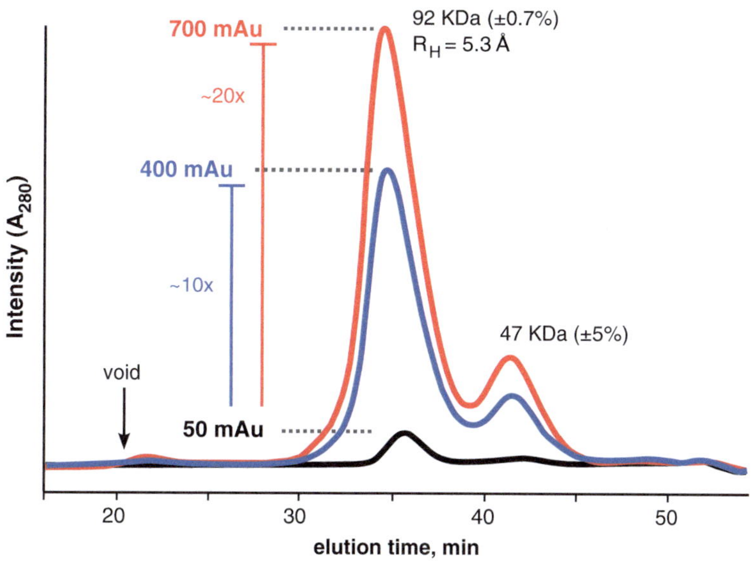


**Fig. 3.7** SEC-MALS analysis of WRN exonuclease in the presence of phosphate and sucrose. The $A_{280}$ signal for the unconcentrated (*black trace*) sample increase nearly tenfold as the initial sample volume was reduced by tenfold (*blue trace*). $A_{280}$ signal increased further (*red trace*) with a further reduction in sample volume. The peak in the void volume (*arrow*) was severely attenuated in the presence of the additives. MALS mass of the main peak (92 kDa) suggests the protein is a dimer with a monomeric mass of 47 kDa

that are less than 5,000 kDa. These resins are compressible and can experience pressure-induced degradation during the initial start-up of the HPLC system. The degradation leads to the loss of material from the column and if connected to a MALS instrument, there will be a large scattering signal during the beginning of the chromatographic run (Fig. 3.9). It is recommended that SEC columns are gradually brought to the operating flow-rate and that the operating flow-rate is maintained continuously during the experimental session.

Alternatively, there is another class of SEC columns growing in popularity. These are silica-based resins that use highly refined porous silica-beads. The beads can withstand greater operating pressures without sacrificing separation resolution but have a narrower operating pH range (pH <8.0). The KW-402.5, -403 and -404 columns (Shodex) offer similar separation ranges as the Superdex/Superose columns. However, the silica-based columns operate with a greater number of theoretical plates and can resolve smaller Stoke's radii differences (Fig. 3.10). The silica-based columns contain negatively-charged silanol groups and can interact with particles differently than the Superdex/Superose columns. It can be expected at low ionic strengths, these interactions may become more influential thus changing the elution characteristics of the column. Similarly, the agarose-based resins are sugars and for carbohydrate or nucleic acid binding proteins, low-ionic strength buffer conditions (<50 mM) may promote particle-column interactions causing a noticeable tailing in the elution peak.

For SEC-SAXS, the choice of column will be determined by the mass of the particle and initial purity of the sample. If the protein elutes too close to the end of the column volume, then there is the risk of poor background subtraction as the differences in small molecules from the injection elute at the same time from the column. Likewise, if the protein elutes too close to the void, then there is the risk of contaminating the SAXS signal with scattering from large aggregates. Therefore, the choice of column should place the particle of interest away from the void volume and the end of the column volume.

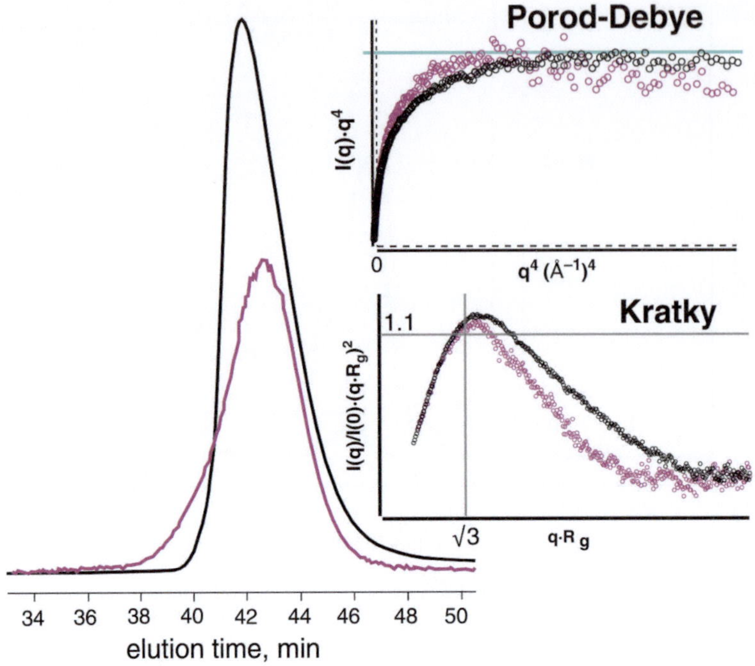


**Fig. 3.8** SEC analysis of a small RNA binding protein. Glycerol is a common reagent used to stabilize proteins against aggregation during freezing or concentrating. While the stability may inhibit material loss of the sample, the glycerol may not promote or inhibit non-ideal behaviour. SEC (*black trace*) and subsequent SAXS of the protein (*black circles*) in 5 % glycerol showed a protein with an asymmetric elution profile. SAXS data indicated the protein was compact with a discrete electron density contrast (plateau in the Porod-Debye plot, *cyan*). However, dimensionless Kratky plot showed a peak that was not globular. Globular proteins exhibit a peak at the Guinier-Kratky point (√3, 1.1). Purifying the protein in 100 mM phosphate and 2% sucrose (substituting glycerol), caused a notable shift in the SEC QELS peak (*magenta*). The peak shows a stable shoulder. SAXS analysis in the new condition revealed a stable Porod-Debye plateau and a shift of the SAXS peak towards the Guinier-Kratky point. The results demonstrate that the thermodynamic state of the protein can be modulated using additives

## 3.6 Summary

A great SAXS sample may make for a good MX sample but the reverse is not always true. Since crystallography optimizes for conditions that promote interparticle interactions, SAXS of samples prepared for crystallography must be assessed for unwanted interactions. These interactions can prevent interpretation of the solution state but can be readily attenuated using small molecule additives. In the RNA world, conditioning screening is employed early in a structural investigation as structured RNAs often require precise concentrations of divalent and monovalent metal ions (Rambo and Tainer 2010a, b; Reyes et al. 2014). Similarly, it is recommended that in the early stages of a SAXS investigation, that additive screening be explored for difficult samples as illustrated with the WRN exonuclease. The choice of buffer condition should be one that minimizes particle-column interactions while optimizing for stability.

SEC-coupled SAXS is available at most synchrotron facilities that focus on SAXS of biological samples in the solution-state (Bizien et al. 2016). These experiments may not be amenable to high-throughput SAXS but offer the most reliable method for collecting quality SAXS data. Since the measurement is under-flow, the resulting SAXS curve will be an accumulation of short exposures that may not be

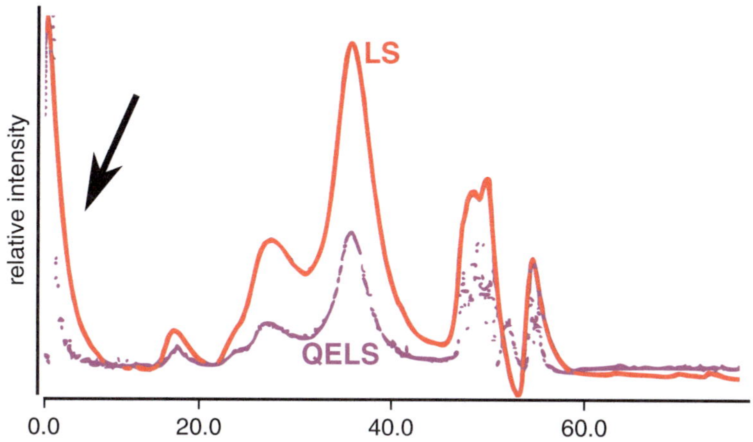


**Fig. 3.9** Rapid pressurization of an SEC column is damaging to the resin. Initial pressurization will degrade the resin and cause debris to elute from the column. This debris can cause considerable light scatter (LS) in the beginning of the column run (*black arrow*) and leads to an elevated baseline. We recommend starting a column at a low flow-rate and incrementing by doubling until the desired flow-rate is achieved. The operating flow-rate should be maintained in a continuous flow-mode until the experimental session ends

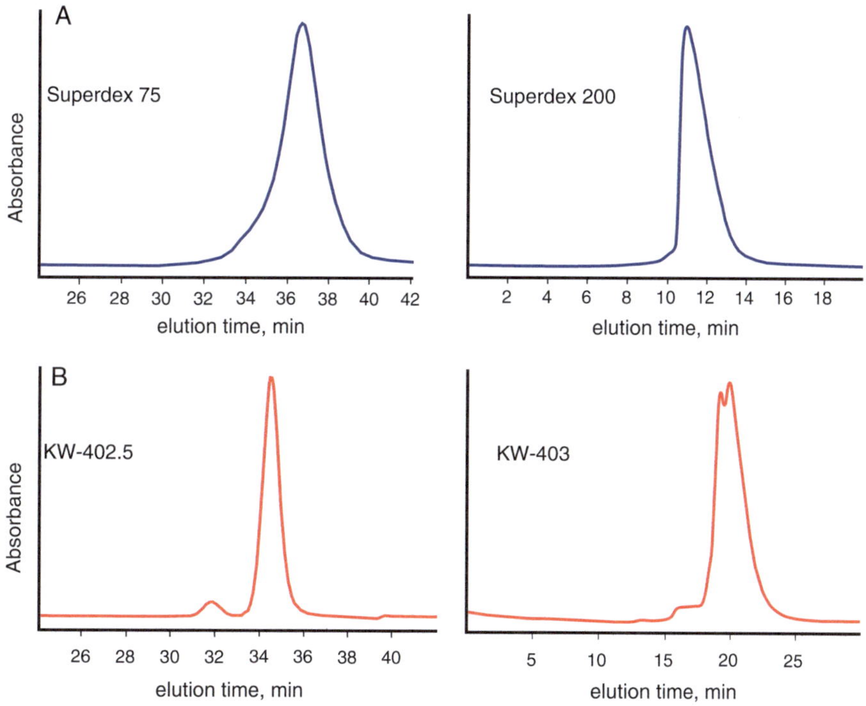


**Fig. 3.10** Comparison of SEC columns. (**a**) SEC analysis of xylanase (*left panel*) using Superdex 75 shows an elution peak with a leading shoulder and likewise, SEC analysis of a larger protein X (*right panel*) using Superdex 200 shows an asymmetric peak that leans towards the void volume. (**b**) SEC Analysis of the same sample on the same day using the Shodex columns shows resolution of the leading shoulder in the xylanase sample and partial resolution of protein X into two distinct peaks. Superdex columns use cross-linked agarose resins that are chemically robust but have fewer theoretical plates available for sample partitioning during SEC separation. For protein X, the peaks are not fully resolved and further analysis may require additive screening to promote a single state

sufficient to capture the SAXS curve at moderately high scattering vectors ($q > 0.2$ Å$^{-1}$). Repeated measurements of the same sample, slower flow-rates or static samples (batch) with increased exposure times would allow for data collection to higher $q$-values. If preparing samples for batch mode (PCR strips or 96-well plates), sample preparation using SEC is optimal but does not guarantee perfect background subtraction. As mentioned previously, collecting samples near the end of the column volume may lead to a buffer mismatch and purifications schemes should be employed that push the particle of interest away from the end of the column volume. More importantly, the buffer that is collected for the background measurement must be treated just as special as the sample containing the protein. Keeping the buffer at a different temperature or exposed to air while the protein sample is stored on ice will allow for different oxidation rates. These difference are noticeable in a reducing environment (DTT, TCEP, BME) and can be a major source of buffer mismatching.

**Acknowledgements** Diamond Light Source is supported by the Wellcome Trust and the Science and Technology Facilities Council (STFC), UK.

## References

Bizien T, Durand D, Roblina P, Thureau A, Vachette P, Perez J (2016) A Brief Survey of State-of-the-Art BioSAXS. Protein Pept Lett 23(3):217–231

Brookes E, Perez J, Cardinali B, Profumo A, Vachette P, Rocco M (2013) Fibrinogen species as resolved by HPLC-SAXS data processing within the UltraScan Solution Modeler (US-SOMO) enhanced SAS module. J Appl Crystallogr 46(Pt 6):1823–1833. doi:10.1107/S0021889813027751

Choi JM, Kang SY, Bae WJ, Jin KS, Ree M, Cho Y (2007) Probing the roles of active site residues in the 3′–5′ exonuclease of the Werner syndrome protein. J Biol Chem 282(13):9941–9951. doi:10.1074/jbc.M609657200

Classen S, Hura GL, Holton JM, Rambo RP, Rodic I, McGuire PJ, Dyer K, Hammel M, Meigs G, Frankel KA, Tainer JA (2013) Implementation and performance of SIBYLS: a dual endstation small-angle X-ray scattering and macromolecular crystallography beamline at the Advanced Light Source. J Appl Crystallogr 46(Pt 1):1–13. doi:10.1107/S0021889812048698

Davies DR, Hol WG (2004) The power of vanadate in crystallographic investigations of phosphoryl transfer enzymes. FEBS Lett 577(3):315–321. doi:10.1016/j.febslet.2004.10.022

Gillis RB, Rowe AJ, Adams GG, Harding SE (2014) A review of modern approaches to the hydrodynamic characterisation of polydisperse macromolecular systems in biotechnology. Biotechnol Genet Eng Rev 30(1–2):142–157. doi:10.1080/02648725.2014.994870

Han Y, Jin BS, Lee SB, Sohn Y, Joung JW, Lee JH (2007) Effects of sugar additives on protein stability of recombinant human serum albumin during lyophilization and storage. Arch Pharm Res 30(9):1124–1131

Henzler-Wildman KA, Lei M, Thai V, Kerns SJ, Karplus M, Kern D (2007) A hierarchy of timescales in protein dynamics is linked to enzyme catalysis. Nature 450(7171):913–916. doi:10.1038/nature06407

Hura GL, Menon AL, Hammel M, Rambo RP, Poole FL 2nd, Tsutakawa SE, Jenney FE Jr, Classen S, Frankel KA, Hopkins RC, Yang SJ, Scott JW, Dillard BD, Adams MW, Tainer JA (2009) Robust, high-throughput solution structural analyses by small angle X-ray scattering (SAXS). Nat Methods 6 (8):606–612. doi:10.1038/nmeth.1353

Jacques DA, Trewhella J (2010) Small-angle scattering for structural biology – expanding the frontier while avoiding the pitfalls. Protein Sci 19(4):642–657. doi:10.1002/pro.351

Jeffries CM, Graewert MA, Blanchet CE, Langley DB, Whitten AE, Svergun DI (2016) Preparing monodisperse macromolecular samples for successful biological small-angle X-ray and neutron-scattering experiments. Nat Protoc 11(11):2122–2153. doi:10.1038/nprot.2016.113

Leibly DJ, Nguyen TN, Kao LT, Hewitt SN, Barrett LK, Van Voorhis WC (2012) Stabilizing additives added during cell lysis aid in the solubilization of recombinant proteins. PLoS One 7(12):e52482. doi:10.1371/journal.pone.0052482

Meisburger SP, Taylor AB, Khan CA, Zhang S, Fitzpatrick PF, Ando N (2016) Domain movements upon activation of phenylalanine hydroxylase characterized by crystallography and chromatography-coupled small-angle X-ray scattering. J Am Chem Soc 138(20):6506–6516. doi:10.1021/jacs.6b01563

Minton AP (2016) Recent applications of light scattering measurement in the biological and biopharmaceutical sciences. Anal Biochem 501:4–22. doi:10.1016/j.ab.2016.02.007

Nishimura N, Hitomi K, Arvai AS, Rambo RP, Hitomi C, Cutler SR, Schroeder JI, Getzoff ED (2009) Structural mechanism of abscisic acid binding and signaling by dimeric PYR1. Science 326(5958):1373–1379. doi:10.1126/science.1181829

Perez J, Nishino Y (2012) Advances in X-ray scattering: from solution SAXS to achievements with coherent beams. Curr Opin Struct Biol 22(5):670–678. doi:10.1016/j.sbi.2012.07.014

Perry JJ, Yannone SM, Holden LG, Hitomi C, Asaithamby A, Han S, Cooper PK, Chen DJ, Tainer JA (2006) WRN exonuclease structure and molecular mechanism imply an editing role in DNA end processing. Nat Struct Mol Biol 13(5):414–422. doi:10.1038/nsmb1088

Rambo RP, Tainer JA (2010a) Bridging the solution divide: comprehensive structural analyses of dynamic RNA, DNA, and protein assemblies by small-angle X-ray scattering. Curr Opin Struct Biol 20 (1):128–137. doi:10.1016/j.sbi.2009.12.015

Rambo RP, Tainer JA (2010b) Improving small-angle X-ray scattering data for structural analyses of the RNA world. RNA 16(3):638–646. doi:10.1261/rna.1946310

Rambo RP, Tainer JA (2011) Characterizing flexible and intrinsically unstructured biological macromolecules by SAS using the Porod-Debye law. Biopolymers 95 (8):559–571. doi:10.1002/bip.21638

Rambo RP, Tainer JA (2013) Super-resolution in solution X-ray scattering and its applications to structural systems biology. Annu Rev Biophys 42:415–441. doi:10.1146/annurev-biophys-083012-130301

Reyes FE, Schwartz CR, Tainer JA, Rambo RP (2014) Methods for using new conceptual tools and parameters to assess RNA structure by small-angle X-ray scattering. Methods Enzymol 549:235–263. doi:10.1016/B978-0-12-801122-5.00011-8

Tarazona MP, Saiz E (2003) Combination of SEC/MALS experimental procedures and theoretical analysis for studying the solution properties of macromolecules. J Biochem Biophys Methods 56(1–3):95–116

Wyatt PJ (1993) Light scattering and the absolute characterization of macromolecules. Analytica Chimica Acta 272(1):1–40:http://dx.doi.org/10.1016/0003-2670(93)80373-S

# How to Analyze and Present SAS Data for Publication

# 4

Martha Brennich, Petra Pernot, and Adam Round

**Abstract**

SAS is a powerful technique to investigate oligomeric state and domain organization of macromolecules, e.g. proteins and nucleic acids, under physiological, functional and even time resolved conditions. However, reconstructing three dimensional structures from SAS data is inherently ambiguous, as no information about orientation and phase is available. In addition experimental artifacts such as radiation damage, concentration effects and incorrect background subtraction can hinder the interpretation of even lead to wrong results. In this chapter, explanations on how to analyze data and how to assess and minimize the influence of experimental artifacts on the data. Furthermore, guidelines on how to present the resulting data and models to demonstrate the data supports the conclusion being made and that it is not biased by artifacts, will be given.

**Keywords**

SAS • X-rays • Neutrons • Contrast • Solution scattering • SEC • BioSAS

## 4.1 Introduction

SAS is not a new technique, the first experiments date back to the 1930s (Guinier 1938), and the technique has been applied to biological macromolecules early on (Hosemann 1939). In recent years the combination of advances in sample production, high flux (X-ray and neutron) sources with rapid access to automated systems and advanced modeling (taking advantage of

M. Brennich (✉)
European Molecular Biology Laboratory, Grenoble, 71 avenue des Martyrs, 38042 Grenoble, France
e-mail: rejma@esrf.fr

P. Pernot
Structural Biology Group, European Synchrotron Radiation Facility, Grenoble, 71 avenue des Martyrs, 38042 Grenoble, France

A. Round
SPB/SFX instrument, European XFEL GmbH, Holzkoppel 4, 22869 Schenefeld, Germany

Faculty of Natural Sciences, Keele University, Keele, Staffordshire ST5 5BG, UK





B. Chaudhuri et al. (eds.), *Biological Small Angle Scattering: Techniques, Strategies and Tips*, Advances in Experimental Medicine and Biology 1009, DOI 10.1007/978-981-10-6038-0_4

modern computing) has made BioSAS a valuable tool for structural biologists.

SAS experiments on biological macromolecules in solution (BioSAS) using both neutrons (SANS) or X-rays (SAXS) provide information on the size and shape of the scattering object. Using Guinier's law (Guinier 1938) the Radius of gyration ($R_g$), a measure of the overall size, can be determined together with the forward scattering intensity ($I_0$), which is proportional to the molecular mass and the macromolecular concentration. Additionally the hydrated volume of the scatterer can be determined using Porod's law (Porod 1982) and the maximum dimension ($D_{max}$) within the scatterer can be estimated through the process of the inverse Fourier transformation (Glatter 1977; Svergun 1992).

As proteins in solution are mobile, all orientations are possible, and the SAS signal only contains orientation-averaged information. Combined with the intrinsic lack of phase information (only intensities can be measured), direct shape reconstruction by inverse Fourier transform is impossible and indirect shape reconstruction by model generation is by nature ambiguous.

Furthermore, the shape (form factor) of the particles of interest, scattering events between particles (structure factor) as well as scattering of the buffer, the sample holder and parasitic scattering of the instrument used contribute to the measured signal. Thus the first step required in data processing is data reduction from a raw 2D pattern to an idealized, artifact-free scattering curve representing the investigated particle only.

In order to aid those wishing to exploit BioSAS experiments this chapter covers the necessary data analysis steps for processing and interpretation of SAS data as well as instructions on how to present data for publication.

## 4.2 Data Reduction with Examples of Common Pitfalls

Data reduction is not a separate part of the experiment which starts once the data acquisition is complete, but an integral part of the data collection as preliminary results of data reduction and analysis provide valuable feedback on data quality. A typical BioSAXS experimental set-up is presented in Fig. 4.1 (Pernot et al. 2013). Many

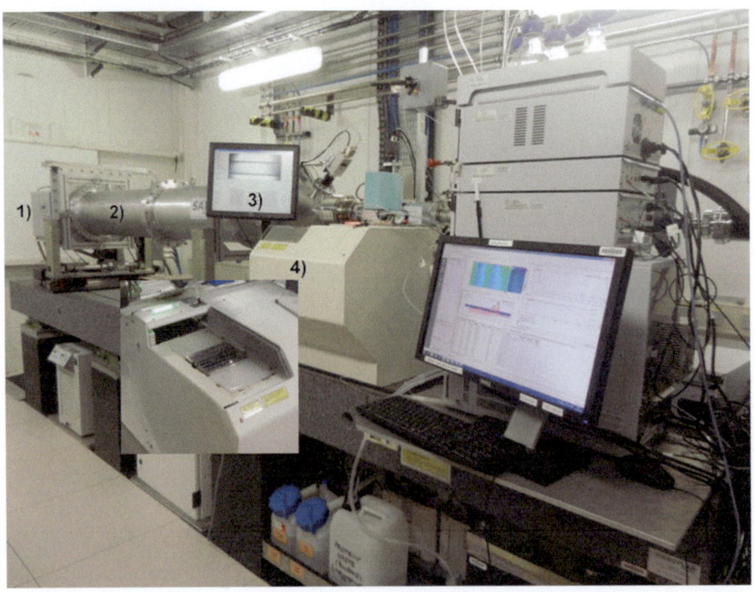


**Fig. 4.1** Experimental setup at the ESRF BioSAXS beamline BM29. This experimental facility is dedicated to SAXS measurements of samples in solution offering both Static (batch) operation and online SEC measurements (both HPLC and FPLC). X-ray scattering images are acquired using a Pilatus 1 M detector *1* Air scattering is avoided by using an evacuated flight tube *2* A touch screen monitor *3* allows easy control of the dedicated sample changer *4* Inset photograph shows the sample changer from the top with sample storage opened

BioSAS instruments, have adopted automated approaches to data reduction as well as preliminary analysis (Brennich et al. 2016; Franke et al. 2012). These tools provide background corrected scattering curves and the useful invariants ($R_g$, $I_0$, Porod Volume and $D_{max}$), which give valuable feedback regarding the sample behavior and data quality (Figs. 4.2, 4.3, and 4.4).

### 4.2.1 Azimuthal Integration

At most modern BioSAS facilities, the scattering signal is detected with area detectors in order detect as many of the scattered photons or neutrons as possible, resulting in 2D scattering images. As in general the scattering of randomly oriented particles in solution is isotropic, these images can be reduced to 1D curves without any loss of information by azimuthal integration (Fig. 4.2a–c). The following information is necessary:

- Type of detector used (its pixel size and geometry)
- Sample-to-detector distance
- Photon or neutron energy
- Direct beam coordinates on the detector
- "integration mask" which lists all pixels to ignore in the integration process (e.g. those hidden by the beamstop, etc.)

A variety of azimuthal integration software packages exist for different operating systems and data formats (Ashiotis et al. 2015; Benecke et al. 2014; Rodriguez-Navarro 2006; Hammersley 1997). The reader can find a listing of commonly used software at http://smallangle.org/content/Software#Reduction-Visualisation. Integrators suited for SANS data also take the distortion of the 2D data due to gravity into account. At most neutron and X-ray facilities suitable (sometimes even automated) software and guidance on how to use it are available.

The results of azimuthal integration are (in some cases already normalized) intensity and its standard deviation versus the scattering vector. Conventions on the scattering vector differ, and it is important to note its units ($nm^{-1}$ or $Å^{-1}$) and whether the scattering vector is equal to $4\pi \sin \theta/\lambda$ or $2 \sin \theta/\lambda$. When comparing data from different instruments, it might be necessary to convert them according to conventions used.

Artifacts appearing after azimuthal integration can be caused by many factors, such as incorrect masking (typically concerns pixels close to the direct beam), integrating anisotropic patterns, 'crazy' pixels (Fig. 4.2d) etc. All such factors, which affect the data, must be identified and corrected before further data processing.

### 4.2.2 Normalization

The normalization of intensities from set-up dependent arbitrary units ($I_{arb}$) to absolute units ($I_{abs}$) by multiplication with a calibration factor is necessary in order to calculate the sample mass from the forward scattering and to correctly compare between different setups. This calibration factor can be determined by measuring the SAXS signal $I_{st}(q)$ of a calibration standard, such as glassy carbon or water, and the conversion to absolute units is given by

$$I_{abs}(q) = \left(\frac{\partial \Sigma}{\partial \Omega}\right)(q) \frac{T_{st} d_{st}}{T_s d_s} \frac{I_{arb}(q)}{I_{st}(q)}$$

where $\left(\frac{\partial \Sigma}{\partial \Omega}\right)(q)$ is the known scattering intensity of the standard, $T_{st}$ and $T_s$ are the X-ray transmissions and $d_{st}$ and $d_s$ the thickness (X-ray path length) of the standard and the sample, respectively (Brian Richard 2013).

Generally, the $q$-dependence of the calibration factor is negligibly small. Therefore it can also be determined from the forward scattering of well-behaved proteins, such as β-amylase from sweet potato. In case of water the flat scattering intensity of 0.01632 $cm^{-1}$ at 20 °C and atmospheric pressure, can be measured at higher angles (Orthaber et al. 2000) as shown in Fig. 4.2e. For protein samples, it can be useful to provide the scattering intensity in units of kDa instead of $cm^{-1}$. The constant scattering intensity of water at 20 °C when scaled for a standard protein

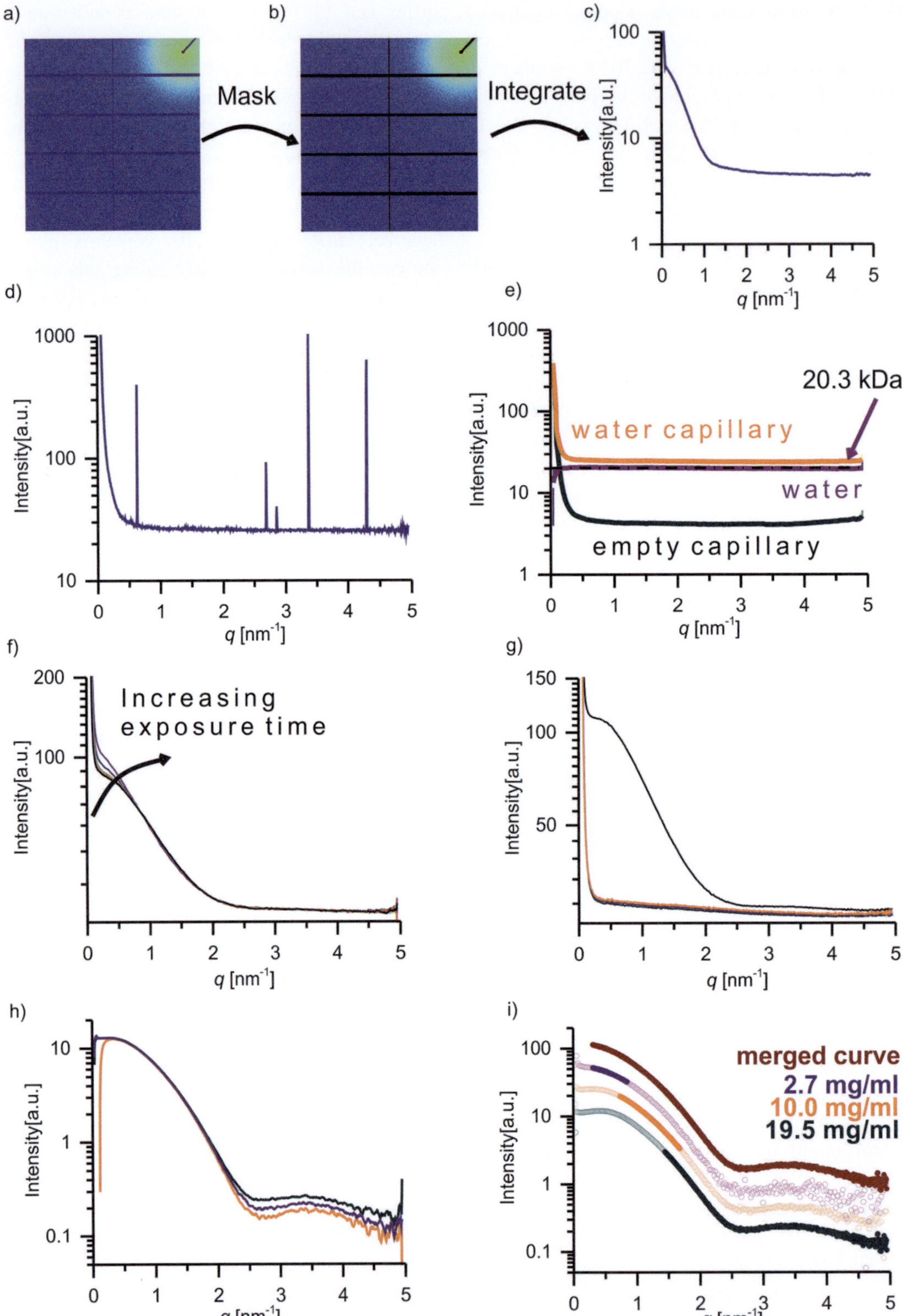


**Fig. 4.2** Data reduction from images to an idealized curve. (**a**) A raw image from a 2D detector. (**b**) The blacked out regions, corresponding to gaps between detector modules, the beamstop, strong parasitic scattering, hot pixels, etc. are neglected ("masked") in further processing. (**c**) Azimuthal integration around the beam

in buffer corresponds to 20.3 kDa (Mylonas and Svergun 2007). Lipids, nucleic acids and protein complexes due to the difference in contrast require modification to this scaling factor, which is dependent on the ratio of protein, lipid and nucleic acid in the investigated particle (see "Guinier approximation" Sect. 3.1 for more details). In practice, different facilities use different calibration methods and one should verify the method and units of the intensity normalization.

### 4.2.3 Averaging of Multiple Frames from Each Sample

Data acquisition is usually split up into several sub-exposures in order to detect (possible) sample degradation corrupting the signal. Especially when using X-rays, radiation often induces sample degradation (radiation damage). To minimize the effect of radiation damage, the sample is typically moved during data acquisition, with the aim that every sub-exposure can be taken on fresh sample. If the sample suffers radiation damage, contains air bubbles or is inhomogeneous, these sub-exposures will not give identical scattering patterns. Therefore, it is necessary to control for outliers. In many cases, this can be done by a qualitative visual control, but in more subtle cases statistical tests such as CORMAP from the ATSAS package need to be used (Franke et al. 2015; Petoukhov et al. 2007). Figure 4.2f shows a case of obvious radiation damage, with at least two frames showing clear radiation induced aggregation. More careful analysis with CORMAP revealed that first systematic changes already occurred as early as in the fifth frame. Only artifact free sub-exposures should be averaged for a better signal to noise ratio and used for subsequent data processing.

### 4.2.4 Background Subtraction for Individual Concentrations

A SAS curve free of contributions of the buffer and set-up can be obtained by simply subtraction of a corresponding buffer measurement from the averaged sample curve. Ideally, the buffer should have been measured in the same sample environment and exposure parameters before and after the sample measurement. If the scattering from these two measurements is not identical (as tested for by e.g. CORMAP (Franke et al. 2015)), the reason has to be investigated (for example inadequate post sample cleaning of the exposure cell, inhomogeneities in the buffer, etc.). In the ideal case both buffer measurements will match and can be averaged (to improve signal to noise ratio) and subtracted from the scattering curve of the sample. In addition, the resulting subtracted curve is often normalized (divided) to the sample concentration (if known).

At this stage, it is necessary to check whether the buffer measurement matches the sample (Jacques and Trewhella 2010). Differences in the chemical composition of buffer and sample affect the transmission of X-rays and thereby the

**Fig. 4.2** (continued) center provides the 1D SAXS curve. (**d**) Failure to mask hot pixels results in characteristic "spikes" in the 1D curve. (**e**) Calibration to absolute units is performed by subtraction of a measurement of an empty capillary (*green*) from that of a water filled capillary (*orange*). The resulting constant signal (*violet*) corresponds to 20.3 kDa or 0.0163 $cm^{-1}$ (when correctly scaled). (**f**) Radiation damage usually results in continuous, systematic changes in signal. Sub-exposures need to be controlled for its onset and affected frames are discarded. (**g, h**) Subtracting a buffer containing too much (*orange*) or too little (*green*) salt mostly affects the high $q$ and very low $q$ regions, as seen in comparison to the ideal subtraction (*violet*). Artefacts like that of the orange curve at very small $q$ are clear warnings. (**i**) To eliminate concentration artefacts while maintaining a good signal to noise ratio, data from different concentrations can be combined. The ideal curve (*red*) was constructed using the filled symbols of the three shown concentrations, while the data corresponding to the open symbols was only used for consistency checks

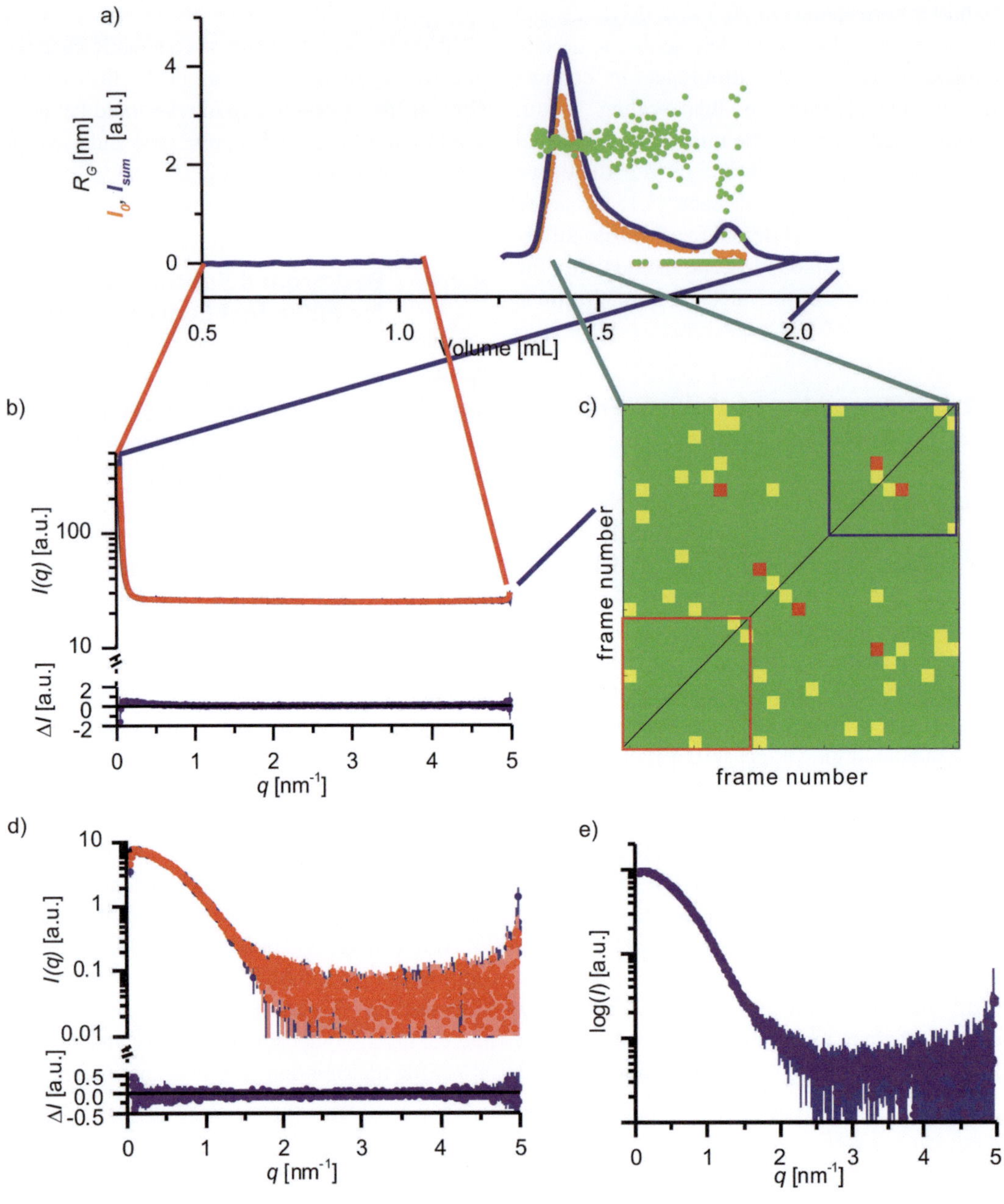


**Fig. 4.3** Data reduction for online SEC-SAXS. (**a**) SEC-SAXS chromatogram presents the total scattering intensity (*blue*) for all frames of the peak and forward scattering (*orange*) and radius of gyration (*green*) for all sample frames. (**b**) Comparison of the buffer collected before (*red*, *upper part*) and after (*blue*, *upper part*) the sample peak. Their difference (*violet*, *lower part*) confirms that the buffer signal stays constant. (**c**) Results of the DATCMP tests for frames in the region of interest. *Green squares* stand for matching frames ($p \geq 0.9$), *red squares* correspond to non-matching frames ($p \leq 0.1$), and *yellow squares* represents all other cases. The *blue* and *red lines* limit the regions compared in Fig. 4.3d). (**d**) Comparison of frames collected at the beginning of the sample peak (*red*, *upper part*) and the end of the sample peak (*blue*, *upper part*). Their difference (*violet, lower part*) confirms that the signal stays constant throughout the peak. (**e**) Final, averaged SAXS curve

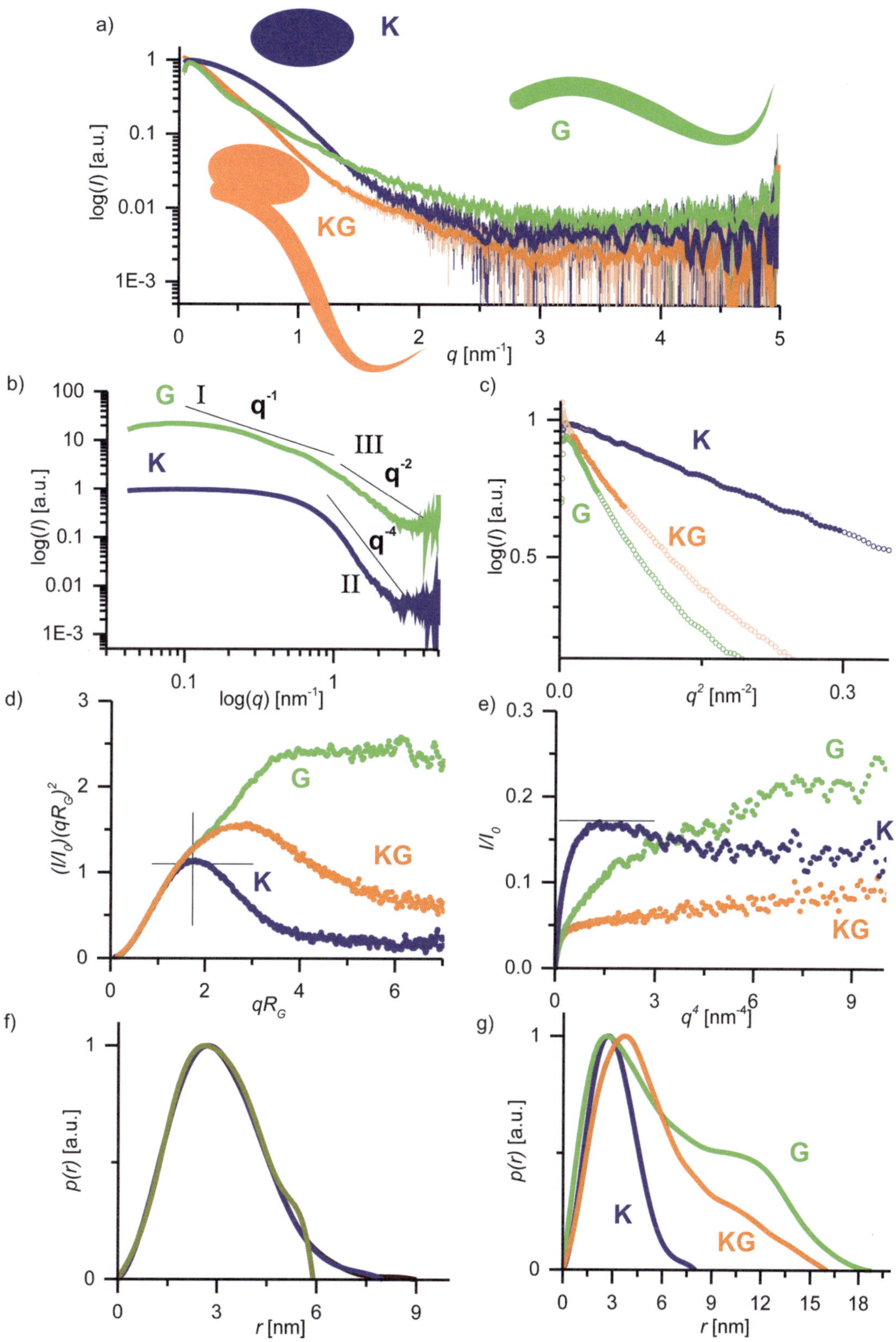


**Fig. 4.4** Primary data analysis of a complex. (**a**) Background corrected SAXS curves of the proteins K, G and their complex KG in semi-logarithmic representation. The curve corresponding to K is flatter at small angles than the curves of G and KG, indicating its smaller size. The curve corresponding to G decays more slowly than K

scaling of the scattering curve. In many cases, the sample contributes only little to the scattering at high angles (scattering vector above 4 $nm^{-1}$), so mismatches can often be identified in this region. Indicators for buffer mismatch are non-matching scattering curves at higher angles, systematically negative regions in the subtracted curve, deviations from Porod's law or differences at high angles between different sample concentrations. However, some of these indicators can also occur for small scatterers at high concentrations and non-globular samples such as intrinsically disordered proteins (IDPs) (Bernado and Svergun 2012).

Figure 4.2g, h show how subtraction of the wrong buffer affects the scattering curve. For both over-subtraction (orange) and under-subtraction (green) the scattering at high angles deviates from the matching case (magenta). In the case of the buffer over-subtraction, the difference in the relative contribution of the capillary scattering additionally results in a sharp downturn of the curve at small angles, which can even affect the determination of the radius of gyration.

### 4.2.5 Merging of Different Concentrations, Extrapolation to Zero-Concentration

BioSAS experiments are almost exclusively carried out with the aim to determine the form factor of scattering particles. However, only the combination of form factor and structure factor is experimentally accessible. Conveniently, the structure factor depends, in contrast to the form factor, on the particle concentration and becomes negligible at sufficiently low concentrations where inter-particle distances are sufficiently large to prevent interactions (Bonneté et al. 1999). Therefore, the effect of the structure factor in BioSAS measurements can be minimized by measuring samples in dilute conditions and crosschecking at multiple concentrations. In addition, other concentration dependent artifacts such as aggregation also diminish at lower concentration.

A concentration which can be assumed to be free of interparticle effects needs to fulfill the following criteria:

**Fig. 4.4** (continued) and KG at high angles, suggesting a higher degree of disorder present. (**b**) SAXS curves of the proteins K and G in double-logarithmic representation. G curve follows a $q^{-1}$ trend at small angles (region I) which is a signature of an elongated form. At high angles, G curves follows a $q^{-2}$ trend (region III), indication of disordered regions. In contrast K curve follows a $q^{-4}$ trend (region II) as expected in the case of a globular protein. (**c**) Guinier plots of proteins K and G and KG complex, with the Guinier region indicated by closed symbols. The small protein K has the longest Guinier region. Increased radius of gyration and anisotropy decrease the length of the Guinier region as observed for protein G. (**d**) Normalized Kratky plot of proteins K and G and KG complex. The symmetric peak of protein K is found at the position predicted for globular proteins. In the case of the KG complex the peak is shifted upwards and right indicating an anisotropic shape. The decay to zero is flatter than for K, suggesting presence of disordered regions. Due to the fact that the area under the peak is not clearly defined, the Porod volume is not reliable. The curve corresponding to protein G has no peak at all but only a plateau, a signature of a high degree of flexibility. The strong up-right shift additionally indicates anisotropy. (**e**) Porod plot of proteins K and G and KG complex. The curve corresponding to globular protein K displays a clear plateau, whereas the signal for the more flexible complex KG and protein G continuously increases. (**f**) Pair distribution $p(r)$ curve of protein K determined using different values of $D_{max}$. A too small $D_{max}$ value results in an abrupt turn of the curve to 0 (*dark yellow curve*). A too large $D_{max}$ (*red curve*) value used for $p(r)$ calculation, results in a long trailing tail close to 0. The curve calculated with $D_{max}$ chosen for further analysis (*dashed blue curve*) approaches 0 gently. (**g**)$p(r)$ curves for proteins K and G and complex KG. The $p(r)$ the small globular K is nearly symmetric. In the case of the more anisotropic KG complex and protein G, the position of the main peak barely moves but the curves becomes more asymmetric and the $D_{max}$ increases. The additional peak at about 12 nm in the curve for G indicates the presence of spatially separated sub-domains

- The signal (subtracted curve) is identical for lower concentrations
- The data at low angles fulfil the Guinier approximation well
- The radii of gyration determined *via* the Guinier approximation and *via* the pair distribution function respectively match each other.

If the signal to noise ratio at this concentration is sufficiently good even at higher angles, it can be used for all further analysis. Otherwise, it is necessary to include data from higher concentrations. For this, one first identifies the point from which on differences between the concentrations are only due to noise, then scales the concentrations to each other in an overlap region and takes the lower angle data from the lower concentration and the higher angle data from the higher concentration. Figure 4.2i presents the creation of an idealized curve based on three concentrations (2.7 mg/ml, 10 mg/ml and 19.5 mg/ml). The regions of each curve used for building the idealized curve are represented by the filled symbols, whereas data represented by open symbols was not included. In the experiment even lower concentration data was collected, however no significant difference to the data at concentration equal to 2.7 mg/ml was found. All initial data points that do not follow Guinier's law are removed for the creation of the idealized curve. At 10 mg/ml and 19.5 mg/ml one can observe strong contributions from inter-particle scattering, which have to be removed before further analysis. The affected regions can be identified by comparison to the next lower concentration. Points at low angles that show significant differences to the data collected at lower concentrations are ignored. The higher $q$ end of the regions that contribute to the idealized curve follows directly from the lower $q$ end of the next higher region and extends just a few data points beyond it. Alternatively, many SAS packages provide routines which allow automatic extrapolation of all measured concentrations to an idealized "concentration zero" curve (Franke et al. 2012).

The result of these approaches is an idealized curve, which should be free from any inter-particle artifacts. This idealized curve is used for further analysis and modeling but the degree of variation observed in the different concentrations should be kept in mind with regards to the confidence in the interpretation.

### 4.2.6 Background Subtraction and Averaging for SEC-SAXS Experiments

A SEC-SAXS experiment typically consists of several hundreds to thousand acquired frames either continuously throughout the elution process or only in regions of interest, e.g. buffer before the elution of the sample from the column, the sample itself and buffer again after the elution of the sample (Watanabe and Inoko 2009; Round et al. 2013; Mathew et al. 2004; Lambright et al. 2013; Grant et al. 2011; Graewert et al. 2015; David and Pérez 2009).

Before further processing, the stability of the buffer baseline needs to be confirmed by comparison of individual measurements. The following effects can negatively affect the baseline:

- Column not completely equilibrated
- Slow drift in the experimental setup
- Sample eluting in column void volume
- Mismatch between running buffer and sample buffer
- Spoiling of the sample environment by additives (or contaminants) in the buffer (typically radiation damage of the buffer and its deposition on walls of the sample exposure cell)
- Spoiling of the sample environment by the sample

Figure 4.3a shows a typical SEC-SAXS chromatogram, providing the total SAXS intensity for each frame as well as the radius of gyration and forward scattering from later processing steps. Between 1.1 and 1.25 mL the shutter was closed to avoid spoiling from the aggregate peak. Between 1.75 and 1.9 mL the total scattering increases due to excess salt injected with the sample (salt peak). When choosing a suitable buffer, one therefore needs to avoid these two regions. Comparing the average of the buffer frames acquired before the aggregate peak (Fig. 4.3b) shows that there is a slight mismatch at small angles, indicative of mild capillary spoiling.

If the buffer signals are matching, measurements can be averaged and subtracted from the individual sample measurements. In some cases, small changes can be interpolated to provide a suitable buffer subtraction for each sample measurement (Brookes et al. 2013). For each individual subtraction, one can then calculate the radius of gyration and forward scattering. If the sample concentration has been measured simultaneously or with a known delay, this information can be combined to estimate the molecular weight. In our example, due to the slight spoiling, we subtracted the buffer recorded before the aggregate peak and obtained the forward scattering and radius of gyration shown in Fig. 4.3a.

Although the aim of performing online size exclusion chromatography (SEC) on mixtures is to separate the different species before collecting SAS data, sometimes the peaks will elute too close to one another. This can lead to overlapping peaks and in these regions the data measured will represent the mixed scattering from the overlapping species, with the proportions contributed to the total observed scattering by each species changing with time. It may still be possible to find regions with only one species, corresponding measurements can be merged together to give the scattering for that species. To identify such regions, one first finds sufficiently large (in the range of one injection volume) regions of stable $R_g$. When such a region is identified, one verifies that the individual (suitably scaled) SAS curves match each other. In our example, a region of 25 frames was identified as potentially stable. To confirm this hypothesis, the subtracted SAXS curves were scaled and a pairwise comparison using CORMAP was performed. The results of the test are visualized in Fig. 4.3c, displaying clearly matching frames ($p \geq 0.9$) as green squares, non-matching frames ($p \leq 0.1$) as red squares and all others as yellow squares. Out of the 300 individual tests in this case, only two gave p-values smaller than 0.1. In the case of low concentration or high noise, it is further advisable, to compare the averages of different sub-regions. If all these tests confirm a stable signal, corresponding measurements can be averaged and used for further processing. A (rare) special case of this scenario appears when protein concentrations are high enough for inter-particle effects to cause a decrease of scattering at low angles. An approach similar to the one described in part 2.5 can be used to combine data from different parts of the chromatogram.

If no stable signal can be found, direct merging of the data is not valid, as the underlying hypothesis of homogeneity and purity does not hold for a mixture of species. Deconvolution using the assumption of overlapping Gaussian peaks can in some cases recover the scattering from the individual species (Brookes et al. 2013). However, it is recommended where possible to re-measure the sample using a better resolving column to separate the peaks experimentally.

## 4.3 1D Curve Analysis

### 4.3.1 Calculation of Model Independent Parameters

#### 4.3.1.1 Initial Assumptions

Interpretation of data from a SAS experiment gives average parameters of the scattering particles. Each model independent parameter provides a single number which is less informative if the sample is not monodisperse, i.e. a single oligomeric species in the same conformation. For mixtures, deconvolution of data to obtain individual curves for the constituents is only possible in special cases (Karlsen et al. 2015) and will not be treated here. Thus not only is validation required in sample preparation but cross checking the expected values with those observed using SAS is essential to avoid misinterpretation.

#### 4.3.1.2 Qualitative Analysis of SAS Curves

Even without any quantitative analysis it is often possible to extract information from SAS curves based on their shape. Figure 4.4a shows the SAXS curves of the proteins K, G and their complex KG, scaled such that their forward scattering matches. Looking at the very small angles, it is obvious that the curve corresponding to G is considerably steeper than those of K and KG, implying that the radius of gyration of this component is actually larger than that of the complex it forms with K. Double logarithmic representation of the results (Fig. 4.4b) highlights some more features: At low angles K flattens off very early whereas as G follows a $q^{-1}$ power law before leveling off (region I in Fig. 4.4b). This $q^{-1}$ behaviour is typical for elongated, rod-like particles (Glatter and Kratky 1982). At high angles, K follows a $q^{-4}$ power law, as expected for well-folded globular proteins (region II in Fig. 4.4b) (Porod 1982), whereas G only decreases as $q^{-2}$, indicating at least some extent of flexibility (region III in Fig. 4.4b) (Reyes et al. 2014; Debye 1947). Hence, even without any advanced analysis, one identifies K as a small globular protein and G as an elongated protein, with at least some highly flexible regions. It can be also noted that their complex is less anisotropic than G, as it seems to have a lower radius of gyration than G protein alone.

#### 4.3.1.3 Guinier Approximation

The SAS signal of any scatterer at small angles can be described by a Gaussian distribution (Guinier 1938)

$$I(q) = I_0 e^{-\frac{(qR_g)^2}{3}}$$

This allows determination of the forward scattering $I_0$ as well as the (average) radius of gyration of the scatterer. In the case of a mixture of similarly sized scatterers, $I_0 = \sum_n f_n I_{0_n}$ and $R_g^2 = \sum_n f_n R_{g_n}^2$, where $f_n$ is the faction, $I_{0_n}$ the forward scattering and $R_{g_n}$ the radius of gyration of the $n$th component, respectively (Segel et al. 1999). The Guinier approximation is only valid for small angles, and therefore when fitting $q \leq 1.3$ needs to be fulfilled for the fit region.

The forward scattering $I_0$ is proportional to the number of scatterers, the square of their mass and electron density (contrast) compared to the surrounding solvent, and more practically to their concentration (in mass/volume). Hence, if the concentration and chemical composition of the sample and buffer is known (including Hydrogen and Deuterium ratio for neutron scattering), it is possible to estimate its mass, and thereby its oligomeric state (assuming it is monodisperse), directly from the Guinier approximation. In the case of X-rays the proportionality factor for protein in water is 1.3 $10^3$ cm kDa, while for nucleic acids (DNA and RNA) it is 2.6 $10^3$ cm kDa due to their higher electron density and thus contrast $\Delta\rho$. For complexes it can be calculated as $N_A/(\Delta\rho v)^2$, where is $N_A$ Avogadro's number, $\Delta\rho$ the contrast and the $v$ partial specific volume.

The quality of the Guinier approximation is best examined in the Guinier plot, log $I$ vs $q^2$. A concave curve in this plot indicates the presence of larger scatterers, often aggregates, while a convex curve indicates repulsion between the scatterers.

If concentration-corrected data are scaled to kDa, the forward scattering is identical to the mass of the scatterer. However, some particles may have an inherently high degree of conformational flexibility. An important consequence of the resulting structural heterogeneity is that the movement of the subunits in relation to each other will not be synchronized across all particles in the X-ray beam. Moreover, it can be assumed that all possible relative positions and orientations will be sampled in the scattering data under the assumption of spherical averaging (all possible orientations are present). This gives rise to an increase in the average size of the scatterers and, moreover, to variation in the particle sizes. These effects cause a deviation from the linear expectation of Guinier's law and, as such, are practically indistinguishable in the 1D data from a small amount of aggregation. This artifact is unlikely to depend on concentration in the dilute concentrations used for SAS experiments. However, in some cases at high concentrations (>10 mg/mL), nearby particles can affect the flexibility (crowding effects), and a concentration dependence can be observed. A convex curve indicated the presence of inter-particle effects ("structure factor"), which typically show a strong concentration dependence and become negligible at sufficiently low concentrations.

Generally, points at very small angles will be ignored for the Guinier analysis. For further analysis, these points should be removed from the curve as they provide no additional information and are prone to be affected by artifacts.

Coming back to our example (Fig. 4.4c), G has the highest radius of gyration and K the lowest, KG one being a bit smaller than that of G. Accordingly, K has the longest Guinier region, going up to over 0.5 $nm^{-1}$. For G, the Guinier region is limited not only by its larger size, but also by its high degree of anisotropy.

For highly anisotropic particles, such as rods or disc additional forms of the Guinier approximation exist. The most relevant for biological macromolecules is the Guinier-approximation for rods, i.e. for particles whose long axis $L$ is much longer than its cross-sectional diameter (Glatter and Kratky 1982):

$I(q) = \frac{I_0}{q} e^{(qR_c)^2/2}$ for $qL \gg 1$ and $qR_C \leq 1.1$.

Analogously to globular particles, it can be used to derive the mass-per-length $M_L$ of a rod and the cross-sectional radius of gyration $R_c$. If the available $q$-range is large enough to determine both $R_g$ and $R_c$. of a macromolecule, its length can be estimated *via*

$$R_g^2 = R_c^2 + \frac{L^2}{12}$$

#### 4.3.1.4 Qualitative Flexibility Analysis

In contrast to the small angle region that only depends on the particles overall size, the high angle region, which corresponds to small distances in real space, corresponds to the flexibility. In this region, unfolded or dis-ordered proteins scatter more strongly than globular proteins of the same size.

These differences can be clearly seen when the data is plotted appropriately: In the normalized Kratky plot $(qR_g)^2\ I(q)/I_0$ vs. $qR_g$. In this representation, globular proteins display a parabolic peak at $\sqrt{3}$. Any anisotropy will move this peak to higher values. On contrast, the signal of completely unfolded proteins will continuously increase in this representation. Flexible proteins can be found between these two extremes (Hammel 2012; Rambo and Tainer 2011).

In the K, G and KG system, the normalized Kratky plot of K displays a symmetric peak exactly at the predicted position, indicating that K is a very globular, well-folded protein with little or no anisotropy (Fig. 4.4d). In contrast, the Kratky plot of G continues to increase well beyond this point, until it finally levels off; a shape corresponding to an elongated, flexible protein. The Kratky plot of the complex KG is found between these two extremes, with its peak shifted towards higher $qR_g$ values, indicating some anisotropy, and its decrease being slower than for K, indicating some remainder of flexibility.

In the Porod-Debye plot, $q^4I(q)$ plotted vs $q^4$, the signal of non-flexible, globular proteins level at a plateau while the signal of unfolded proteins continue to increase linearly. Flexible proteins display a decrease in slope, but do not level off completely.

In Fig. 4.4e K shows a typical Porod-plateau, while the more flexible KG complex just levels off. The very flexible G on its own continues to increase.

In addition to being helpful for determining whether a protein is flexible, these representations also can highlight problems with the buffer subtraction due to their emphasis on the higher angles.

The interested reader can find excellent discussions on flexibility assessment in (Hammel 2012; Rambo and Tainer 2011).

#### 4.3.1.5 Indirect Fourier Transform

In most cases, the calculation of pair distance distribution function *p*(*r*) of a SAS curve is ambiguous, as there are several free parameters (Svergun 1992; Semenyuk and Svergun 1991):

- The region of the curve used for its calculation: Typically, its lower limit is given by the Guinier approximation, whereas the upper limit depends on what the *p*(*r*) is calculated for: For bead modeling with DAMMIF, DENFERT etc. normally $8/R_g$ is sufficient, while for GASBOR modeling the range should extend to at least 3.5 $nm^{-1}$. In addition, most algorithms work best when the cut-off lies in a region of decreasing intensity.
- The maximum distance $D_{max.}$ A good starting point for $D_{max}$ is the value of $3R_g$. From there, it needs to be adjusted in such a way that the resulting distribution approaches zero non-abruptly without strong oscillations, a trailing tail or even negative data points. In addition, a wrong $D_{max}$ results in a *p*(*r*) based radius of gyration that deviates significantly from the one determined *via* the Guinier approximation.
- The smoothing factor. It needs to be decreased if the fit no longer matches the data and increased if either the *p*(*r*) function of the fit show oscillations.

In the case of flexible proteins, it can be difficult to find a suitable *p*(*r*) function and to determine $D_{max}$. Similar problems are observed if the background is not well corrected.

Figure 4.4f shows how an incorrect $D_{max}$ affects the *p*(*r*) function. A too small value results in a sharp down-turn of the curve and often a near perpendicular approach towards 0 (green curve), while a too large value results in a long and extended tail (blue curve). Figure 4.4g shows how different particle shapes affect the *p*(*r*) function: For the small globular protein K (blue curve), it has a sharp, nearly symmetric peak. The more anisotropic shape of the KG complex (orange curve) is reflected in the asymmetric shape of the peak, with a slower descent towards larger distances. For G alone (green), the asymmetric shape of the peak is even more pronounced. The shoulder at larger distances indicates the possibility of two (or more) separated domains.

For elongated, rod-like objects, the cross-sectional pair-distribution function $p_c(r)$ can be determined instead.

#### 4.3.1.6 Porod Analysis

For globular scatterers, another helpful SAS invariant is the Porod invariant

$$Q = \int_0^\infty I(q)q^2 dq = 2\pi I_0/V_p$$

which allows to determine the Porod volume $V_p$ of the scatterer (Porod 1982). In both cases of proteins and nucleic acids, the Porod volume (in units of $nm^3$) corresponds to about 1.5–2 times the molecular weight (in kDa) (Petoukhov et al. 2007). This determination does not depend on the absolute scaling of the SAS curve, therefore this method for mass determination can be applied even if the concentration of the sample or/and the absolute scaling of the data are unknown.

Note that as the above integral only converges for mostly globular proteins, the Porod volume tends to deviate strongly from volume expected based on the mass if the sample is highly flexible or even disordered. This can be easily understood when one considers that *Q* matches the volume under the peak in the Kratky plot, which is only finite for globular objects.

For calculating the Porod volume, one needs to extrapolate the data to infinity by fitting the higher angle data around the Porod-Plateau in the Porod-Debye plot by a power law, whose exponent (the *Porod* exponent) needs to be smaller than −3. Based in this extrapolation, *Q* and thereby the volume can be easily calculated.

In this example, the Porod volume of the flexible protein G is not well defined, as its SAXS curve only decreases as $q^{-2}$ (as noticed in its Kratky plot of Fig. 4.4d). On the other hand, the globular protein K shows a well-defined Porod plateau (Fig. 4.4e), which permits to determine its Porod volume. In the case of the KG complex, the power law fit of the higher $q$ region gives a Porod exponent of $-3$, prohibiting the determination of the Porod volume.

Some algorithms for indirect Fourier transform provide the Porod volume as an additional result. It should also be noted, that most software tools will provide a result for the Porod volume, even when the conditions for determination are not met. These results are typically not related to the actual volume of the macromolecule in question.

#### 4.3.1.7 Correlated Volume

Another approach for estimating the mass of non-flexible macro-molecules is provided by the so-called volume-of-correlation (Rambo and Tainer 2013) given by

$$V_c = \frac{I_0}{\int_0^\infty I(q)q\mathrm{d}q}$$

The advantage of this approach is that the above integral usually converges well in the available data range and no extrapolation of the data is necessary. Additionally, it converges as long as the Porod exponent is smaller than $-2$. In our example, this implies that while the Porod volume of the KG complex is not well defined, the correlated volume can be determined.

In this case, the mass (in kDa) of proteins is roughly equal to $8V_c^2/R_g$ and of RNA to $\left(107V_c^2/R_g\right)^{0.8}$ if the scattering vector is provided in $\mathrm{nm}^{-1}$. For DNA no data are available to our knowledge.

### 4.3.2 Comparison to Predicted Scattering Curves of Atomistic Models

The calculation of theoretical scattering from known atomic coordinates is a direct problem with no inherent ambiguity i.e. one always obtains the same SAS curve for any set of known atomic coordinates. However, macromolecules in solution are surrounded by a hydration shell that contributes to the SAS signal. Atomistic models generally do not account for these highly dynamic structures (Zhang et al. 2007). Different implementations are available to account for these water molecules and their effect on the SAS signal. Some programs use molecular dynamics to describe the hydration shell (Chen and Hub 2015; Knight and Hub 2015), but more often it is modeled as a fixed-width shell around the macromolecule. In order to perform comparisons with actual data, the parameters of the hydration shell are generally adjusted to provide the best fit between model and data (Svergun et al. 1995; Schneidman-Duhovny et al. 2010). Particular attention should be paid to the values used for the hydration layer when comparing the resulting fits of multiple structures to the same data to avoid misinterpretation.

## 4.4 Building and Interpretation of 3D Models

If no or incomplete structural information is available, it is often possible to build 3D models of the macromolecule based on the SAS data. Approaches to model building span from the construction of bead models to fully fledged molecular dynamics simulations. Here, we will only discuss a few of them and refer the readers to Chap. 7 for more details.

### 4.4.1 *Ab-initio* Modeling

*Ab-initio* modeling techniques allow construction of three dimensional bead models optimized to the SAS signal. This is also mathematically possible in the case of corrupted data (mixture, non-matching buffer, etc.), therefore the data needs to be validated before attempting a reconstruction.

Most algorithms assume a monodisperse system and uniform contrast (i.e. all beads are identical) (Svergun et al. 2001; Svergun 1999; Franke and Svergun 2009), but some specialized programs can model oligomeric mixtures (Petoukhov et al. 2007), DNA-protein complexes (Petoukhov et al. 2007; Svergun 1999), hydration layers (Koutsioubas and Perez 2013) etc. Other programs can include known partial structures in the modeling (Petoukhov and Svergun n.d.).

As the reconstruction of three dimensional models from the one dimensional SAS curve is intrinsically ambiguous and the high number of free parameters makes stochastic approaches to model building necessary, it is essential to repeat the modeling several times and to compare the results. The similarity between two resulting models can be quantified by the normalized spatial discrepancy (NSD). If two models systematically differ from each other, their NSD exceeds 1, for identical objects, it is 0. Therefore, large NSD values are indicative of ambiguity in the modeling.

### 4.4.2 Rigid-Body Modeling

In the case of complexes or multi-domain proteins, the structures of the individual components or homologues thereof are often known. The relative positions of the individual "rigid bodies" can be modeled to fit the SAS data (Petoukhov et al. 2007; Petoukhov and Svergun n.d.). Inclusion of additional constraints to conserve known connectivity between domains is also often required especially if the domain is symmetric. However, inclusion of any constraint will bias the model thus if the constraint is false, being based on an incorrect assumption the resulting model may still fit the data and lead to false conclusions. As for *ab-initio* modeling, the results are often ambiguous and repeated reconstructions are necessary. As for *ab initio* modeling, the robustness of the reconstruction can be assessed by calculating the NSD between the resulting models. If different rigid-body modeling approaches are applied, $\chi^2$ can be used to identify which model describes the available data best. As the absolute value of $\chi^2$ depends mostly on the signal quality it cannot be used to assess the absolute quality of the resulting fit.

## 4.5 Presentation of SAS Data for Publication

When presenting BioSAS data for publication, it is generally recommended to follow the IUCr guidelines (Jacques et al. 2012)

- Scattering data should be presented either in logarithmic intensity scaling or in double-logarithmic representation. Linear representation of the intensity hides key features of the curves and should be avoided.
- Guinier fits should be shown in the Guinier representation, showing a sufficient data range to evaluate the quality of the fit.
- If $p(r)$ functions are used for further modelling, these must be shown. When comparing different functions, it is not uncommon to scale their respective maxima to 1.
- The (normalized) Kratky plot should be shown to allow assessment of flexibility.
- When presenting models, the variability of the results needs to be illustrated. In the case of bead models, this means that averaged and filtered models need to be shown.
- The instrument and conditions used for data acquisition as well as numeric primary processing results have to be presented (see an example in Table 4.1).
- Data and models should be made publicly available *via* an appropriate venue such as the SAS Biological Data Bank (SASBDB).

## 4.6 Conclusions

The analysis and interpretation of small angle scattering data rely on correctly collected and reduced data as well as on sample properties such as monodispersity or compactness.

This chapter describes the necessary steps of data reduction and the most common approaches to data analysis. While the majority of these steps can be automatized, it is still necessary to understand the underlying assumptions and possible sources of error so that one can verify the validity

**Table 4.1** Data-collection and scattering-derived parameters. The data in this table belongs to the data shown in Fig. 4.2i. Software listed in *italics* provides examples for its class, but was not used in this analysis

| | |
|---|---|
| Data-collection parameters | |
| Instrument: | ESRF BM29 |
| Beam geometry | 0.7 mm × 0.7 mm |
| Wavelength (Å) | 0.99 |
| q-range ($Å^{-1}$) | 0.0032–0.49 |
| Exposure time (sec) | 1 per frame, 10 frames |
| Concentration range (mg/ml) | 0.22–19.22 |
| Temperature (K) | 293 |
| Structural parameters | |
| $I_0$ ($cm^{-1}$) [from P(r)] | 0.01184 |
| $R_g$ ($Å^{-1}$) [from P(r)] | 15.6 |
| $I_0$ ($cm^{-1}$) [from Guinier] | 0.01152 ± 0.0004 |
| $R_g$ ($Å^{-1}$) [from Guinier] | 15.4 ± 2.5 |
| $D_{max}$ (Å) | 58 |
| Porod volume $V_p$ ($Å^3$) | $22 \cdot 10^3$ |
| Molecular-mass determiantion | |
| Partial specific volume ($cm^3\ g^{-1}$) | 0.724 |
| Contrast ($\Delta\rho \times 10^{12}\ cm^{-2}$) | 3.047 |
| Molecular mass $M_{r.}$ [from $I_0$] | 14.4 ± 0.10 |
| Calculated monomeric $M_{r.}$ from sequence (kDa) | 14.3 |
| Software employed | |
| Primary data reduction | BM29 online data analysis, pyFAI, Primus |
| 1D data processing | Primus, Gnom |
| *Ab initio* analysis | n.a. (e.g. *DAMMIF*) |
| Validation and averaging | n.a. (e.g. *SUPALM, DAMAVER*) |
| Rigid-body modeling | n.a. (e.g. *SASREF*) |
| Computation of model intensities | n.a. (e.g. *WAXSiS*) |
| Three-dimensional graphics representations | n.a. (e.g. *PyMOL*) |

of results and does not need to blindly trust them. Any conclusions on the shape and behaviour of a sample in solution drawn from SAS data should take into account the quality and reliability of the data.

Data analysis should be viewed as part of an exploratory, interactive process designed to test hypotheses and learn more about the system under study. Cross checking with complementary information is a highly valuable part of the analysis procedure enabling any differences to be highlighted and investigated. In this way artifacts and biased conclusions can be avoided, novel insights can be discovered and more in depth interpretation of the data with greater confidence can be obtained compared to using SAS alone.

Tips

- Show the data: scattering curve in log-log or log-lin is the minimum requirement. For models, show the fitted curves!
- Garbage in = garbage out: Most tools will give you an answer even if the prerequisites are not met at all – always check before!
- Units matter: In particular, the scattering vector $q$ can be reported in either $nm^{-1}$ or $Å^{-1}$

(continued)

- Crosschecks: Utilise information from complimentary techniques wherever possible to allow validation
- Easy to fit noisy data: always be critical
- Don't over-interpret models: Results are not atomic models, just a filled volume! Your solution might not be unique!

**Acknowledgments** The authors would like to thank Dr. Bárbara Calisto Machado for providing the SAXS data of the K and G proteins and KG complex.

## References

Ashiotis G, Deschildre A, Nawaz Z, Wright JP, Karkoulis D, Picca FE, Kieffer J (2015) The fast azimuthal integration python library. J Appl Crystallogr 48(Pt 2):510–519. doi:10.1107/S1600576715004306

Benecke G, Wagermaier W, Li C, Schwartzkopf M, Flucke G, Hoerth R, Zizak I, Burghammer M, Metwalli E, Muller-Buschbaum P, Trebbin M, Forster S, Paris O, Roth SV, Fratzl P (2014) A customizable software for fast reduction and analysis of large X-ray scattering data sets: applications of the new DPDAK package to small-angle X-ray scattering and grazing-incidence small-angle X-ray scattering. J Appl Crystallogr 47(5):1797–1803. doi:10.1107/S1600576714019773

Bernado P, Svergun DI (2012) Structural analysis of intrinsically disordered proteins by small-angle x-ray scattering. Mol BioSyst 8(1):151–167. doi:10.1039/c1mb05275f

Bonneté F, Finet S, Tardieu A (1999) Second virial coefficient: variations with lysozyme crystallization conditions. J Cryst Growth 196(2–4):403–414. doi: http://dx.doi.org/10.1016/S0022-0248(98)00826-4

Brennich ME, Kieffer J, Bonamis G, De Maria AA, Hutin S, Pernot P, Round A (2016) Online data analysis at the ESRF bioSAXS beamline, BM29. J Appl Crystallogr 49:203–212. doi:10.1107/S1600576715024462

Brian Richard P (2013) Everything SAXS: small-angle scattering pattern collection and correction. J Phys Condens Matter 25(38):383201

Brookes E, Pérez J, Cardinali B, Profumo A, Vachette P, Rocco M (2013) Fibrinogen species as resolved by HPLC-SAXS data processing within the UltraScan Solution Modeler (US-SOMO) enhanced SAS module. J Appl Crystallogr 46(6):1823–1833

Chen PC, Hub JS (2015) Interpretation of solution x-ray scattering by explicit-solvent molecular dynamics. Biophys J 108(10):2573–2584. doi:10.1016/j.bpj.2015.03.062

David G, Pérez J (2009) Combined sampler robot and high-performance liquid chromatography: a fully automated system for biological small-angle X-ray scattering experiments at the Synchrotron SOLEIL SWING beamline. J Appl Crystallogr 42(5):892–900. doi:10.1107/S0021889809029288

Debye P (1947) Molecular- weight determination by light scattering. J Phys Colloid Chem 51(1):18–32. doi:10.1021/j150451a002

Franke D, Svergun DI (2009) DAMMIF, a program for rapid ab-initio shape determination in small-angle scattering. J Appl Crystallogr 42(2):342–346. doi:10.1107/S0021889809000338

Franke D, Kikhney AG, Svergun DI (2012) Automated acquisition and analysis of small angle X-ray scattering data. Nucl Instrum Methods Phys Res Sect A 689:52–59. doi:http://dx.doi.org/10.1016/j.nima.2012.06.008

Franke D, Jeffries CM, Svergun DI (2015) Correlation Map, a goodness-of-fit test for one-dimensional X-ray scattering spectra. Nat Methods 12(5):419–422. doi:10.1038/nmeth.3358

Glatter O (1977) A new method for the evaluation of small-angle scattering data. J Appl Crystallogr 10(5):415–421

Glatter O, Kratky O (1982) Small angle x-ray scattering. Academic, London

Graewert MA, Franke D, Jeffries CM, Blanchet CE, Ruskule D, Kuhle K, Flieger A, Schäfer B, Tartsch B, Meijers R, Svergun DI (2015) Automated pipeline for purification, biophysical and x-ray analysis of biomacromolecular solutions. Sci Rep 5:10734. doi:10.1038/srep10734

Grant TD, Luft JR, Wolfley JR, Tsuruta H, Martel A, Montelione GT, Snell EH (2011) Small angle x-ray scattering as a complementary tool for high-throughput structural studies. Biopolymers 95(8):517–530

Guinier A (1938) La diffusion des rayons X sous les très faibles angles appliquée à l'étude de fines particules et de suspensions colloïdales. Compte Rendu de l'Académie des Sciences de Paris 206:1374–1376

Hammel M (2012) Validation of macromolecular flexibility in solution by small-angle X-ray scattering (SAXS). Eur Biophys J EBJ 41(10):789–799. doi:10.1007/s00249-012-0820-x

Hammersley A (1997) FIT2D: an introduction and overview. European synchrotron radiation facility internal report ESRF97HA02T 68

Hosemann R (1939) Neues röntgenographisches Verfahren zur Bestimmung des submikroskopischen Feinbaues eines Stoffes. Diskussion des Zellulosemodells. Z Phys 114(3–4):133–169

Jacques DA, Trewhella J (2010) Small-angle scattering for structural biology – expanding the frontier while avoiding the pitfalls. Protein Sci 19(4):642–657. doi:10.1002/pro.351

Jacques DA, Guss JM, Svergun DI, Trewhella J (2012) Publication guidelines for structural modelling of small-angle scattering data from biomolecules in solution. Acta Crystallogr D Biol Crystallogr 68 (6):620–626

Karlsen ML, Thorsen TS, Johner N, Ammendrup-Johnsen I, Erlendsson S, Tian X, Simonsen JB, Hoiberg-Nielsen R, Christensen NM, Khelashvili G, Streicher W, Teilum K, Vestergaard B, Weinstein H, Gether U, Arleth L, Madsen KL (2015) Structure of dimeric and tetrameric complexes of the BAR domain protein PICK1 determined by small-angle X-ray scattering. Structure 23(7):1258–1270. doi:10.1016/j.str.2015.04.020

Knight CJ, Hub JS (2015) WAXSiS: a web server for the calculation of SAXS/WAXS curves based on explicit-solvent molecular dynamics. Nucleic Acids Res 43 (W1):W225–W230. doi:10.1093/nar/gkv309

Koutsioubas A, Perez J (2013) Incorporation of a hydration layer in the 'dummy atom' ab initio structural modelling of biological macromolecules. J Appl Crystallogr 46(6):1884–1888. doi:10.1107/S0021889813025387

Lambright D, Malaby AW, Kathuria SV, Nobrega RP, Bilsel O, Matthews CR, Muthurajan U, Luger K, Chopra R, Irving TC (2013) Complementary techniques enhance the quality and scope of information obtained from SAXS. ACA Trans:1–12

Mathew E, Mirza A, Menhart N (2004) Liquid-chromatography-coupled SAXS for accurate sizing of aggregating proteins. J Synchrotron Radiat 11 (4):314–318. doi:10.1107/S0909049504014086

Mylonas E, Svergun DI (2007) Accuracy of molecular mass determination of proteins in solution by small-angle X-ray scattering. J Appl Crystallogr 40(s1): s245–s249. doi:10.1107/S002188980700252X

Orthaber D, Bergmann A, Glatter O (2000) SAXS experiments on absolute scale with Kratky systems using water as a secondary standard. J Appl Crystallogr 33(2):218–225. doi:10.1107/S0021889899015216

Pernot P, Round A, Barrett R, De Maria AA, Gobbo A, Gordon E, Huet J, Kieffer J, Lentini M, Mattenet M (2013) Upgraded ESRF BM29 beamline for SAXS on macromolecules in solution. J Synchrotron Radiat 20 (4):660–664. doi:10.1107/S0909049513010431

Petoukhov MV, Svergun DI Global rigid body modeling of macromolecular complexes against small-angle scattering data. Biophys J 89(2):1237–1250. doi:10.1529/biophysj.105.064154

Petoukhov MV, Konarev PV, Kikhney AG, Svergun DI (2007) ATSAS 2.1-towards automated and web-supported small-angle scattering data analysis. Applied crystallography

Porod G (1982) General theory. In: Glatter O, Kratky O (eds) Small angle X-ray scattering. Academic, London

Rambo RP, Tainer JA (2011) Characterizing flexible and intrinsically unstructured biological macromolecules by SAS using the Porod-Debye law. Biopolymers 95 (8):559–571. doi:10.1002/bip.21638

Rambo RP, Tainer JA (2013) Accurate assessment of mass, models and resolution by small-angle scattering. Nature 496(7446):477–481. doi:10.1038/nature12070

Reyes FE, Schwartz CR, Tainer JA, Rambo RP (2014) Methods for using new conceptual tools and parameters to assess RNA structure by small-angle X-ray scattering. Methods Enzymol 549:235–263. doi:10.1016/B978-0-12-801122-5.00011-8

Rodriguez-Navarro A (2006) XRD2DScan: new software for polycrystalline materials characterization using two-dimensional X-ray diffraction. J Appl Crystallogr 39(6):905–909. doi:10.1107/S0021889806042488

Round A, Brown E, Marcellin R, Kapp U, Westfall CS, Jez JM, Zubieta C (2013) Determination of the GH3.12 protein conformation through HPLC-integrated SAXS measurements combined with X-ray crystallography. Acta Crystallogr D Biol Crystallogr 69 (10):2072–2080. doi:10.1107/S0907444913019276

Schneidman-Duhovny D, Hammel M, Sali A (2010) FoXS: a web server for rapid computation and fitting of SAXS profiles. Nucleic Acids Res 38(suppl 2): W540–W544. doi:10.1093/nar/gkq461

Segel DJ, Bachmann A, Hofrichter J, Hodgson KO, Doniach S, Kiefhaber T (1999) Characterization of transient intermediates in lysozyme folding with time-resolved small-angle X-ray scattering. J Mol Biol 288(3):489–499

Semenyuk AV, Svergun DI (1991) GNOM – a program package for small-angle scattering data processing. J Appl Crystallogr 24(5):537–540. doi:10.1107/S002188989100081X

Svergun D (1992) Determination of the regularization parameter in indirect-transform methods using perceptual criteria. J Appl Crystallogr 25(4):495–503

Svergun DI (1999) Restoring low resolution structure of biological macromolecules from solution scattering using simulated annealing. Biophys J 76 (6):2879–2886. doi:10.1016/S0006-3495(99)77443-6

Svergun D, Barberato C, Koch M (1995) CRYSOL – a program to evaluate X-ray solution scattering of biological macromolecules from atomic coordinates. J Appl Crystallogr 28(6):768–773

Svergun DI, Petoukhov MV, Koch MH (2001) Determination of domain structure of proteins from X-ray solution scattering. Biophys J 80(6):2946–2953. doi:10.1016/S0006-3495(01)76260-1

Watanabe Y, Inoko Y (2009) Size-exclusion chromatography combined with small-angle X-ray scattering optics. J Chromatogr A 1216(44):7461–7465. doi:10.1016/j.chroma.2009.02.053

Zhang L, Wang L, Kao YT, Qiu W, Yang Y, Okobiah O, Zhong D (2007) Mapping hydration dynamics around a protein surface. Proc Natl Acad Sci U S A 104 (47):18461–18466. doi:10.1073/pnas.0707647104

# 5 Designing and Performing Biological Solution Small-Angle Neutron Scattering Contrast Variation Experiments on Multi-component Assemblies

Susan Krueger

**Abstract**

Solution small-angle neutron scattering (SANS) combined with contrast variation provides information about the size and shape of individual components of a multi-component biological assembly, as well as the spatial arrangements between the components. The large difference in the neutron scattering properties between hydrogen and deuterium is key to the method. Isotopic substitution of deuterium for some or all of the hydrogen in either the molecule or the solvent can greatly alter the scattering properties of the biological assembly, often with little or no change to its biochemical properties. Thus, SANS with contrast variation provides unique information not easily obtained using other experimental techniques.

If used correctly, SANS with contrast variation is a powerful tool for determining the solution structure of multi-component biological assemblies. This chapter discusses the principles of SANS theory that are important for contrast variation, essential considerations for experiment design and execution, and the proper approach to data analysis and structure modeling. As sample quality is extremely important for a successful contrast variation experiment, sample issues that can affect the outcome of the experiment are discussed as well as procedures used to verify the sample quality. The described methodology is focused on two-component biological complexes. However, examples of its use for multi-component assemblies are also discussed.

**Keywords**

Small-angle neutron scattering • Contrast variation • Isotopic substitution • Biological assemblies

S. Krueger (✉)
NIST Center for Neutron Research, NIST, Gaithersburg, MD 20899, USA
e-mail: susan.krueger@nist.gov



B. Chaudhuri et al. (eds.), *Biological Small Angle Scattering: Techniques, Strategies and Tips*, Advances in Experimental Medicine and Biology 1009, DOI 10.1007/978-981-10-6038-0_5

## 5.1 Introduction

Small-angle neutron scattering (SANS) is able to provide the size, molecular mass and shape of a macromolecular complex in solution on length scales between approximately 10 Å to about 1,000 Å (Jacrot 1976; Svergun and Koch 2003; Svergun 2010; Jacques and Trewhella 2010; Zaccai 2012; Ankner et al. 2013). SANS combined with contrast variation allows for the unique retrieval of the internal structure and organization of chemically distinct components in a biological assembly. Recent reviews (Neylon 2008; Whitten and Trewhella 2009; Heller 2010; Gabel 2015; Zaccai et al. 2016) as well as classic papers (Engelman and Moore 1975; Ibel and Stuhrmann 1975; Jacrot 1976) describe the contrast variation technique as it applies to biological systems in detail.

Contrast variation takes advantage of the large difference in neutron scattering properties between hydrogen and deuterium. By systematically varying the ratio of $H_2O$ to $D_2O$ ($H_2O$:$D_2O$ ratio) of the solvent, conditions are found in which one component in a multi-component assembly has the same scattering properties as the solvent. Under these conditions, the "matched" component is essentially invisible, much like a transparent material, e.g., a glass rod, can be optically invisible when in a solution that matches its index of refraction. In theory, a systematic series of measurements can be made allowing the scattering from each component to be determined as well as their positions with respect to each other in the complex. Substitution of deuterium for hydrogen in one or more of the components of the complex is another form of contrast variation.

This chapter introduces contrast variation experiments, outlining pertinent principles of SANS theory, essential considerations for experiment design and execution, and the proper approach to data analysis and structure modeling. The discussions on experiment planning and sample quality verification provide a guide for preparing samples that will lead to a successful contrast variation experiment. While the methodology described here is focused on two-component complexes, its use for multi-component assemblies is also discussed.

## 5.2 Theory

### 5.2.1 Scattering Intensity

The measured SANS intensity from a macromolecule consists of a coherent and an incoherent contribution such that

$$I(\vec{q}) = I_{coh}(\vec{q}) + I_{inc}. \tag{5.1}$$

The coherent contribution is dependent on the scattering vector, $\vec{q}$, which has a magnitude defined as

$$q = \frac{4\pi \sin(\theta)}{\lambda}, \tag{5.2}$$

where $2\theta$ is the scattering angle (typically in degrees), measured from the axis of the incoming neutron beam, and $\lambda$ is the neutron wavelength. The wavelength is usually expressed in nm or Å, such that $q$ is stated in units of $\text{nm}^{-1}$ or $\text{Å}^{-1}$. The shape of the coherent scattering intensity profile depends on the shape of the molecule. On the other hand, the incoherent contribution is not $q$-dependent and contributes mainly to the noise level or "background". Thus, the scattering intensity will refer to only the coherent term unless otherwise specified. The incoherent term will be further discussed in the Experimental Method, Data Analysis and Structure Modeling section.

Since SANS does not provide information on the length scale of atomic bonds, the strength of the scattering interaction can be described in terms of a uniform scattering length density of the entire molecule, $\rho$, within the molecular volume, $V$ (in $\text{cm}^3$ or $\text{Å}^3$). The scattering length density is usually expressed in units of $\text{cm}^{-2}$ or $\text{cm Å}^{-3}$, but can be found stated in units of $\text{Å}^{-2}$. The SANS intensity from the molecule can be written in terms of the scattering length density as

$$I(\vec{q}) = \rho^2 V \left| \frac{1}{V} \int_V e^{i\vec{q}\cdot\vec{r}} d\vec{r} \right|^2, \qquad (5.3)$$

or

$$I(\vec{q}) = \rho^2 V \left| F(\vec{q}) \right|^2, \qquad (5.4)$$

where $\left| F(\vec{q}) \right|^2$ is the form factor of the molecule.

### 5.2.2 Contrast

The coherent scattering from a biological molecule in a solvent can be thought of in terms of the cartoon shown in Fig. 5.1. Each fish represents one of $N$ randomly-oriented molecules in the solvent. The molecules are identical in composition, size and shape and they are not interacting with each other. In other words, the solution is monodisperse, containing $N$ identical particles and dilute, such that the particles do not sense one another.

The coherent scattering from any one of the molecules in the solution equals that from the molecule with scattering length density, $\rho$, in the molecular volume, $V$, plus that from the solvent with scattering length density, $\rho_s$, in an effectively infinite volume, minus that from a "ghost" molecule with scattering length density of the solvent, $\rho_s$, in the volume, $V$. The coherent scattering from a solvent of infinite extent is a delta function at $q = 0$ because the solvent has no long range-order; thus, it is not observed in practice. The scattering is therefore defined only in terms of the difference in the scattering length densities of the molecule and the solvent within the same molecular volume. Thus, Eq. 5.4 becomes

$$I(\vec{q}) = (\rho - \rho_s)^2 V \left| F(\vec{q}) \right|^2. \qquad (5.5)$$

The difference in scattering length densities is known as the contrast,

$$\Delta\rho = \rho - \rho_s. \qquad (5.6)$$

Figure 5.2 shows the scattering length density of water as a function of the percent $D_2O$ (% $D_2O$) in the solvent, along with scattering length densities of some typical biological molecules and compounds. The large slope of the water line as compared to the others is due to the large difference in scattering length density between hydrogen and deuterium. The lines are horizontal for $CH_2$ and $CD_2$ because there is no exchange of deuterium for hydrogen as the % $D_2O$ increases in the solvent. However, for proteins and nucleic acids, labile hydrogen atoms, i.e., those bound to nitrogen and oxygen, will exchange with deuterium in the solvent, causing an increase in scattering length densities for increased % $D_2O$.

The vertical double-arrowed line at 10% $D_2O$ represents the contrast between protein and water at that % $D_2O$ value. A similar line can be drawn at any % $D_2O$. Note that the protein and water lines cross near 40% $D_2O$ indicating that the scattering length densities of the molecule and solvent are the same, i.e., $\Delta\rho = 0$. This is called a contrast match point and occurs at approximately 40% $D_2O$ for typical proteins. The contrast match points for DNA, RNA, lipid head groups, and $CH_2$ can be found in the same manner. Note that the scattering length densities for perdeuterated molecules, in which all of the non-exchangeable hydrogen atoms, i.e., those bound to carbon, have been chemically replaced

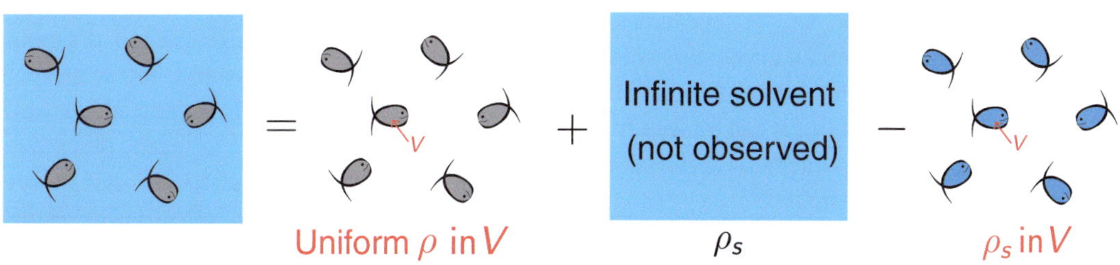


**Fig. 5.1** Contributions to the coherent scattering from a macromolecule in solution

**Fig. 5.2** Neutron scattering length density as a function of percentage of $D_2O$ (% $D_2O$) in the solvent for some typical biological molecules and compounds (Zaccai et al. 2016)

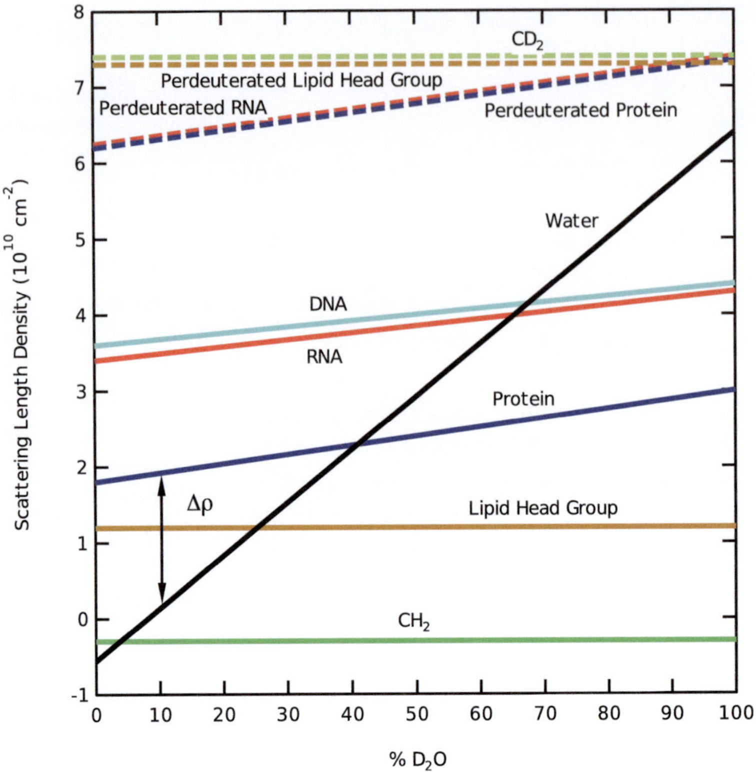


by deuterium, do not cross the water line. Thus, the theoretical contrast match points for these molecules would be greater than 100% $D_2O$, which is not obtainable in practice.

The SANS intensity from all $N$ monodisperse, randomly-oriented biological macromolecules in a dilute solution can be written in terms of the contrast as

$$\begin{aligned} I(q) &= n\,(\Delta\rho)^2\,V^2\left\langle\left|F(\vec{q})\right|^2\right\rangle \\ &= n\,(\Delta\rho)^2\,V^2 P(q), \end{aligned} \tag{5.7}$$

where $n$ is the number density ($N$ per unit volume, $v$), of molecules (in $cm^{-3}$). The brackets represent an averaging over all orientations of the molecule. The rotationally averaged form factor is sometimes also called $P(q)$. It can be seen from Eq. 5.7 that the scattering intensity is zero at the contrast match point, $\Delta\rho = 0$.

A typical contrast variation experiment involves measuring a complex consisting of two components that have different scattering length densities in solvent consisting of mainly water with perhaps a small amount of salts and buffering compounds. When water mixtures with different $H_2O$:$D_2O$ ratios are used, the contrast of each component will change as a function of the concentration of $D_2O$ in the solvent. Thus, contrast match points exist for each of the components as well as the entire complex. As the $H_2O$:$D_2O$ conditions in the solvent are varied, different parts of the complex scatter more strongly depending on their individual scattering length densities. By varying the amount of $D_2O$ in the solvent, one component can be essentially transparent at its contrast match point while the others are still visible. If enough different contrast conditions are measured, the scattering intensities for the individual components as well as their positions relative to each other are obtained. It is this feature of SANS that makes the method so powerful for selective measurement of individual components within a complex.

From Fig. 5.2, it is clear that proteins and nucleic acids have different contrast match points. The protein contrast match point is around 40% $D_2O$, meaning that only the DNA or RNA is visible at this contrast. The DNA and RNA contrast match points are around 65% $D_2O$ such that only the protein is visible under these conditions. Therefore, complexes consisting of proteins and nucleic acids are ideal candidates for contrast variation experiments (Gabel 2015).

For a complex consisting of two proteins, replacement of some or all of the non-exchangeable hydrogen atoms with deuterium in one of the components is required in order for the two contrast match points to be different. Since the contrast match point of perdeuterated proteins is above 100% $D_2O$, partially deuterated proteins are generally used for contrast variation experiments so that the contrast match point of the deuterated component is somewhere between 60% $D_2O$ and 100% $D_2O$ (Jacques et al. 2011). The exact contrast match point of a deuterated component is dependent on the amount of deuteration achieved. The contrast variation experiment can be used to verify this parameter, especially if a reliable determination cannot be made by other methods such as NMR or mass spectrometry. The method can be extended to larger assemblies consisting of multiple copies of the two different components (Appolaire et al. 2014).

Contrast variation methods also exist for solving the structures of multi-component assemblies. The triple isotopic substitution method (TISM) (Serdyuk and Zaccai 1996) allows for the determination of the scattering from one component at a time. Each determination requires three measurements with different deuteration levels of the component of interest. The label triangulation method (LTM) (Engelman and Moore 1972; Hoppe 1973) allows for the structure of a complex to be obtained by determining the distances between pairs of components. Each distance determination requires the selective labeling of the two components of interest. After determining a requisite number of distances, a 3D model of the structure can be determined by triangulation. The main applications of this method have been for the determination of the location of various components of the small (30S) and large (50S) subunits of the ribosome. These historic works are referenced in (May and Nowotny 1989).

Both the TISM and LTM methods require a large number of measurements. In practice, measurements of complexes consisting of three or more components are often made under solvent conditions that match each of the individual components. While some components still must be deuterated, this approach limits the number of measurements needed. An attempt is then made to find model structures that best fit all of the contrast variation data. This approach often works well when atomistic level starting structures of individual components are already available or can be built using homology modeling techniques, distance constraints are available from other techniques and/or docking software can be used to build structure models. The value of any such model is determined by its ability to reproduce the SANS data.

### 5.2.3 Radius of Gyration and Forward Scattering Intensity

The radius of gyration, $R_g$, which is similar to the moment of inertia with respect to the scattering center of mass, and the forward scattering intensity, $I(0)$, which is the scattering intensity at $q = 0$, are two important model-independent parameters that are obtained from SANS data. $R_g$ provides information about the size of the molecule whereas $I(0)$ provides information about its molecular weight. Both parameters depend on the contrast, which can provide important clues about the spatial distribution of contrast within the molecule.

By definition, $P(q)$ in Eq. 5.7 is equal to 1 at $q = 0$. Thus,

$$I(0) = n\,(\Delta\rho)^2\,V^2. \tag{5.8}$$

The number density, $n$ ($cm^{-3}$), can be written in terms of the concentration of the molecule, $c$ ($g\ cm^{-3}$), as

$$n = \frac{cN_A}{M_w}, \tag{5.9}$$

where $M_w$ is the molecular weight of the molecule (in Da, where 1 Da = 1 g mole$^{-1}$) and $N_A$ is Avogadro's number. In addition, the molecular volume, $V$, can be written in terms of the partial specific volume, $\bar{v}$ (cm$^3$ g$^{-1}$), as

$$V = \frac{\bar{v}M_w}{N_A}. \tag{5.10}$$

Equations 5.8, 5.9 and 5.10 can be used to relate $I(0)$ to the $M_w$ of the molecule if the SANS data are on an absolute scale, usually in units of cm$^{-1}$. While $I(0) = 0$ at the contrast match point, this is not true at larger angles that correspond to length scales on the order of the internal scattering length density fluctuations that were ignored by assuming a uniform scattering length density in Eq. 5.3.

The Guinier approximation (Guinier and Fournet 1955),

$$I(q) = I(0)\exp\left(-q^2\frac{R_g^2}{3}\right), \tag{5.11}$$

can be used on the low-$q$ portions of the data to obtain values for $R_g$ and $I(0)$. This low-$q$ analysis is valid only in the region where $qR_g \lesssim 1.3$, so the valid $q$ range depends on the size of the molecule. A shape must be assumed for the molecule to relate $R_g$ to the molecular dimensions. $R_g$ and $I(0)$ are found by plotting the natural log of Eq. 5.11 such that

$$\ln(I(q)) = \ln(I(0)) - q^2\frac{R_g^2}{3}. \tag{5.12}$$

A linear fit of $\ln(I(q))$ vs $q^2$ (Eq. 5.12) to the low-$q$ portion of the data allows the determination of $R_g$ from the slope and $I(0)$ from the intercept.

Another method to obtain $R_g$ and $I(0)$, which makes use of all of the data rather than a limited data set at small $q$ values, is to use the distance distribution function, $P(r)$ vs $r$ (Glatter and Kratky 1982). This function represents the probability distribution of distances, $r$, between all pairs of atoms in the molecule. The result is a smooth histogram-like plot with peaks at the most probable distances in the molecule. Thus, the shape of the $P(r)$ vs $r$ curve depends strongly on the shape of the molecule and can vary as a function of contrast.

$P(r)$ is typically obtained from the SANS data using an indirect Fourier transformation method (Glatter 1977; Moore 1982; Semenyuk and Svergun 1991) using the relation

$$I(q) = 4\pi V\int_0^{D_{max}} P(r)\frac{\sin(qr)}{qr}\,dr. \tag{5.13}$$

This analysis requires a stipulation by the user of a maximum dimension, $D_{max}$, beyond which $P(r) = 0$. Typically, several values of $D_{max}$ are explored in order to find the range over which the $P(r)$ function doesn't change as a function of $D_{max}$. Typically, the condition that $P(r) = 0$ at $r = 0$ is also assumed. $R_g$ and $I(0)$ can be derived from $P(r)$, as described in (Glatter 1982).

## 5.3 Experimental Method, Data Analysis and Structure Modeling

The successful SANS contrast variation experiment requires a number of steps from experiment planning and sample preparation to data collection and analysis, culminating with structure modeling. A flow chart showing these different steps is shown in Fig. 5.3 and each is described in detail in this section.

### 5.3.1 Sample Considerations

To obtain the most information possible from a contrast variation experiment, the main requirement is that the biological complex must be measured under dilute and monodisperse conditions in all $H_2O$:$D_2O$ solvents. In other

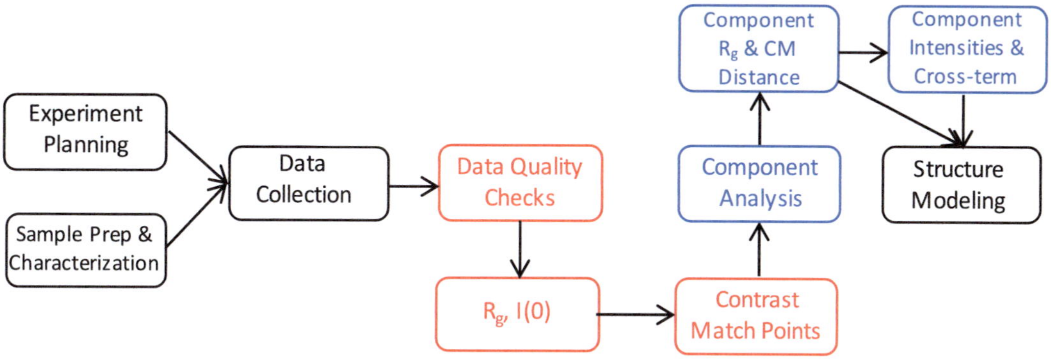


**Fig. 5.3** Flow chart showing the steps necessary for performing a successful SANS contrast variation experiment

words, the sample must be of high purity. Ideally, this means that all the molecules have the same stoichiometry under all contrast conditions and there are no "free" components in the solution. Furthermore, all of the complexes are in identical conformations and there are no interactions between the complexes. The above equations describing the scattering intensity from macromolecules in solution assume these conditions are met.

In practice, there are many reasons why a sample might not meet all of these conditions. Sample preparation for contrast variation experiments involves making biological complexes in solvents containing $D_2O$. In some cases, the complexes themselves must be made using deuterated components, which is typically accomplished by expressing one of the protein components using bacteria grown in deuterium-enriched media.

Whenever deuterium is introduced into the molecule or solvent, there is an increased chance for undesirable aggregation, even under conditions where the complex would not aggregate were deuterium absent. SANS is a volume-weighted technique. Since larger aggregates have larger volumes, a small amount of larger aggregates contribute more to the scattering intensity than a larger amount of small aggregates. While the signature of aggregation is most obvious at lower $q$ values, the contribution to the total scattering persists over a much wider $q$ range. Another issue when using deuterated components is that the complex might not form as readily or stably. This can lead to a situation for which the complex is associating and dissociating at a constant rate. In such situations, it may not always be possible to avoid having an excess of the constituent components in the solution.

The amount of sample needed for a contrast variation experiment is on the order of 0.5 mL at a concentration of 1–5 mg mL$^{-1}$ for complexes with a typical $M_w$ between 50 and 100 kDa. Under solvent conditions where the contrast, and thus $I(q)$ is low, higher concentrations must sometimes be used. This may introduce interparticle interference or aggregation effects that need to be mitigated as much as possible, e.g., by adding salts to the solvent to screen electrostatic interactions or by adding compounds that inhibit hydrophobic interactions. The mitigation of interparticle interference and aggregation effects is a difficult problem since every complex is different and measures that work in one case may not work in another.

Given the amount of material needed for a contrast variation experiment, the care that must be taken to prepare the samples, and the cost and time involved in getting access to a SANS beamline, it is important that such experiments are well planned and that these issues are considered and tested well before committing to making samples for a complete set of contrast variation measurements. Steps that can be taken to insure sample purity are discussed in detail in (Jacques and Trewhella 2010). Thorough sample characterization under the conditions that are being used for SANS should be performed prior to the SANS experiments using not only biochemical assays such as SDS-PAGE and gel filtration, but also complementary physical characterization techniques such as size exclusion chromatography with multi-angle light scattering (SEC-MALS), dynamic light scattering (DLS), or analytical ultracentrifugation (AUC). The amount of deuteration in a deuterated component can be assessed by mass spectroscopy or deuterium NMR. Small-angle X-ray scattering (SAXS) measurements can also be very helpful to detect concentration or $D_2O$-dependent aggregation as will be discussed in more detail below.

### 5.3.2 Experiment Planning

For two component systems, the contrast of each component as well as that for the complex can be calculated from the chemical composition of each component and the stoichiometry of the complex as a function of the solvent composition. Two applications for performing this calculation are the MULCh (Whitten et al. 2008) software and the Contrast Calculator (Sarachan et al. 2013) module of the SASSIE (Curtis et al. 2012) software. For proteins and nucleic acids, the chemical composition is defined in terms of the amino acids or nucleotide bases, respectively, including the amount of H-D exchange and deuteration for each component. Non-protein or nucleic acid components can be entered based on their chemical formulas. Non-water solvent components such as salts and buffering compounds can be included in the calculation of the solvent scattering length densities to determine if they significantly alter the contrast match points. This is usually not the case if they are present in typical mmoles $L^{-1}$ (mM) quantities.

Both contrast calculating programs provide a web interface for easy access and use. While MULCh is limited to two-component complexes, it also provides options for data analysis that will be described further below. The Contrast Calculator module allows for an unlimited number of components in the complex and additionally calculates expected $I(0)$ values as a function of solvent composition. The predicted $I(0)$ values are invaluable in determining the concentration needed to obtain a measurable signal for each solvent condition. If atomic coordinates are available, such as from X-ray crystallography or NMR structures, model scattering intensity curves can be calculated at each contrast to perform the SANS experiment *in silico* prior to the actual experiment. Software such as CRYSON (Svergun et al. 1998) or the SasCalc (Watson and Curtis 2013) module of SASSIE can be used for this purpose. The experimental SANS data may not match that calculated from the model structure if, for instance, disordered residues are missing or if the structure doesn't accurately represent the complex in solution. Model structures must be complete if they are to be used as starting points for further modeling once SANS data are obtained. Thus, missing N or C terminal residues, internal loops, domains and/or disordered linkers must be added if necessary.

The calculations of the scattering length densities, contrast, $I(0)$ and model scattering intensity curves used to plan an experiment apply only to the coherent scattering component in Eq. 5.1. The hydrogen and deuterium in the biological complex also have an incoherent scattering component, which contributes a small amount to the total scattering in dilute solutions. On the other hand, the incoherent scattering from a solvent containing mostly $H_2O$:$D_2O$ is significant. In fact, it can often be greater than the calculated coherent intensity of a typical complex in a dilute solution. For example, the calculated coherent $I(0)$ value for a 1 mg $mL^{-1}$ protein-protein complex consisting of a trimer of the chaperone, Skp, bound to 50% deuterated outer membrane protein, OmpA, is 0.086 $cm^{-1}$ in $H_2O$ (i.e., 0% $D_2O$) and 0.04 $cm^{-1}$ in 98% $D_2O$ (Sarachan et al. 2013). However, the incoherent scattering intensity for 98% $D_2O$ is $\approx$ 0.07 $cm^{-1}$ and that of $H_2O$ is $\approx$ 1.0 $cm^{-1}$ (Rubinson et al. 2008). Thus, the incoherent component is the dominant contributor to the total scattering in $H_2O$, and remains a significant contributor in 98% $D_2O$.

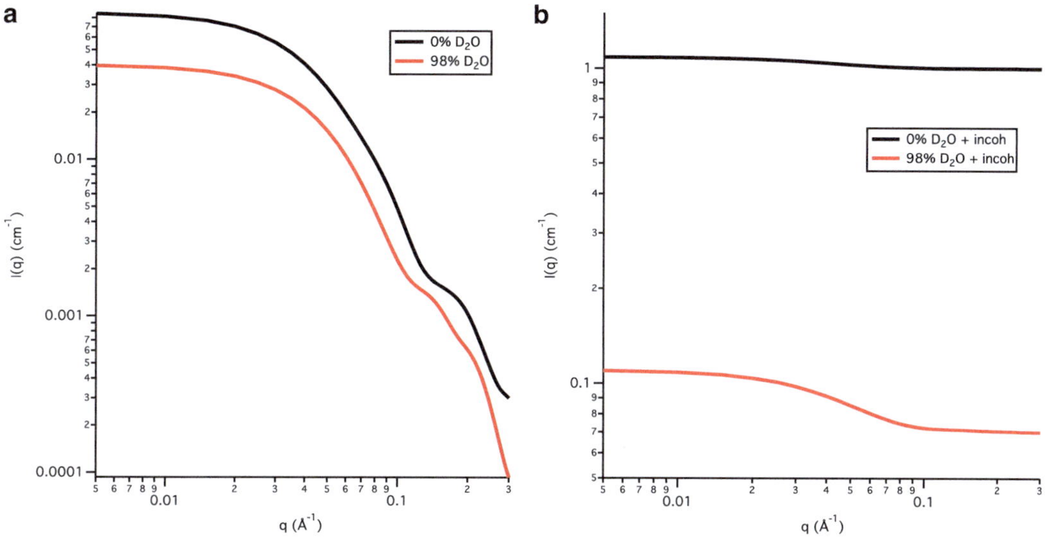


**Fig. 5.4** Calculated I(q) values for a Skp-OmpA complex in $H_2O$ (i.e., 0% $D_2O$) and 98% $D_2O$ solvents (**a**) without incoherent scattering and (**b**) with incoherent scattering. The coherent scattering curves were calculated using the SasCalc module in SASSIE

The calculated scattering curves for the Skp-OmpA complex are shown in Fig. 5.4 with and without the addition of the incoherent scattering component. The scattering that is actually measured is that shown in Fig. 5.4b. Under ideal conditions where there are no error bars on the data points in Fig. 5.4b and the incoherent background is measured equally well, the scattering curves in Fig. 5.4a could be recovered. However, this is not the case in practice. Features such as that shown at $q \approx 0.2$ Å$^{-1}$ for the complex in $H_2O$ (Fig. 5.4a) generally are not observable in practice since the incoherent component already significantly influences the total scattering at $q \approx 0.1$ Å$^{-1}$ (Fig. 5.4b). On the other hand, features such as those shown at 0.15 Å$^{-1}$ $\leq q \leq$ 0.25 Å$^{-1}$ for the complex in 98% $D_2O$ typically are observable after subtraction of the incoherent scattering component. This subtraction is accomplished by measuring the incoherent scattering from a solvent that matches that of the sample as closely as possible. Thus, dialysis is an ideal method for exchanging the biological complex into buffers with different $H_2O$:$D_2O$ ratios, as the dialysate solution can then be used to measure the incoherent background scattering.

The incoherent scattering from the solvent must be considered when planning data collection times. For a given calculated coherent *I* (0) value, it will take longer to obtain data in $H_2O$ than in $D_2O$ if the data are to have similar statistics after the incoherent background is subtracted. Furthermore, since the incoherent background is much higher in $H_2O$, the maximum *q* value obtained after buffer subtraction will be lower for data obtained in $H_2O$. Thus, when planning an experiment, samples prepared in solvents containing less than 50% $D_2O$ often must be measured a higher concentration so that the coherent *I*(0) is higher to make up for the higher incoherent background under these conditions. It should be noted that the scattering from water can contain both incoherent multiple scattering and inelastic scattering contributions and thus can differ depending on instrument conditions and sample geometry (Carsughi et al. 2000; Rubinson et al. 2008; Do et al. 2014). As the instrument scientists are most familiar with their instruments, it is important to rely on their advice for the *I*(0) values that can reasonably be measured at a given contrast at their facility.

Prior to performing a series of contrast variation measurements, two series of sample concentration measurements should be performed, one in $H_2O$ and one in >95% $D_2O$, to determine the maximum concentration at which the sample is dilute and monodisperse. These measurements also identify whether there are any effects due to the presence of $D_2O$ in the solvent. Discussions with instrument scientists, who are familiar with the range of $I(0)$ values that are measurable at their facility, provide the best resource in planning the range of sample concentrations to examine. This will avoid wasting time measuring the complex at sample concentrations that would result in $I(0)$ values that are too low to be measured. It should also be noted that the accuracy of the concentration is extremely important, as it is the main source of error for determining the $M_w$ from $I(0)$ (Eqs. 5.8 and 5.9).

The SANS curves from a complex often are significantly different in $H_2O$ and $D_2O$ solvents. Indeed, this is the reason for performing a contrast variation experiment in the first place. Therefore, the measurements described above are for comparison of the behavior of the complex as a function of concentration separately in $H_2O$ and $D_2O$. The concentration needed to satisfy the dilute and monodisperse conditions may be lower in $D_2O$ than in $H_2O$. If a SAXS instrument is available, SAXS measurements are a good option to determine whether there are any effects from either deuterium in the complex (when it consists of a deuterated and non-deuterated protein, for example) or $D_2O$ in the solvent. Since X-rays are not sensitive to the difference between hydrogen and deuterium, the scattering curves should ideally be the same regardless of sample or solvent deuteration. Another advantage is that SAXS measurements require smaller samples, on the order of 50 μL, requiring less sample preparation time for these important feasibility measurements. The SAXS measurement can also serve as another contrast point that can assist in modeling the structure of the complex.

Once optimal concentrations in $H_2O$ and $D_2O$ solvents have been found, then the feasibility of performing the contrast variation experiment can be further assessed by referring back to the predicted $I(0)$ values obtained for the complex as a function of $D_2O$ in the solvent. For instance, is there a conflict between the concentration needed to obtain a reasonable signal and that needed to obtain a dilute and monodisperse sample at a given solvent condition? Is there a way to mitigate the issue, perhaps by altering the solvent conditions slightly? Again, an experienced instrument scientist can be a valuable resource when considering these types of issues.

### 5.3.3 Data Collection and Reduction to $I(q)$ vs $q$

Data collection consists of measuring both the transmission of neutrons through the sample and the scattering of neutrons from the sample. Typically, these two different types of measurements require different instrument configurations and are performed separately. Since the sample consists of a biological complex in a solvent, these same measurements are also made for the solvent alone. Recall from Fig. 5.1 that the coherent scattering from the solvent is not observable. However, Fig. 5.4b shows that the incoherent scattering from a solvent containing mostly $H_2O$:$D_2O$ is significant and therefore must be subtracted from the total sample scattering, which also contains this incoherent component from the solvent. Since both the sample and solvent alone are measured in a sample holder, often a quartz cuvette or a demountable cell with quartz windows, the transmission and scattering from the empty holder (i.e, the empty cell) should also be measured.

Typical windows used for SANS such as quartz, aluminum and titanium have significant scattering in the forward direction. Subtracting the scattering of the empty cell from both the sample and the solvent alone will eliminate the scattering contribution from the windows so that the shape of the scattering from the sample and solvent can be accurately observed. The solvent scattering should then be approximately flat since it is mainly incoherent and, hence, not $q$-dependent (Eq. 5.1). Furthermore, the scattering observed from the sample is then attributed to the biological complex of interest plus the solvent (as in Fig. 5.4b) with no contribution from the sample holder. The transmission measurement is important for proper subtraction of the empty cell and for placing the data on an absolute scale as described, for example, in (Glinka et al. 1998).

An additional consideration when collecting data is the time spent measuring each sample. Counting times for transmission measurements from the sample, solvent and empty cell are typically just a few minutes. On the other hand, counting times for scattering measurements from the sample (complex plus solvent) are concentration, contrast and instrument dependent and should be determined in consultation with the instrument scientist. For dilute samples, the counting statistics on the sample and solvent alone should be equivalent since the incoherent scattering from the solvent dominates the total scattering at most $q$ values measured (Fig. 5.4). Thus, counting times for the sample and solvent alone should be the same. This is especially critical for measurements made under low contrast conditions. An exception can sometimes be made for a high $M_w$ complex, or for measurements made under high contrast conditions, if the scattering is significantly higher than that of the solvent alone in the low-$q$ region of the $I(q)$ vs $q$ curve. However, the counting statistics of the sample and solvent scattering should still be equal in the higher-$q$ portion of the curves, where the signal from the complex is weak. Since the empty cell scatters predominately in the forward direction, the counting times for these measurements can be shorter, matching only the statistics of the sample in the low-$q$ region.

In a dilute solution, the incoherent scattering from the hydrogen and deuterium in the biological complex itself is negligible compared to that from the solvent. While the background levels of both sample and solvent alone should be the same in this case, mismatches can occur for a variety of reasons. Some possible sources for these mismatches include statistical differences in transmission values, the influence of incoherent multiple scattering, micro- or macroscopic bubbles in the solution, and mismatch in $H_2O$:$D_2O$ ratio between the sample and solvent alone. However, under dilute conditions, the incoherent scattering from the solvent alone can be removed by scaling it to match the high-$q$ region of the total sample scattering and then subtracting it. The acceptability of the solvent subtraction can be assessed by not requiring that $P(r) = 0$ at $r = 0$ when calculating $P(r)$ from the SANS data (Jacques and Trewhella 2010). If $P(0)$ is positive or negative, then the solvent has been under- or over-subtracted, respectively.

This approximation, used to remove the solvent scattering by scaling it to match the sample scattering, does not apply to concentrated solutions (concentrations $\approx$ 10 mg mL$^{-1}$ and above for samples in a $D_2O$ buffer), where the incoherent scattering from the non-exchangeable hydrogen atoms in the complex itself is significant. As an alternative, it is tempting to simply subtract a constant representing the incoherent scattering from the solvent rather than subtract the measured scattering from the solvent alone, but this only works if the scattering from the solvent is truly flat. Typically the solvent scattering is not exactly flat due to incoherent multiple scattering, inelastic scattering, and other contributions(Carsughi et al. 2000; Rubinson et al. 2008; Do et al. 2014), with the shape depending on the $H_2O$:$D_2O$ conditions of the solvent. Given these issues, in practice it is quite difficult to accurately subtract the solvent scattering from the sample scattering. Thus, it is not uncommon to find a systematic background mismatch when comparing the calculated SANS curves to the measured SANS data during the structure modeling stage.

Methods of reducing data from the raw data collected on a 2D position-sensitive detector to the 1D $I(q)$ vs $q$ data vary depending on the facility. However, if the end result is a set of separate $I(q)$ vs $q$ curves on an absolute scale for both the sample and the solvent alone, the scattering from the solvent can be successfully subtracted. If the solvent subtraction is performed as part of the 2D data reduction, then it is more difficult to diagnose problems that often arise during the solvent subtraction step.

### 5.3.4 Data Quality Checks

#### 5.3.4.1 $R_g$ and $I(0)$

Once contrast variation data have been reduced and the contribution from the solvent has been subtracted, the basic parameters, $R_g$ and $I(0)$ should be obtained at all contrasts using both the Guinier and $P(r)$ analyses (Eqs. 5.11–5.13). Guinier analysis is usually provided by scattering facilities as part of the data reduction software. If $P(r)$ analysis isn't also included, a program such as GNOM (Semenyuk and Svergun 1991) or BayesApp (Hansen 2014) can be used. The experimental values of $I(0)$ are then compared to the calculated values at each contrast in order to confirm the integrity of the samples. The number density, $n$, is defined as in Eq. 5.9, but it is now in terms of the concentration and $M_w$ of the entire complex. Similarly, $\Delta\rho$ refers to the mean contrast of the entire complex and $V$ is the volume of the complex. The $M_w$ values should match the calculated values to within 10% if the data are on an absolute scale, $I(0)$ has an error of 1-2%, and the concentrations are measured to within 5%.

If the solution consists of a mixture of monomers and dimers of the complex or another combination of low-order oligomers, a valid Guinier region may still exist, but the $I(0)$ value will be larger than that of the monomeric complex. If there are larger aggregates of the molecule in the solution, the Guinier region will occur at smaller $q$ values than expected and Eq. 5.12 will not be linear in the $q$-range related to the expected size of the complex. Rather it will have some curvature and the fit to a straight line in the expected $q$-range will be poor. The effects can be subtle or very obvious depending on the severity of the aggregation (Jacques and Trewhella 2010). If aggregation is present, whether subtle or severe, the $R_g$ and $I(0)$ values no longer represent that of a monomer complex in a monodisperse solution. Rather, they are influenced by the larger aggregates present in the solution. The $P(r)$ function also changes in response to aggregation, primarily in the high-$r$ region. In cases of minor aggregation, it may still be possible to calculate a $P(r)$ corresponding to the monomer but the more severe the aggregation, the more difficult it is to determine $D_{max}$ (Jacques and Trewhella 2010).

#### 5.3.4.2 Contrast Match Point Analysis

To further confirm the integrity of the sample at all contrasts, the contrast match point of the of the complex can be determined from the experimental data. Expanding on Eqs. 5.8–5.10$I(0)$ can be written in terms of a two-component complex as,

$$I(0) = \frac{cM_w}{N_A}\left(f_1\Delta\rho_1\bar{v}_1 + f_2\Delta\rho_2\bar{v}_2\right)^2, \quad (5.14)$$

where $f_1$ and $f_2$ are the respective mass fractions of the first and second components in the complex. Now, $\Delta\rho_1$ and $\Delta\rho_2$ are respectively the scattering contrasts of the first and second components and $\bar{v}_1$ and $\bar{v}_2$ are respectively the partial specific volumes of the first and second components (Sarachan et al. 2013).

Because $I(0)$ is proportional to $c$ and to $(\Delta\rho)^2$, which is in turn dependent on the fraction of $D_2O$ in the solvent, $f_{D2O}$, the contrast match point of the complex can be determined from the x-intercept of a linear fit to $\sqrt{I(0)/c}$ vs $f_{D_2O}$. An example of this analysis is shown in Fig. 5.5, where experimental values of $\sqrt{I(0)/c}$ vs $f_{D_2O}$ are plotted for the aforementioned Skp-OmpA complex that was measured in 0%, 15%, 30% and 98% $D_2O$ buffer. An unweighted linear fit to the data resulted in a line with a slope of $-0.51 \pm 0.01$ and a y-intercept of $0.294 \pm 0.006$, where the errors represent one standard deviation. The x-intercept corresponding to the match point of the complex was calculated to be $0.576 \pm 0.014$, or $57.6\% \pm 1.4\%$ $D_2O$. This analysis can also be performed using MULCh (Whitten et al. 2008) and requires the experimental $I(0)$ values, the concentration of the complex and the fraction of $D_2O$ in the buffer at each contrast.

Comparison of this experimentally determined contrast match point to that calculated with $\Delta\rho_1$ and $\Delta\rho_2$ determined from the chemical composition of each component provides another quality assurance test on the data in that the

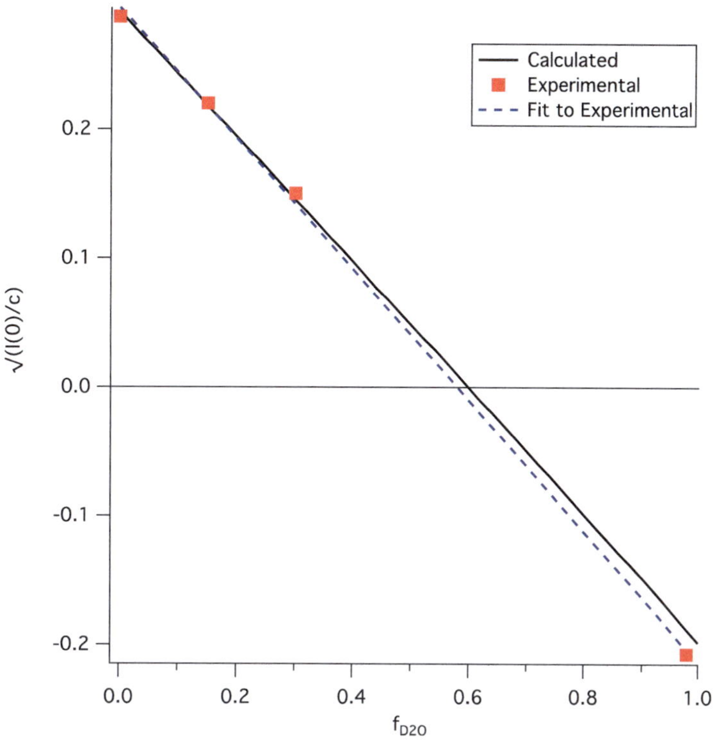


**Fig. 5.5** Experimental $\sqrt{I(0)/c}$ values as a function of $f_{D_2O}$ for a Skp-OmpA complex measured in 0%, 15%, 30% and 98% $D_2O$ buffers (*red squares*). The *blue dashed line* is the linear fit to the data while the *black solid line* is the calculated curve assuming 50% deuteration for OmpA

calculated and experimentally determined contrast match points should agree with each other. If that is the case, then there is confidence that the calculated match points of the individual components are correct. The match point calculated from the chemical composition of the Skp-OmpA complex assuming 50% deuteration for OmpA using the Contrast Calculator module of SASSIE (Sarachan et al. 2013) was 59.8% $D_2O$, in excellent agreement with the experimental result. It should be noted that the experimental and calculated complex match points can agree even if the slopes of the experimental and theoretical lines are quite different. The fit to the experimental data points should be compared to the theoretical curve as well. In this case, the data point for the 98% $D_2O$ sample falls slightly below the calculated line, which could mean that the sample concentration was lower than the value used in the analysis. Thus, even if the concentration is measured to a precision of ±5% prior to the experiment, the actual concentration of material in the beam can be affected by micro- or macro-bubbles in the sample, a small amount of precipitate in the sample (not uncommon for samples in $D_2O$ buffers), pipetting errors while diluting the sample and other such issues. Therefore, it is a good idea to recheck the sample concentrations after the SANS experiment if possible.

### 5.3.5 Component Analysis

#### 5.3.5.1 Component $R_g$ and Center of Mass (CM) Distance

The $R_g$ values obtained at each contrast are related by the relationship (Ibel and Stuhrmann 1975),

$$R_g^2 = R_m^2 + \frac{\alpha}{\Delta\rho} - \frac{\beta}{\Delta\rho^2}, \tag{5.15}$$

where $R_m$ is the $R_g$ value of the equivalent complex with a homogeneous scattering length density, $\alpha$ is the second moment of the density fluctuations and $\beta$ is the square of the first moment of the density fluctuations. For two component systems with different scattering length densities, the term $\alpha$ relates to the distribution of scattering length densities relative to the center of mass (CM) of the complex, and the term $\beta$ provides the separation of the scattering CM of the two components (Moore 1982). A Stuhrmann plot (Ibel and Stuhrmann 1975) of $R_g^2$ vs $(\Delta\rho)^{-1}$ (Eq. 5.15) is used to determine $R_m$, $\alpha$, and $\beta$. If the plot is linear, then $\beta = 0$ and the CM of the two components are concentric. In this case, the sign of the slope of the line, $\alpha$, is an indication of whether the component with the higher scattering length density is on the interior (negative) or exterior (positive) of the complex. Examples of parabolic and linear Stuhrmann plots are shown in Fig. 5.6. In practice, it is not always easy to distinguish between a parabolic and linear Stuhrmann plot, especially if $R_g$ doesn't change appreciably as a function of contrast ($\alpha$ is close to zero) or if $R_g$ values are not available close to the individual contrast match points of the two components.

Similar information can be obtained from the parallel axis theorem

$$R_g^2 = \frac{\Delta\rho_1 V_1}{\Delta\rho V} R_1^2 + \frac{\Delta\rho_2 V_2}{\Delta\rho V} R_2^2 + \frac{\Delta\rho_1 \Delta\rho_2 V_1 V_2}{(\Delta\rho V)^2} D_{CM}^2, \quad (5.16)$$

where $R_1$ and $R_2$ are the radii of gyration of the components and $D_{CM}$ is the distance between the scattering CM of the two components (Engelman and Moore 1975). Here, $\Delta\rho_1$, $\Delta\rho_2$, $V_1$ and $V_2$ refer to the individual components and $\Delta\rho$ and $V$ refer to the complex. The parallel axis theorem provides the radii of gyration of the components and the CM distance between them directly, whereas they are calculated from the definitions of $\alpha$ and $\beta$ when the Stuhrmann analysis is used. In both cases, $R_1$, $R_2$ and $D_{CM}$ are contrast-independent values that can be used in the structure modeling process. The MULCh software provides both of these analyses (Whitten et al. 2008). The inputs are the measured $R_g$ values as a function of $f_{D2O}$ in the solvent. The contrasts, $\Delta\rho$, $\Delta\rho_1$ and $\Delta\rho_2$, are determined from the chemical composition of each component based on the sequence information input earlier in the analysis.

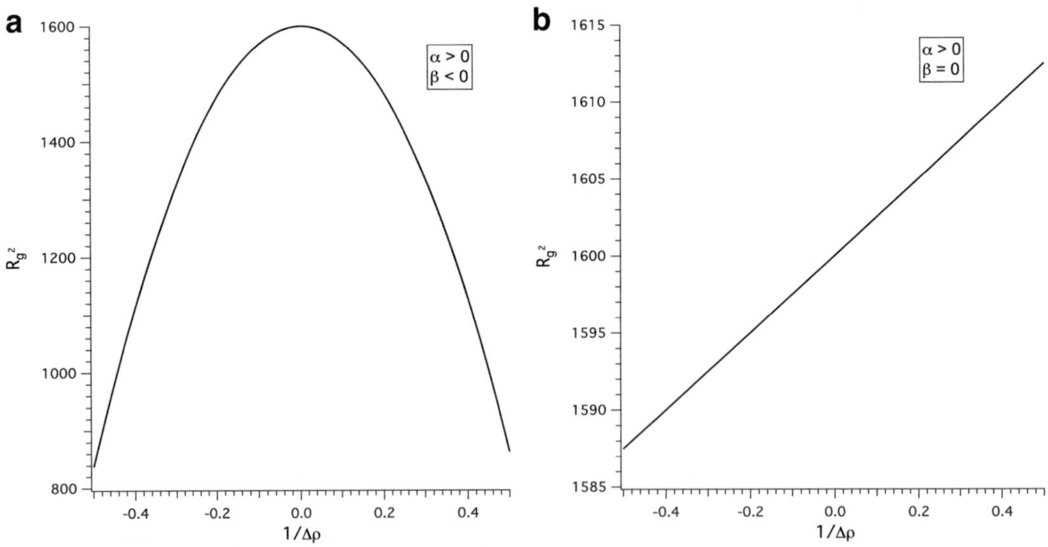


**Fig. 5.6** Examples of Stuhrmann plots from a two-component complex in which **a** the centers of mass are not concentric and **b** the centers of mass are concentric and the component with the higher scattering length density is on the exterior of the complex

#### 5.3.5.2 Component Scattering Intensities and Cross-Term

The scattering intensity from a two-component system with different scattering length densities can be written as (Whitten et al. 2008)

$$I(q) = \Delta\rho_1{}^2 I_1(q) + \Delta\rho_1\Delta\rho_2 I_{12}(q) + \Delta\rho_2{}^2 I_2(q), \quad (5.17)$$

where $I_1(q)$ and $I_2(q)$ are the scattering intensities of components 1 and 2, respectively, and $I_{12}(q)$ is the scattering intensity due to the interference between the two components. $I_1(q)$ and $I_2(q)$ are related to the shapes of the two components and $I_{12}(q)$ is related to their spatial distribution. For a given set of measured contrast variation intensities, $I(q)$, and known values for the contrasts, $\Delta\rho_1$ and $\Delta\rho_2$, the three unknowns, $I_1(q)$, $I_2(q)$ and $I_{12}(q)$, are found by solving the resultant set of linear equations at each $q$ value. Thus, data must be obtained at a minimum of three contrasts to solve for the three unknowns. The MULCh software (Whitten et al. 2008) provides this analysis. The measured $I(q)$ vs $q$ SANS curves and complex concentrations are required for each contrast and $\Delta\rho_1$ and $\Delta\rho_2$ are determined from the chemical composition of each component based on the sequence information input earlier in the analysis.

In practice, successful contrast variation studies that have resulted in the determination of $I_1(q)$, $I_2(q)$ and $I_{12}(q)$ have employed at least five strategically chosen contrast points including the contrast match points of the individual components, where high quality data were obtained (Whitten et al. 2008). An example of a successful application of this method can be found in a study of a kinase, KinA, in complex with an inhibitor, Sda (Whitten et al. 2007). Based on this study and an analysis of the corresponding theoretical scattering curves with noise added, a data collection strategy was recommended that includes a minimum of two contrast points on either side of the average match point of the entire complex (Whitten et al. 2008). This results in a well-spread range of contrast points that allows the individual $R_1$, $R_2$ and $D_{CM}$ to be obtained with good accuracy and precision.

For a complex consisting of a protein and nucleic acid or a protein and a deuterated protein (with a deuteration level such that the match point is between 60% $D_2O$ and 80% $D_2O$), a typical contrast variation data set would include 0% $D_2O$ and 100% $D_2O$, along with 20% $D_2O$, 40% $D_2O$ (the protein match point) and a fifth contrast point between 70% $D_2O$ and 80% $D_2O$ (near the match point of the second component). To obtain accurate scattering intensities of the components (Eq. 5.17), high quality data are needed at all measured contrasts. This analysis can be a useful tool to model the components separately and then arrange them with respect to each other in their proper position within the complex. However, structure modeling can proceed on the basis of the individual $R_1$, $R_2$ and $D_{CM}$ information alone, as is shown by example below.

### 5.3.6 Structure Modeling

Both SAXS and SANS are being used for structural determination of large biological complexes and for complexes containing flexible regions in solution. Many options are available for modeling multimeric biological complexes using a combination of rigid body and atomistic approaches, as described in recent reviews (Putnam et al. 2007; Rambo and Tainer 2010; Schneidman-Duhovny et al. 2012; Boldon et al. 2015). The SASSIE software suite (Curtis et al. 2012) is one tool that is available to assist in the atomistic and rigid body structure modeling of biological molecules for comparison to SAXS and SANS data. SASSIE provides users access to molecular dynamics, Monte Carlo, docking and rigid body modeling methods to assist in generating structure models and assessing how well models match the data. Constraints can be incorporated from other techniques such as NMR and AUC. SASSIE has been used for the structure modeling of many biological systems, including intrinsically disordered monomeric proteins (Curtis et al. 2012), large protein complexes (Krueger et al. 2011, 2014) and single-stranded nucleic acids (Peng et al. 2014). It has also been applied to the study of monoclonal antibodies using free energy analysis (Clark et al. 2013). A web version is available (https://sassie-web.chem.utk.edu/sassie2/) for ease of access and to handle the intensive computational requirements of the structural modeling and data analysis.

For a two-component complex, SANS and contrast variation experiments provide the added structural information from the individual components as constraints for modeling the entire complex. If obtainable, the scattering intensities of the separate components (Eq. 5.17) can be helpful for the modeling of the individual components and for construction of the model structure for the entire complex (Whitten et al. 2008). However, the contrast-independent $R_1$, $R_2$ and $D_{CM}$ distance constraints found by the Stuhrmann (Eq. 5.15) and parallel axis theorem (Eq. 5.16) analyses add unique information that can be used in the modeling process even in the absence of the component scattering intensities. Often, structural information for one or both of the components alone in solution is used as a starting point for their structures in the complex. Whether or not models are constructed from the scattering intensities of the separate components, the model SANS curves should always be judged against the entire contrast variation data set.

The first step in the structure modeling process is to construct starting models that satisfy the $R_1$, $R_2$ and $D_{\mathrm{CM}}$ (as well as $I_1(q)$, $I_2(q)$ and $I_{12}(q)$, if available) constraints found from the data analysis. Since these parameters have errors associated with them, several starting model structures that encompass the range of these parameters may be needed. Model structures that are consistent with the SANS data at all contrast conditions take full advantage of the information content of the contrast variation data set and provide the most robust representation of the data. Working model structures are first tested against the data by calculating both $R_g$ and the theoretical SANS curves from the model structures at each contrast measured. The theoretical SANS curves are then compared to the measured SANS curves, usually using a goodness of fit criterion such as the reduced $\chi^2$ equation

$$\chi^2 = \frac{1}{(N_{\mathrm{p}} - 1)} \sum_q \frac{\left(I_{\mathrm{exp}}(q) - I_{\mathrm{calc}}(q)\right)^2}{\sigma_{\mathrm{exp}}(q)^2}, \tag{5.18}$$

where $I_{\mathrm{exp}}(q)$ is the experimentally determined SANS intensity curve, $I_{\mathrm{calc}}(q)$ is the calculated intensity curve from the model structure and $\sigma_{\mathrm{exp}}(q)$ is the $q$-dependent error of the $I_{\mathrm{exp}}(q)$ values. The sum is taken over $N_{\mathrm{p}}$ independent data points.

The model structures producing the best-fit curves to the SANS data are then evaluated to verify that the $R_g$ values at each contrast match the values obtained from the experimental data. This process can be partially automated by plotting $\chi^2$ vs $R_g$ at each contrast to identify the structures that match the experimental $R_g$ values as well as give the best fits to the data. The chi-square filter module in SASSIE (Curtis et al. 2012) provides this analysis. The inputs are the SANS data at a given contrast and the calculated SANS curves from the model structures at that same contrast.

The subset of best-fit model structures at each contrast are then further filtered to obtain the best overall agreement with the entire SANS contrast variation data set. The global best-fit structures can be evaluated using a parameter such as the average reduced $\chi^2$ value

$$\chi^2(\mathrm{avg}) = \frac{1}{N_c} \sum_i \chi_i^2, \tag{5.19}$$

where $N_{\mathrm{c}}$ is the number of contrast variation $I(q)$ vs $q$ scattering curves and $\chi_i^2$ is the reduced $\chi^2$ value for the $i^{\mathrm{th}}$ scattering curve. An example of this procedure is found in a recent methods paper (Zaccai et al. 2016), in which complexes of the Skp chaperone with two different unfolded Omp proteins (uOmps) were studied, Skp-OmpA and Skp-OmpW.

Structure modeling is an iterative process, especially for complexes since the structures of two or more components and their spatial arrangement need to be determined. This can be illustrated by examining the recent Skp-OmpA and Skp-OmpW study (Zaccai et al. 2016) more closely. For both complexes, the uOmp component was determined to be 50% ± 5% deuterated from the contrast match point analysis. First, hybrid model structures for Skp-OmpW were tested in which the Skp component was modeled by an all-atom structure and OmpW was modeled by an ellipsoid of revolution. Once a working model structure was found that agreed with all of the contrast variation data, the structure of the Skp component was adopted for the Skp-OmpA complex as well.

In the Skp-OmpA complex, the OmpA component contained a periplasmic domain connected to the transmembrane domain by a flexible linker. The Complex Monte Carlo module of SASSIE (Curtis et al. 2012) was used to create an ensemble of possible conformations of the periplasmic domain that best match the entire contrast variation data set. In this case, both all-atom and ellipsoid models were used for the OmpA component. However, in both cases, there was a mismatch in the 98% $D_2O$ data at higher $q$ values, as shown in Fig. 5.7a. Since the scattering from the deuterated OmpA is weak at this contrast, the mismatch in the data was attributed to the Skp component, which was not varied from that found for Skp-OmpW (Zaccai et al. 2016). In fact, the 98% $D_2O$ scattering curves from Skp-OmpW and Skp-OmpA clearly differed in shape at the higher $q$ values. It was evident that the Skp trimer is more symmetric, with respect to the position of the three monomers relative to each other, in the Skp-OmpA complex than in the Skp-OmpW complex. This was an important finding that provided support to the notion that Skp changes its conformation to accommodate different uOmps (Zaccai et al. 2016).

To determine if a more symmetric Skp structure could be found to better agree with the 98% $D_2O$ data for the Skp-OmpA complex, additional structures of Skp were further examined. The $R_g$ values were calculated for 60 Skp structures that were recorded during a biased MD simulation

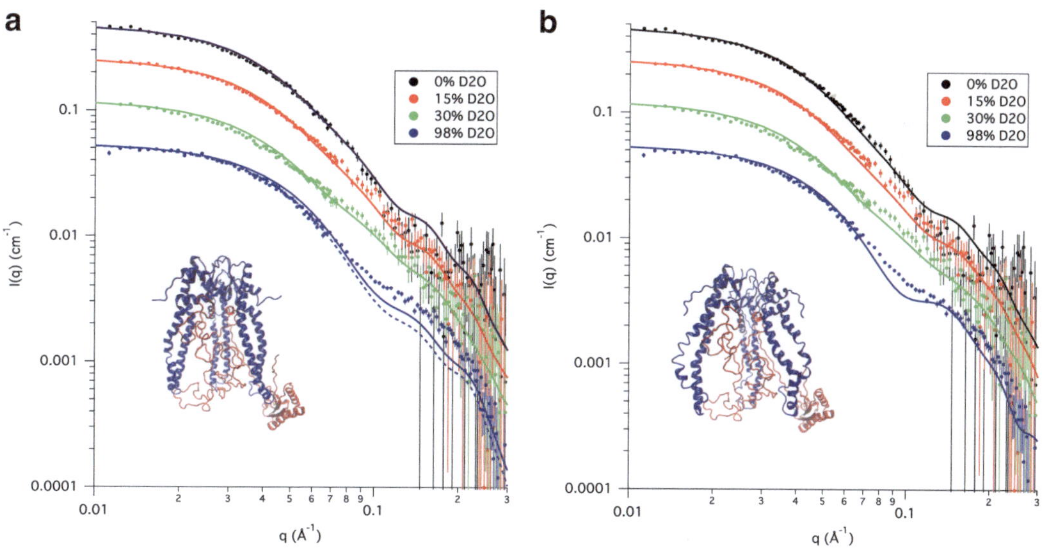


**Fig. 5.7** SANS contrast variation data for a Skp-OmpA complex. Solid lines represent the calculated SANS curves from the model structures in the inset assuming 60% deuteration for OmpA. Skp is shown in *blue* and OmpA is shown in *red*. **a** The model curves were calculated from the Skp-OmpA structure described in (Zaccai et al. 2016). The *dashed line* for the 98% $D_2O$ contrast was calculated assuming 50% deuteration for OmpA for comparison. **b** Best-matched model curve calculated using the same OmpA structure as in **a** paired with a more symmetric Skp structure. *Error bars* represent the standard error of the mean with respect to the number of pixels used in the data averaging

(Zaccai et al. 2016) in which the Skp monomers were splayed out to specific separations constrained to specific $R_g$ values. The ten structures that were in the best agreement with the $R_g$ value of Skp from the Stuhrmann analysis of Skp-OmpA were paired with the OmpA component from the best-matched Skp-OmpA structure. The new Skp-OmpA structures were checked for overlap between the basis CA atoms and then energy minimized using NAMD (Phillips et al. 2005). The SasCalc (Watson and Curtis 2013) module of SASSIE (Curtis et al. 2012) was then used to calculate their scattering curves. Two sets of calculations were performed assuming both 50% and 60% deuteration for the OmpA component. The model SANS curves were then compared to the SANS data at all contrasts using the Chi-Square Filter module of SASSIE (Curtis et al. 2012) to identify four structures that were in better agreement with the 98% $D_2O$ data. The model SANS curve from the overall best-matched structure is shown in Fig. 5.7b for each contrast. The Skp structures from the complexes shown in Fig. 5.7 have been superimposed in Fig. 5.8 for comparison. While the Skp structure in Fig. 5.7b is more symmetric with respect to the location of the three monomers as expected, there is still a mismatch in the range $0.06\ \text{Å}^{-1} \leq q \leq 0.1\ \text{Å}^{-1}$.

Given the flexibility of the monomers in the Skp trimer, it is not surprising that a single structure does not match the 98% $D_2O$ data. Mismatches of the type observed in Fig. 5.7b often are an indication of polydispersity. Since a large part of the OmpA structure is disordered, including the entire transmembrane (TM) domain that is encapsulated by Skp, it is also likely that the TM domain of OmpA exists in many different forms. Therefore, Skp likely takes on different structures to encapsulate these different OmpA TM structures. This is in addition to the conformations that can be assumed by the periplasmic domain of OmpA. While Skp and OmpA presumably exist in multiple different conformations in the Skp-OmpA complex, the SANS data reveal that an overall symmetry with respect to the configuration of the Skp monomers exists in the best-matched structural ensemble.

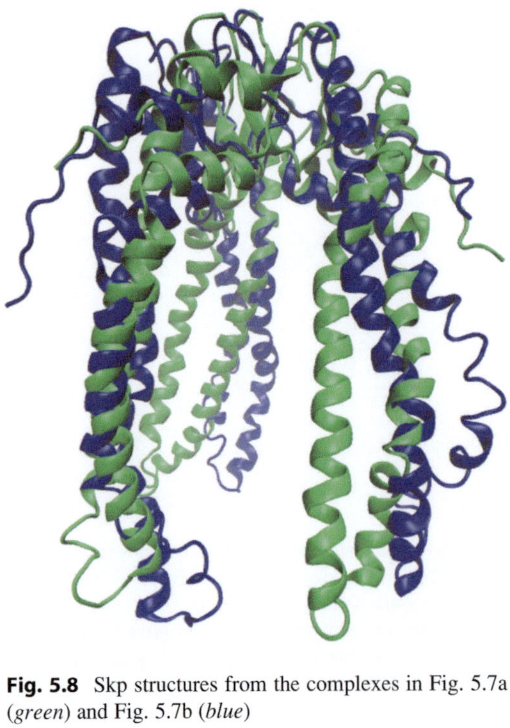

**Fig. 5.8** Skp structures from the complexes in Fig. 5.7a (*green*) and Fig. 5.7b (*blue*)

## 5.4 Concluding Remarks

Contrast variation combined with small-angle neutron scattering is a powerful tool for determining the structure of biological assemblies in solution. Contrast variation can be easily applied using neutrons due to the different scattering properties between hydrogen and deuterium. Through isotopic substitution of deuterium for hydrogen in both the molecule and/or solvent, the structures of individual components in a complex can be determined as well as their spatial arrangement. This represents unique information that cannot be easily obtained using other experimental techniques.

Experiments should be well-planned and the quality of the samples verified in advance to make the best use of beam time at neutron scattering facilities. Software tools are available to assist in experiment planning and facilities employ experienced instrument scientists who can offer assistance as well. Sample quality is extremely important for a successful contrast variation experiment. The integrity of the sample must be verified under all contrast conditions to have confidence in the data analysis and structure modeling.

After quality samples have been measured and the data reduced, model structures can provide valuable insight into the experimental system. Calculated SANS curves from model structures should be consistent with the SANS data at all contrasts. A wide variety of structure modeling software is becoming available to assist in developing and testing models to identify those that agree with SANS data.

**Tips for Performing a Successful SANS Contrast Variation Experiment**

**Sample Preparation**

- Prepare highly pure samples
  - SDS-PAGE, gel filtration to remove larger $M_w$ species
  - A280:260 nm to detect nucleic acid
  - DLS, AUC to assess aggregation
  - SEC-MALS for monodispersity
- Buffer and sample should match as closely as possible
- Dialyze into final buffer
- Measure sample concentration as accurately as possible
- Re-check after experiment
- Measure amount of deuteration in deuterated components as accurately as possible
  - Mass spectroscopy
  - Deuterium NMR

**Preliminary Calculations and Measurements**

- Measure sample at multiple concentrations in 0% and 100% $D_2O$ buffers before contrast variation experiment
- Assess concentration effects and $D_2O$-dependent aggregation
- Use SAXS if possible to also assess more subtle $D_2O$ effects
- Plan the experiment ahead of time
  - Calculate contrast match points of complex and components
  - Calculate expected I(0) values vs fraction of $D_2O$ in the solvent
  - Determine contrast conditions for measurement, sample concentrations and counting times (with help from beam line instrument scientist)

**Data Collection**

- Measure data on an absolute scale
- Measure the buffers for the same counting time as the samples

**Data Reduction and Analysis**

- Calculate $I(q)$ vs $q$ for samples and buffers
- Subtract buffer scattering
- Perform preliminary Guinier and $P(r)$ analysis to obtain $R_g$ and $I(0)$ at each contrast
- Verify calculated complex match point from experimental data
- Perform Stuhrmann and Parallel Axis Theorem analyses to obtain $R_1$, $R_2$, and $D_{CM}$
- Perform component analysis to obtain $I_1(q)$, $I_2(q)$ and $I_{12}(q)$ (if possible)

**Structure Modeling**

- Verify that model structures are complete (no missing residues, loops, domains, linkers)
- Verify that starting model structures satisfy $R_1$, $R_2$, $D_{CM}$ (and $I_1(q)$, $I_2(q)$ and $I_{12}(q)$)

(continued)

- Choose modeling method (s) appropriate for complex
  - Rigid body
  - Monte Carlo for flexible regions
  - Molecular Dynamics
- Best-fit structures must fit the entire contrast variation data set
- Narrow choices at each contrast and find subset that matches entire data set

## References


Ankner JF, Heller WT, Herwig KW et al (2013) Neutron scattering techniques and applications in structural biology. In: Coligan JE, Dunn BM, Speicher DW, Wingfield PT (eds) Current protocols in protein science. Wiley, Hoboken

Appolaire A, Girard E, Colombo M et al (2014) Small-angle neutron scattering reveals the assembly mode and oligomeric architecture of TET, a large, dodecameric aminopeptidase. Acta Crystallogr D Biol Crystallogr 70:2983–2993. doi:10.1107/S1399004714018446

Boldon L, Laliberte F, Liu L (2015) Review of the fundamental theories behind small angle X-ray scattering, molecular dynamics simulations, and relevant integrated application. Nano Rev. doi:10.3402/nano.v6.25661

Carsughi F, May RP, Plenteda R, Saroun J (2000) Sample geometry effects on incoherent small-angle scattering of light water. J Appl Crystallogr 33:112–117. doi:10.1107/S0021889899013643

Clark NJ, Zhang H, Krueger S et al (2013) Small-angle neutron scattering study of a monoclonal antibody using free-energy constraints. J Phys Chem B 117:14029–14038. doi:10.1021/jp408710r

Curtis JE, Raghunandan S, Nanda H, Krueger S (2012) SASSIE: a program to study intrinsically disordered biological molecules and macromolecular ensembles using experimental scattering restraints. Comput Phys Commun 183:382–389. doi:10.1016/j.cpc.2011.09.010

Do C, Heller WT, Stanley C et al (2014) Understanding inelastically scattered neutrons from water on a time-of-flight small-angle neutron scattering (SANS) instrument. Nucl Instrum Methods Phys Res Sect Accel Spectrometers Detect Assoc Equip 737:42–46. doi:10.1016/j.nima.2013.11.030

Engelman DM, Moore PB (1972) A new method for the determination of biological quarternary structure by neutron scattering. Proc Natl Acad Sci U S A 69:1997–1999

Engelman D, Moore PB (1975) Determination of quaternary structure by small-angle neutron scattering. Q Rev Biophys 4:219–241

Gabel F (2015) Small-angle neutron scattering for structural biology of protein–RNA complexes. Methods in enzymology. Elsevier, 391–415

Glatter O (1977) A new method for the evaluation of small-angle scattering data. J Appl Crystallogr 10:415–421. doi:10.1107/S0021889877013879

Glatter O (1982) Data treatment. In: Small-angle x-ray scattering. Academic Press, New York, pp 119–165

Glatter O, Kratky O (1982) Small-angle x-ray scattering. Academic Press, New York

Glinka CJ, Barker JG, Hammouda B et al (1998) The 30 m small-angle neutron scattering instruments at the national institute of standards and technology. J Appl Crystallogr 31:430–445. doi:10.1107/S0021889897017020

Guinier A, Fournet G (1955) Small-angle Scattering of X-rays. Wiley, New York

Hansen S (2014) Update for *BayesApp* : a web site for analysis of small-angle scattering data. J Appl Crystallogr 47:1469–1471. doi:10.1107/S1600576714013156

Heller WT (2010) Small-angle neutron scattering and contrast variation: a powerful combination for studying biological structures. Acta Crystallogr D Biol Crystallogr 66:1213–1217. doi:10.1107/S0907444910017658

Hoppe W (1973) The label triangulation method and the mixed isomorphous replacement principle. J Mol Biol 78:581–585. doi:10.1016/0022-2836(73)90480-4

Ibel K, Stuhrmann HB (1975) Comparison of neutron and x-ray scattering of dilute myoglobin solutions. J Mol Biol 93:255–265

Jacques DA, Trewhella J (2010) Small-angle scattering for structural biology – expanding the frontier while avoiding the pitfalls. Protein Sci Publ Protein Soc 19:642–657. doi:10.1002/pro.351

Jacques DA, Langley DB, Hynson RMG et al (2011) A novel structure of an antikinase and its inhibitor. J Mol Biol 405:214–226. doi:10.1016/j.jmb.2010.10.047

Jacrot B (1976) The study of biological structures by neutron scattering from solution. Rep Prog Phys 39:911–953. doi:10.1088/0034-4885/39/10/001

Krueger S, Shin J-H, Raghunandan S et al (2011) Atomistic ensemble modeling and small-angle neutron scattering of intrinsically disordered protein complexes: applied to minichromosome maintenance protein. Biophys J 101:2999–3007. doi:10.1016/j.bpj.2011.11.006

Krueger S, Shin J-H, Curtis JE et al (2014) The solution structure of full-length dodecameric MCM by SANS and molecular modeling: structure of dodecameric MCM helicase. Proteins Struct Funct Bioinf 82:2364–2374. doi:10.1002/prot.24598

May RP, Nowotny V (1989) Distance information derived from neutron low- $Q$ scattering. J Appl Crystallogr 22:231–237. doi:10.1107/S0021889888014281

Moore PB (1982) Small-angle scattering techniques for the study of biological macromolecules and macromolecular aggregates. In: Ehrenstein G, Lecar H (eds)

Methods of experimental physics. Academic, New York, pp 337–390

Neylon C (2008) Small angle neutron and x-ray scattering in structural biology: recent examples from the literature. Eur Biophys J 37:531–541. doi:10.1007/s00249-008-0259-2

Peng Y, Curtis JE, Fang X, Woodson SA (2014) Structural model of an mRNA in complex with the bacterial chaperone Hfq. Proc Natl Acad Sci 111:17134–17139. doi:10.1073/pnas.1410114111

Phillips JC, Braun R, Wang W et al (2005) Scalable molecular dynamics with NAMD. J Comput Chem 26:1781–1802. doi:10.1002/jcc.20289

Putnam CD, Hammel M, Hura GL, Tainer JA (2007) X-ray solution scattering (SAXS) combined with crystallography and computation: defining accurate macromolecular structures, conformations and assemblies in solution. Q Rev Biophys 40:191–285. doi:10.1017/S0033583507004635

Rambo RP, Tainer JA (2010) Bridging the solution divide: comprehensive structural analyses of dynamic RNA, DNA, and protein assemblies by small-angle x-ray scattering. Curr Opin Struct Biol 20:128–137. doi:10.1016/j.sbi.2009.12.015

Rubinson KA, Stanley C, Krueger S (2008) Small-angle neutron scattering and the errors in protein structures that arise from uncorrected background and intermolecular interactions. J Appl Crystallogr 41:456–465. doi:10.1107/S0021889808004950

Sarachan KL, Curtis JE, Krueger S (2013) Small-angle scattering contrast calculator for protein and nucleic acid complexes in solution. J Appl Crystallogr 46:1889–1893. doi:10.1107/S0021889813025727

Schneidman-Duhovny D, Kim S, Sali A (2012) Integrative structural modeling with small angle X-ray scattering profiles. BMC Struct Biol 12:17. doi:10.1186/1472-6807-12-17

Semenyuk AV, Svergun DI (1991) GNOM – a program package for small-angle scattering data processing. J Appl Crystallogr 24:537–540. doi:10.1107/S002188989100081X

Serdyuk IN, Zaccai G (1996) The triple isotopic substitution method in small-angle neutron scattering: application to studying macromolecular complexes. J Mol Struct 383:197–200. doi:10.1016/S0022-2860(96)09286-1

Svergun DI (2010) Small-angle X-ray and neutron scattering as a tool for structural systems biology. Biol Chem. doi:10.1515/bc.2010.093

Svergun DI, Koch MHJ (2003) Small-angle scattering studies of biological macromolecules in solution. Rep Prog Phys 66:1735–1782. doi:10.1088/0034-4885/66/10/R05

Svergun DI, Richard S, Koch MHJ et al (1998) Protein hydration in solution: experimental observation by x-ray and neutron scattering. Proc Natl Acad Sci 95:2267–2272

Watson MC, Curtis JE (2013) Rapid and accurate calculation of small-angle scattering profiles using the golden ratio. J Appl Crystallogr 46:1171–1177. doi:10.1107/S002188981301666X

Whitten AE, Trewhella J (2009) Small-angle scattering and neutron contrast variation for studying bio-molecular complexes. In: Foote RS, Lee JW (eds) Micro and nano technologies in bioanalysis. Humana Press, Totowa, pp 307–323

Whitten AE, Jacques DA, Hammouda B et al (2007) The structure of the KinA-Sda complex suggests an allosteric mechanism of histidine kinase inhibition. J Mol Biol 368:407–420. doi:10.1016/j.jmb.2007.01.064

Whitten AE, Cai S, Trewhella J (2008) *MULCh* : modules for the analysis of small-angle neutron contrast variation data from biomolecular assemblies. J Appl Crystallogr 41:222–226. doi:10.1107/S0021889807055136

Zaccai G (2012) Straight lines of neutron scattering in biology: a review of basic controls in SANS and EINS. Eur Biophys J EBJ 41:781–787. doi:10.1007/s00249-012-0825-5

Zaccai NR, Sandlin CW, Hoopes JT, et al (2016) Deuterium labeling together with contrast variation small-angle neutron scattering suggests how Skp captures and releases unfolded outer membrane proteins. In: Kelman Z (ed) Methods in enzymology. Elsevier, pp 159–210

# SAS-Based Structural Modelling and Model Validation

6

Maxim V. Petoukhov and Anne Tuukkanen

**Abstract**

Small angle scattering of X-rays (SAXS) and neutrons (SANS) is a structural technique to study disordered systems with chaotic orientations of scattering inhomogeneities at low resolution. An important example of such systems are solutions of biological macromolecules. Rapid development in the methodology for solution scattering data interpretation and model building during the last two decades brought the analysis far beyond the determination of just few overall structural parameters (which was the only possibility in the past) and ensured SAS a firm position in the methods palette of the modern life sciences. The advances in the methodology include *ab initio* approaches for shape and domain structure restoration from scattering curves without *a priori* structural knowledge, classification and validation of the models, evaluation of potential ambiguity associated with the reconstruction. In rigid body and hybrid modelling approaches, solution scattering is synergistically used with other structural techniques utilizing the complementary information such as atomic models of the components, intramolecular contacts, subunits orientations etc. for the reconstruction of complex systems. The usual requirement of the sample monodispersity has been loosed recently and the technique can now address such systems as weakly bound oligomers and transient complexes. These state-of-the-art methods are

M.V. Petoukhov (✉)
Hamburg Unit, European Molecular Biology Laboratory, EMBL c/o DESY, Notkestrasse 85, 22607 Hamburg, Germany

Federal Scientific Research Centre “Crystallography and Photonics”, RAS, Leninsky prospect 59, 119333 Moscow, Russia

A.N. Frumkin Institute of Physical Chemistry and Electrochemistry, RAS, Leninsky prospect 31, 119071 Moscow, Russia
e-mail: maxim.petoukhov@embl-hamburg.de

A. Tuukkanen
Hamburg Unit, European Molecular Biology Laboratory, EMBL c/o DESY, Notkestrasse 85, 22607 Hamburg, Germany
e-mail: anne.tuukkanen@embl-hamburg.de





B. Chaudhuri et al. (eds.), *Biological Small Angle Scattering: Techniques, Strategies and Tips*, Advances in Experimental Medicine and Biology 1009, DOI 10.1007/978-981-10-6038-0_6

described together with the examples of their applications and the possible ways of post-processing of the models.

**Keywords**
Solution scattering • *Ab initio* reconstruction • Model validation • Hybrid modelling • Ambiguity assessment • Multisubunit complexes • Modular proteins

## 6.1 Introduction

Small-angle scattering (SAS) is a universal low-resolution structural tool to analyze individual biological macromolecules and complex systems and processes permitting one to assess broad range of sizes and molecular weights. The method can be used on its own but is mostly effective when synergistically employed with other biophysical, biochemical and structural techniques. SAS is not only applied for determination of the overall parameters of the solute such as molecular mass (*MM*), radius of gyration ($R_g$), hydrated volume ($V$) maximum diameter ($D_{max}$) and distance distribution function *p(r)* but most importantly it allows one three-dimensional (3D) analysis of the quaternary structure from one-dimensional scattering profiles collected from dilute and preferably monodisperse solutions of macromolecules. This chapter will describe various modelling approaches to analyze the SAS data developed in the last two decades, ranging from *ab initio* methods to rigid body and hybrid modelling techniques and the tools for validation of the obtained results. The developed approaches ensured SAS a deserved place in the methods portfolio of a modern structural biologist.

The general principle of the SAS-based structural modelling is to find optimal values of the parameters describing the model (the number of parameters may vary from few to thousands depending on the approach) that yield the computed scattering intensity fitting the experimental solution scattering data. In a generic modelling approach, this search is performed to minimize discrepancy $\chi^2$ between the experimental profile $I_{exp}(s)$ and that computed from the model $I_{calc}(s)$:

$$\chi^2 = \frac{1}{N-1}\sum_{j=1}^{N}\left[\frac{I_{\exp}(s_j) - cI_{calc}(s_j)}{\sigma(s_j)}\right]^2 \qquad (6.1)$$

where $c$ is a scaling factor, $N$ is the number of angular points and $\sigma$ denotes experimental errors. Acceptable $\chi^2$ value from a correct model should be around 1.0 provided that the experimental errors are correctly estimated.

The reconstruction of a three-dimensional model of an object from its one-dimensional scattering pattern is an ill-posed problem, as different structures may provide one and the same scattering profile. In practice to reduce the ambiguity, a target function $E$ is minimized containing, in addition to the discrepancy term, a set of weighted penalties formulating additional requirements to the model:

$$E = \chi^2 + \sum \alpha_i P_i \qquad (6.2)$$

The penalty terms $P_i$ can impose both "natural" physical constraints (e.g. the requirement of interconnectivity of the entire model) and restrictions accounting for additional structural information from complementary structural, computational and biochemical techniques.

## 6.2 *Ab Initio* Bead Modelling

A classical task in SAS-based structural analysis is *ab initio* restoration of the particle shape, which can be done in the absence of any *a priori* structural information. A simplification reducing the ambiguity of shape reconstruction consists in representing the particle at low resolution as a homogeneous body. This simplification yields the valid s-range for shape reconstruction up to 0.2–0.3 $Å^{-1}$. Originally the reconstructions were performed by 3-parameters geometrical bodies (Feigin and Svergun 1987; Glatter and Kratky 1982) or using an angular envelope function (Svergun et al. 1996) based on the spherical harmonics. The latter approach, although allowing one to construct more complicated bodies, still provided models with relatively simple shapes (i.e. with no internal holes). Nowadays, shape determination utilizes models represented by finite volume elements (for example densely packed beads on a regular hexagonal grid first proposed by Chacon et al. 1998). Given a large number of the beads (typically, several thousands), a Monte-Carlo-based search is applied for the target function minimization which allows one to avoid the pitfalls (trapping in the local minima) of the conventional gradient methods. In a popular approach called DAMMIN (Svergun 1999), simulated annealing (SA) is employed to assign beads in the search volume (typically, a sphere with diameter $D_{max}$) either to the particle or to the solvent to yield a configuration with the optimal $E$. The search starts with a random distribution of the phase indices solvent/particle and at each step the phase of a randomly selected bead is toggled, whereby all the beads positions remain fixed (see Fig. 6.1). The penalty terms used here ensure that a compact interconnected ensemble of beads is built at the end of the procedure as a molecule cannot consist of separate fragments, and the modelling is done at low resolution in homogeneous approximation such that reconstruction of fine details is not justified. Non-spherical search volumes including e.g. those from electron microscopy (EM) may also be used. In a newer method DAMMIF which is now available (Franke and Svergun 2009), an adaptive search volume is employed and permanent interconnectivity of the model is required from the start of the modelling. This makes the algorithm more efficient and allows to gain the speed by 20–40 times. In both approaches, symmetry can be taken into account, which is very useful in reconstructions of oligomeric particles like (homo-) dimers, tetramers etc. From practical point of view both DAMMIN and DAMMIF require as an input a special file with pair distance distribution function $p(r)$ produced by the indirect Fourier transform package GNOM (Svergun 1992, 1993).

For the macromolecular assemblies consisting from parts with different contrasts (e.g. nucleoproteins or protein complexes studied by SANS contrast variation with selective labeling of specific subunits), a multiphase bead modelling approach (Svergun and Nierhaus 2000) can be applied. In this case the beads are assigned to a solvent or to a specific component, whereby the components may have different scattering densities. Typically, multiple scattering curves fitting is performed, e.g. from contrast variation series or when individual scattering profiles of the subunits are also available in addition to the scattering from the entire particle.

## 6.3 Dummy Residues Modelling

An *ab initio* modelling method developed specifically for the structural analysis of proteins utilizes a concept of so called dummy residues (DR) (Svergun et al. 2001). In this approach, implemented in the program GASBOR, each amino acid of a protein disregarding of its size is represented by an averaged DR with an effective average scattering form factor and centered at the approximate $C_{\alpha}$ atom position. Therefore the approach does not assume particle homogeneity providing more adequate representation of the internal structure. The amount of DRs equals to the number of amino acids in the primary sequence of the protein. The modelling procedure starts from a configuration with random positions of DRs, and a SA search is performed of the spatial distribution of DRs inside a

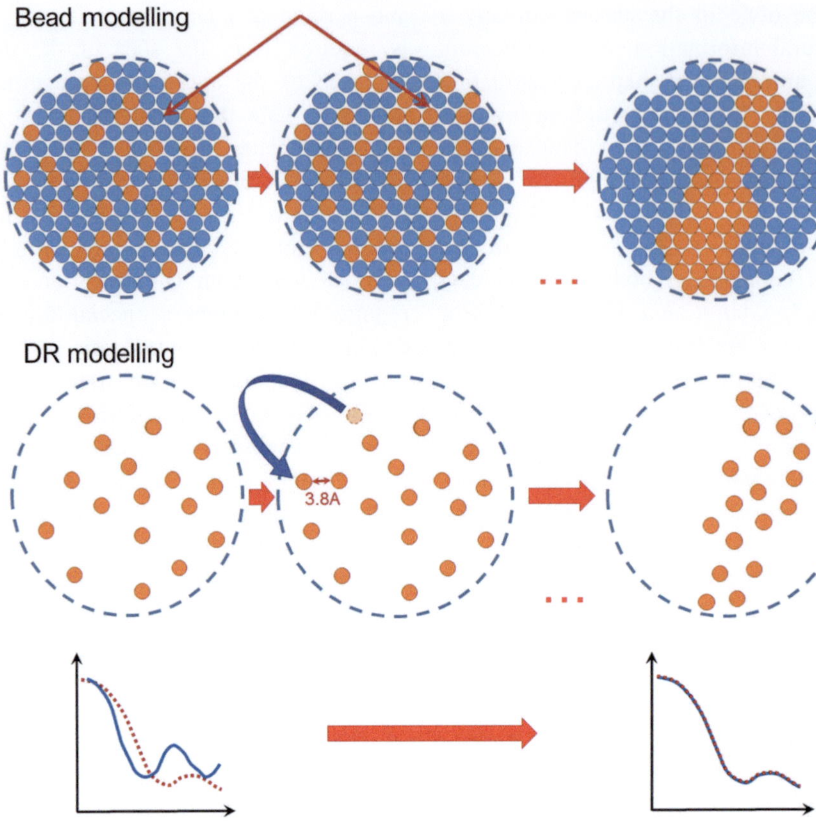


**Fig. 6.1** Comparison of *ab initio* methods. In bead modelling approach the positions of the beads remain fixed on a hexagonal grid and only their phases are toggled (between solvent and particle) during SA procedure. In dummy residues approach, at each SA step a DR is moved to a new location at 3.8 Å from another DR. Target functions being minimized in both scenarios ensure that a physically meaningful model is obtained at the end of the procedure that fits the experimental SAXS data

spherical search volume with diameter $D_{max}$ equal to the maximum particle size. At each step a randomly selected DR is moved to a distance of 3.8 Å from another randomly selected DR, so that the parameters being optimized are the coordinates of DRs (Fig. 6.1). The resulting assembly is required to be locally compatible with a typical distribution of backbone $C_\alpha$ atoms in proteins, and the scattering computed from this assembly should fit the experimental data. The protein-like distribution of DRs in an acceptable model is ensured by penalty-driven restraint that controls the histogram of nearest neighbors of a DR within ten concentric shells of 1 Å width.

## 6.4 Evaluation of the Modelling Convergence and Averaging

Modelling path in Monte-Carlo-based minimization depends on the random seed and multiple runs of the algorithm are typically performed producing potentially different solutions. The models obtained in individual runs are then compared with each other to assess the convergence of the reconstruction and to average the models. The post-processing of multiple *ab initio* results employs a normalized spatial discrepancy NSD (Kozin and Svergun 2001) criterion as a measure of dissimilarity between individual models:

$$NSD(S_1, S_2) = \left\{ (1/2) \left[ \left( \frac{1}{N_1 d_2^2} \right) \sum_{i=1}^{N_1} \rho^2(s_{1i}, S_2) + \left( \frac{1}{N_2 d_1^2} \right) \sum_{i=1}^{N_2} \rho^2(s_{2i}, S_1) \right] \right\}^{1/2} \tag{6.3}$$

where $N_i$ is the number of points $s_i$ in the structure $S_i$, the fineness $d_i$ is the average distance between the neighbouring points in $S_i$ and ρ is the minimum value among the distances between the given point and all points in the other structure. Two similar structures typically yield NSD values below or around unity, if NSD exceeds 2.0 the structures are substantially dissimilar.

The most typical model (having the lowest average NSD to all the others) is selected and used as a reference for the generation of the average model (Volkov and Svergun 2003) whereas possible outliers (with a significantly higher average NSD) are also identified and discarded. After the alignment of all the models but the outliers with the reference one, the entire collection of beads is remapped onto a densely packed hexagonal grid so what each grid point is characterized by its occupancy factor depending on the beads density in its vicinity. The ensemble of grid points with non-zero occupancies forming a total spread region is filtered and a portion of beads with higher occupancies is selected to yield the volume equal to the average excluded volume of all the reconstructions. Such most populated volume represents the averaged model of multiple *ab initio* reconstructions.

The obvious limitation of this procedure is that the excluded outliers, which also fit the scattering data and fulfil the modelling restraints, may in fact be the true positives. In a more versatile approach (Petoukhov et al. 2012) none of the models are discarded and instead all are clustered into groups, such that each group contains similar models. The evaluation of the effective distance (dissimilarity) between the models and division into clusters is performed based either on NSD criteria or using root-mean-square-deviation (RMSD) in case of the models with one-to-one correspondence between the volume elements. Within each cluster, the most typical model is selected and the averaged shape is built in the same way as described above. Comparison between the representatives of individual clusters provides the possibility to evaluate the non-uniqueness of the reconstruction.

## 6.5 Evaluation of an Inherent Ambiguity of Scattering Curve

It is generally not known beforehand, what is the adequate number of Monte-Carlo runs to reliably cover the conformational space in order not to miss the true positive. Recently a tool has been developed (Petoukhov and Svergun 2015) which addresses the ambiguity of shape reconstruction associated with a given scattering profile from ideal monodisperse solution. The approach is based on a pre-generated comprehensive library of scattering patterns from shape skeletons describing a manifold of low-resolution particle shape topologies. The shape skeletons are represented by several (up to seven) beads closely packed on a hexagonal grid. The corresponding scattering profiles are mapped onto a normalized scale $I/I_0$ vs $sR_g$ (Fig. 6.2) eliminating the size information and keeping the topology information only. The amount of the skeleton patterns similar to the given SAS experimental profile provides a measure of ambiguity associated with the data set and thus of potential non-uniqueness of the *ab initio* shape restoration from it. For profiles with high ambiguity measure a larger number of reconstructions (possibly above hundred) is necessary to cover the solution space whereas less ambiguous profiles can be evaluated in just few modelling attempts. This approach called AMBIMETER thus allows one to meaningfully define the shape reconstruction strategy. For highly ambiguous scattering profiles,

**Fig. 6.2** A density map of normalized scattering profiles computed from shape skeletons made of up to seven densely packed beads and describing a manifold of low-resolution particle shape topologies. Experimental scattering profiles passing through the more populated areas have higher propensity to ambiguous shape reconstructions

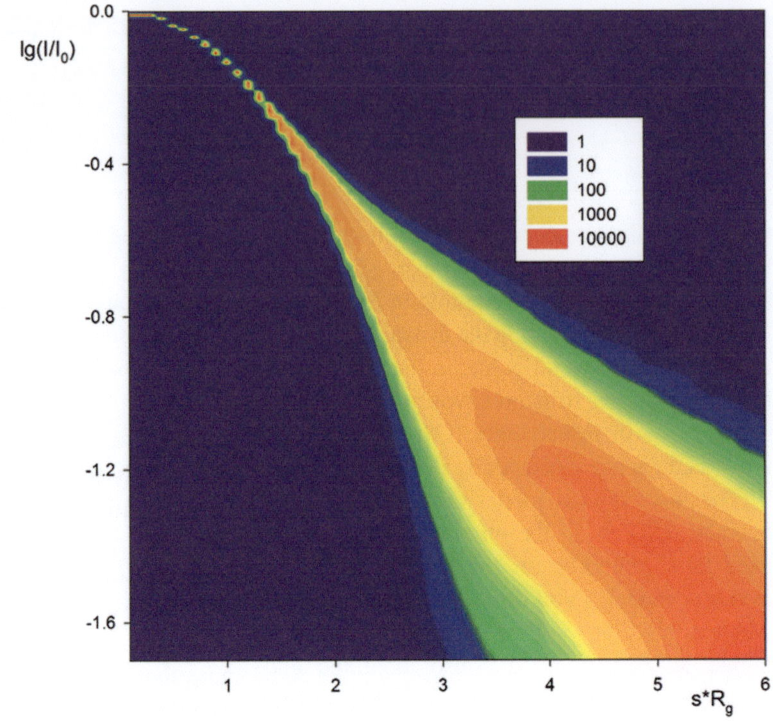


reconstructions with symmetry restrictions, if applicable, is recommended.

## 6.6 Quality Assessment and Resolution of *Ab Initio* Models

Quality assessment of SAS-based *ab initio* models is essential for their biological interpretation and further use. The previously described averaging procedure and evaluation of ambiguity provide indirect quality estimates of the reconstructed model ensemble but not quantitative assessment of the quality in terms of resolution. The amount of structural information that can be extracted from a SAS profile and, thus, the accuracy of produced *ab initio* models depend on several other factors. The sources of systematic and random errors include the signal-to-noise ratio of the employed SAS profile, the measured data range, and the selected modeling approach. In X-ray crystallography, resolution is defined based on the highest diffraction peak that can be distinguished from the background in the reciprocal space. The maximum momentum transfer value $s_{max}$ of a SAS profile gives however only the nominal theoretical limit of resolution that can be achieved because SAS data is inherently ambiguous as a result of the rotational averaging. The employed *ab initio* modelling method and its level of coarse-graining (use of dummy beads or dummy residues) also affect the highest resolution that can be reached. Several methods are available for MX and NMR to assess the structural quality of models with knowledge-based scores which evaluate how models fit with the known stereo-chemical features of proteins such as the atom distance distributions found in high-resolution structures (e.g. programs like CING (Doreleijers et al. 2012) or MolProbity (Chen et al. 2010)). As SAS *ab initio* models do not contain information at atomic detail, such a validation-based approach is not applicable for solution scattering. For NMR models, the structure quality is also assessed based on the root-mean-square-deviation (RMSD) of the reconstructed atomic ensembles compatible with the data. In contrast to NMR ensembles, multiple SAS models do not have a one-to-one bead correspondence and the calculation of the RMSD is not possible. The resolution of Electron cryo-Microscopy (EM) models is commonly estimated by Fourier Shell Correlation (FSC) method, where the electron density maps reconstructed from two distinct sets of experimental images are compared in reciprocal space (Penczek 2010). This method is not directly applied to SAS given that a single experimental scattering profile cannot be split into two independent datasets. Until recently, no objective resolution criteria such as existing for X-ray crystallographic or EM structures were available for SAS-based models. To address this problem, a novel FSC based approach was developed to evaluate the resolution of an *ab initio* reconstruction using an ensemble of models (Tuukkanen et al. 2016). The variability of multiple SAS models derived from a single scattering dataset is utilized to estimate the model resolution. It was demonstrated that the average FSC function over an ensemble that reflects the variability of models is directly related to the resolution of the individual models in the shape reconstruction. If $A$ and $B$ are two *ab initio* models and $A(\mathbf{s})$ and $B(\mathbf{s})$ are their three-dimensional scattering amplitudes (here, $\mathbf{s}$ is the scattering vector in reciprocal space), a one-dimensional FSC is defined as the function of the momentum transfer

$$FSC(s) = \frac{\sum_{[s,\Delta s]} \boldsymbol{A}(\boldsymbol{s}) \cdot \boldsymbol{B}^*(\boldsymbol{s})}{\sqrt{\sum_{[s,\Delta s]} |\boldsymbol{A}(\boldsymbol{s})|^2 \sum_{[s,\Delta s]} |\boldsymbol{B}(\boldsymbol{s})|^2}} \quad (6.4)$$

where $[s, \Delta s]$ are the radius and width of the spherical shell in reciprocal space. In general, FSC decreases with increasing momentum transfer reflecting the loss of structural similarity with increasing resolution. The variability measure $\Delta_{ens}$ is defined as $2\pi/s_{ens}$, where $s_{ens}$ is the momentum transfer value at which the average FSC drops below 0.5. This approach was benchmarked against a set of proteins with known high-resolution structures. The cross-correlated resolution of an ensemble $\Delta_{SAS}$ (representing the actual resolution of the *ab initio* models) was determined from the averaged pairwise FSC functions between SAS models and the high-resolution structure at the same cut-off value for benchmarking purposes. The variability measure $\Delta_{ens}$ demonstrated excellent correlation with the cross-validated resolution $\Delta_{SAS}$ in the benchmark protein set. This allows one to directly estimate the resolution of *ab initio* models from the variation of the ensembles using linear regressions models. Figure 6.3 shows the averaged FSC function between the *ab initio* bead models based on simulated SAXS data from myoglobin and its crystal structure (PDB code: 1WLA) (blue) and the averaged pairwise FSC between the *ab initio* models (red). The FSC approach implemented in the program SASRES consists of four steps: First, the *ab initio* models reconstructed from a given SAS profile are structurally aligned pairwise using the programs SUPCOMB or SUPALM (Kozin 2001; Konarev 2001). Then scattering amplitudes of the aligned models are computed and the pairwise FSC functions evaluated using spherical harmonics. Finally, the model variability and resolution are determined based on the average of the pairwise FSC functions (Fig. 6.4). The FSC-based resolution analysis is suggested to become a standard step of *ab initio* modeling and the resolution to be reported in publications and depositions of SAS data and models.

## 6.7 *Ab Initio* Modelling Workflow

As a summary, we propose the following steps in the workflow of SAS-based *ab initio* shape reconstruction: (1) The pair distance distribution function $p(r)$ which is the starting input of the *ab initio* modelling process is determined for the SAS data in question e.g. with the program GNOM, (2) The ambiguity of the SAS data and the number of independent *ab initio* modelling runs required can be estimated using the AMBIMETER approach, (3) The appropriate modelling program is selected depending on the desired level of details in models and the available data range, either a dummy bead (DAMMIN/DAMMIF) or a dummy residue

**Fig. 6.3** Resolution assessment for *ab initio* models of myoglobin. *Red*, the averaged FSC between the *ab initio* bead models and the high-resolution X-ray crystal structure (PDB code: 1WLA) yielding cross-validated resolution $\Delta_{cc}$; *blue*, the averaged pairwise FSC between the *ab initio* models providing the variability of the ensemble $\Delta_{ens}$

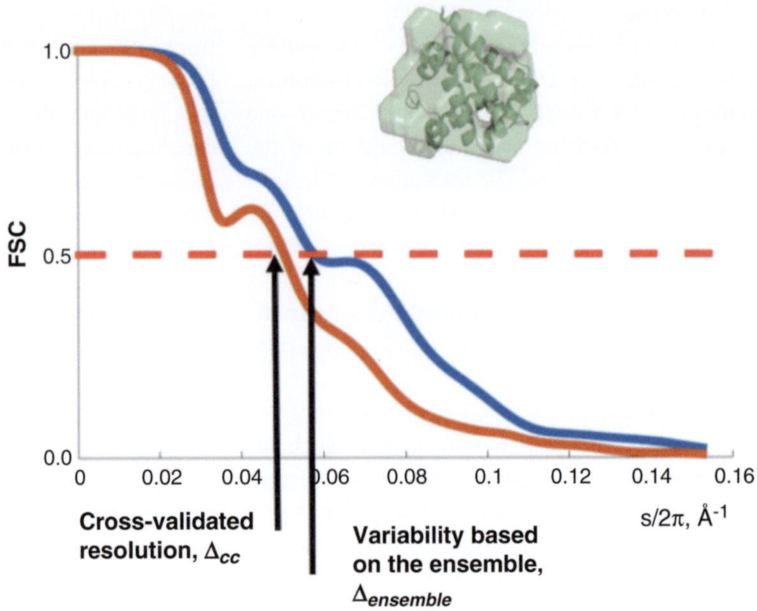


representation (GASBOR), and several independent modelling runs are performed, (4) The independently generated models are compared by structurally aligning and clustering them using the NSD-metric (e.g. with the program DAMAVER), (5) The most-typical model having the lowest NSD value to all other models can be used to generate an average model which presents the conformational space filled by all models as an occupancy map (outlier models with high NSD-value to the most-typical model are excluded), (6) In order not to discard any models that might be correct despite a high NSD-value against the central model, the results of the clustering analysis and the representative structures from each cluster should be studied, (7) The variability of the all models can be used to assess the resolution of the reconstruction using the program SASRES.

## 6.8 Validation of Available Atomic Structures Against SAS Data

When no structural information about the particle is available, *ab initio* modeling is the only applicable method. Possibilities of SAS are significantly widened in the presence of atomic models of the entire macromolecule or of its individual fragments. The sources of the high-resolution structures are typically X-ray crystallography, NMR or computational homology modelling approaches. A natural question here is, what are the benefits of having SAS data for an object whose atomic coordinates are known? In fact, SAS can help to analyze possible differences between the quaternary structure in crystal and solution (due to crystal packing forces), identify the oligomeric state of the macromolecule in solution, select biologically active conformation among alternative and even to perform a screening of a multitude of computational models to pick the one(s) compatible with the scattering data. Hence, validation of the experimental and predicted high-resolution structures and selection between alternative models is a frequent and straightforward applications of SAS.

An accurate evaluation of the scattering patterns in solution from atomic coordinates is a non-trivial task (Svergun et al. 1995, 1998) due to the fact that in a SAS experiment the scattering from the blank (pure buffer) is subtracted from sample signal. Moreover, the resulting profile is influenced by the solvent effects, in particular by

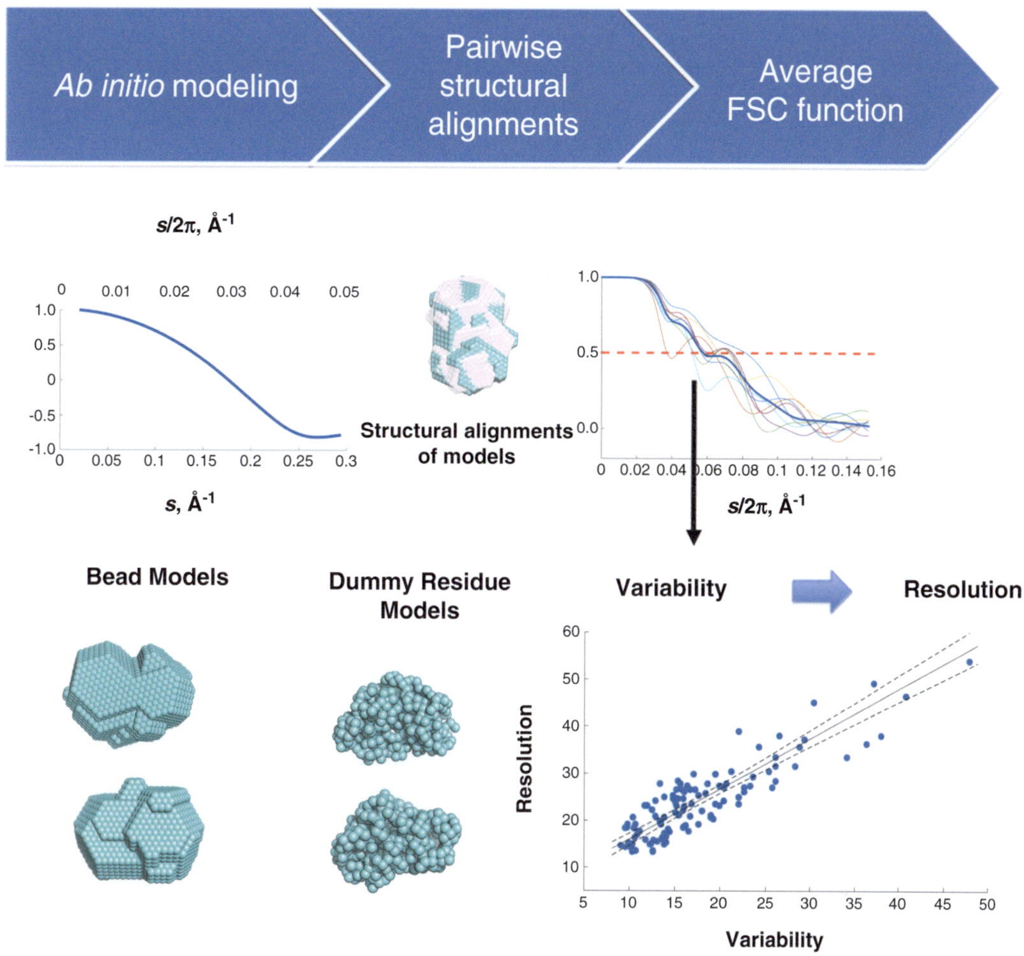


**Fig. 6.4** Overview of the FSC approach for estimating variability of structural ensembles. The reconstructed bead or dummy-residue models are structurally aligned and their pairwise FSC functions are computed. The average of all pairwise FSC functions is used to determine the variability estimate $\Delta_{ens}$ as $2\pi/s_{ens}$, where $s_{ens}$ is the momentum transfer value at which the average FSC drops below 0.5. The corresponding resolution is estimated based on the variability

the presence of a hydration shell whose average density is about 10 % higher than that of the bulk solvent. Therefore, the equation for scattering by a macromolecule in solution takes generally the form:

$$I(s) = \left\langle |A_a(\mathbf{s}) - \rho_s A_s(\mathbf{s}) + \delta\rho_b A_b(\mathbf{s})|^2 \right\rangle_\Omega \tag{6.5}$$

where $A_a(s)$ is the scattering amplitude of the particle in a vacuum, $A_s(s)$ and $A_b(s)$ are, respectively, the normalized scattering amplitudes from the excluded volume inaccessible to the solvent and from the hydration shell with the increased density. $\rho_s$ and $\delta\rho_b$ are, respectively, the solvent density and the excess density of the shell. Multipole expansion utilizing spherical harmonics (Stuhrmann 1970) is typically employed to speed up the calculations by expressing the three terms by partial amplitudes such that the spherical averaging is performed analytically.

The traditional algorithms available for two decades (Svergun et al. 1995, 1998) for computing X-ray and neutron small-angle scattering profiles based on atomic structures of macromolecules can either predict the theoretical curve with the default

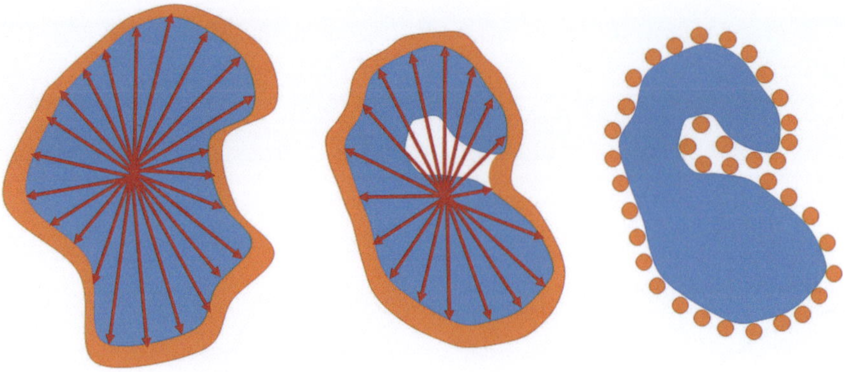

**Fig. 6.5** Evaluation of the hydration shell for the calculation of solution scattering profile from an atomic model. For a globular shape (*left panel*), an envelope function based on a quasi-uniform angular grid provides reasonable description of the particle's hydration shell. For more complex shapes (*middle*) this approach provides limited reliability. A more adequate shell representation in this case can be obtained by dummy solvent beads surrounding the surface of the particle (*right panel*)

values for all parameters or fit the provided experimental data by adjusting the hydration shell contrast, the averaged atomic group radius, and the total excluded volume. They utilize the angular envelope function to evaluate the hydration layer (Fig. 6.5). These tools were applied in a number of studies and in particular helped in distinguishing between alternative NMR models proposed for the Josephin domain of ataxin-3 involved in the ubiquitin/proteasome pathway (Nicastro et al. 2006) and for validation of a rather unusual 42-meric assembly of E2 catalytic core of an archaeal 2-oxoacid dehydrogenase multienzyme complex in the crystal structure (Marrott et al. 2012) (Fig. 6.6). Representation of the hydration shell with an angular envelope function performs well for the particles with simple globular shapes but has drawbacks in the case of a complex shapes (Fig. 6.5). Recently, a novel algorithm with improved features for hydration shell modeling has been developed for X-ray case and incorporated into the original method (Svergun et al. 1995). In the new approach verified by molecular dynamics simulations, protein hydration shell is modeled with explicit dummy water molecules attached to the molecular surface. It is especially beneficial for structures with complex geometries, for example containing cavities or holes.

## 6.9 Rigid Body Modelling of Quaternary Structure of Oligomeric Proteins and Complexes

In the studies of macromolecular complexes and modular multidomain proteins it is common that the atomic structures of individual subunits or distinct domains are available from X-ray crystallography on NMR while the structure of the entire particle is unknown. Models of such complex systems can be built by rigid body assembly of the components based on experimental information from complementary methods such as SAS. The SAS rigid body modeling is data-driven, i.e., the spatial arrangement of the subunits is sought by a direct fitting of available experimental scattering data from the complex. In rigid body modelling one neglects the possible small changes in the ternary structure of the subunits upon binding (these are hardly revealed at low resolution). The entire particle is reconstructed by moving and rotating the building blocks (subunits or domains) with fixed structures with respect to one another (Fig. 6.7). For each rigid body its position and orientation is described by six rotational and translational parameters whose optimal values are to be

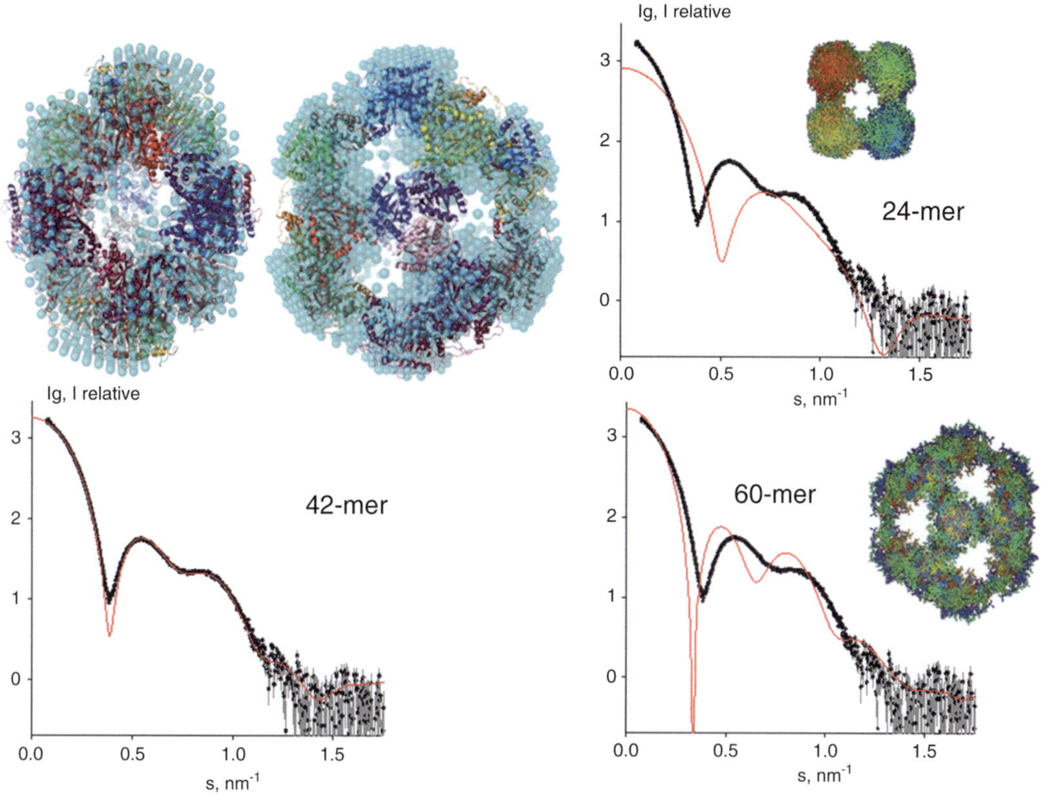


**Fig. 6.6** Validation of an unusual 42-meric assembly of E2 catalytic core of an archaeal 2-oxoacid dehydrogenase by SAXS. Left panel shows the comparison between the symmetric bead model (*cyan beads*) reconstructed *ab initio* from the SAXS profile of E2 catalytic core with the 42-meric assembly observed in the crystal structure (*$C_\alpha$-ribbon*). Right panel demonstrates that the alternative models consisting of 24 and 60 monomers can be ruled out based on the SAXS profile. Experimental data are denoted by *black dots*, the fits from atomic models as *red solid lines*

found while fitting the SAS data. The scattering from the complex can be analytically expressed using spherical harmonics from the partial scattering amplitudes of individual components (Petoukhov and Svergun 2005; Svergun 1991):

$$I(s) = 2\pi^2 \sum_{l=0}^{\infty} \sum_{m=-l}^{l} \left| \sum_{k=1}^{K} A_{lm}^{(k)}(s) \right|^2 \quad (6.6)$$

The partial scattering amplitudes $A_{lm}{}^{(k)}(s)$ of the $k^{th}$ rigid body depend on its scattering amplitudes in the reference position and orientation, and on three rotation angles and three shift values. A number of rigid body modelling algorithms are available allowing to address various use cases (Petoukhov and Svergun 2005, 2006). For oligomeric particles, symmetry can be taken into account as a rigid constraint such that the changes are applied to the asymmetric part only and the symmetry mates are generated automatically. In contrast variation measurements or in the studies where partial constructs are also available (e.g. subcomplexes or deletion mutants), multiple SAXS/SANS patterns can be fitted simultaneously. The use of multiple data sets increases the reliability of the reconstruction as not only the information on the overall shape but also about internal organization is employed. An interesting example combining

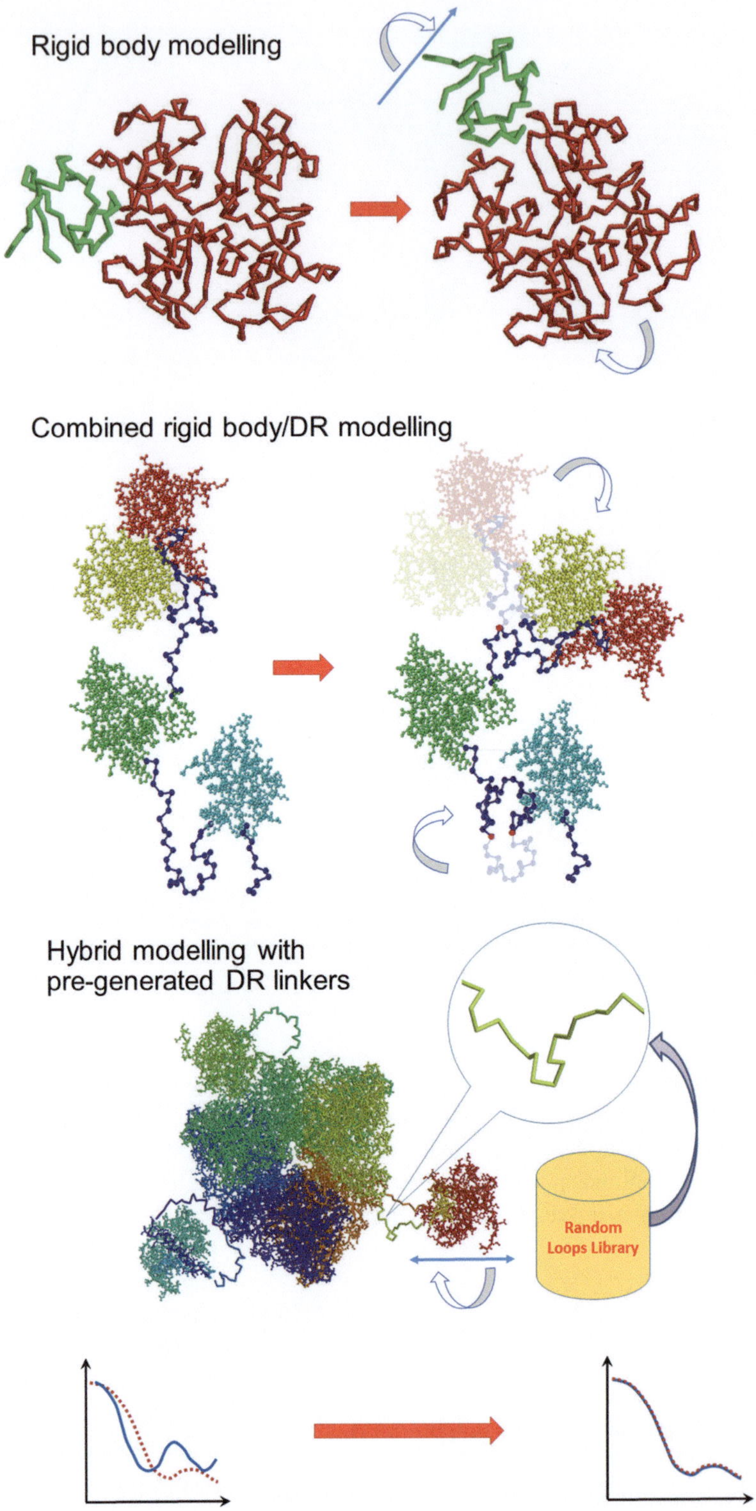


**Fig. 6.7** 3D–modelling algorithms utilizing available atomic models of the fragments of complex particles. In conventional rigid body modelling (*top*), the position and orientations of individual subunits are adjusted to fit the experimental SAS data. For modular proteins with interdomain linkers of unknown structure, hybrid modelling approach is applied combining rigid body positioning of the domains with conformational analysis of the linkers using DR-chain representation (*middle*). For multichain particles with missing fragments, a pre-generated library of loops conformations is employed in the hybrid modelling (*bottom*)

SAXS and SANS on a complex particle with partial constructs is given by the study of the Met receptor and the *Listeria monocytogenes* invasion protein InlB (Niemann et al. 2008), where more than 40 scattering profiles were fitted simultaneously. Combination of specific deuteration of InlB with the use of truncation of the C-terminal Ig-like domains of Met allowed the authors to reliably determine the location of the ligand in the complex from the solution scattering data. The SAS model was later confirmed by a crystallographic study of a short Met construct in complex with InlB (Fig. 6.8).

For the complex particles with a reasonable initial guess, an interactive local refinement can be applied (Konarev et al. 2001), otherwise automated SA-based global modeling is performed without user intervention. Rigid body modelling approaches typically apply physical restraints (for instance, the requirements of interconnectivity and the absence of steric clashes) formulated as penalty terms in the target function. In addition, capabilities are provided to incorporate additional information from other methods (Petoukhov and Svergun 2005) including distances between specific residues (from mutagenesis (Gherardi et al. 2006), Fourier transform infrared spectroscopy or NMR chemical shifts (Xu et al. 2008)) and relative subunit orientations (from NMR residual dipolar coupling (Marino et al. 2006)).

## 6.10 Hybrid Modelling

In many practical applications the high-resolution models of the subunits are only partially available, while the structures of some fragments remain completely unknown. Quite often it is the case for multidomain proteins consisting of globular domains linked by flexible loops. The high-resolution structures or homology models available for the individual domains usually do not cover the linkers. For such cases, a hybrid modelling approach is proposed combining rigid-body and *ab initio* modeling techniques for structural characterization of the entire assembly. The idea is to simultaneously find optimal positions and orientations of the domains/subunits moved as rigid bodies and probable conformations of the flexible linkers attached to the appropriate terminal residues of the domains. These linkers are represented by DRs described in the *ab initio* modelling section. The notable difference however is that instead of the gas of DRs that condenses into a shape, a linker or protein fragment with unknown structure is substituted by an always interconnected chain of DRs such that the spacing of 0.38 nm between the subsequent residues is maintained during the modelling process. In one modelling approach a single modification of the system is performed by a random rotation of the model portion about an axis passing through a random DR or connecting two randomly selected DRs (Fig. 6.7). As additional restraints in this approach, proper bond and dihedral angle distributions and the absence of steric clashes are required. Multiple scattering data sets from deletion mutants, if available, can be fitted simultaneously whereby the appropriate portion of the structure is used for scattering intensity calculation for each data set. Such hybrid modelling has been employed e.g. in a study of the conformation of polypyrimidine tract binding protein consisting of four RRM domains connected by the linkers, whereby the experimental scattering intensities from the full-length protein and its deletion mutants composed by all possible combinations of the domains were fitted by a single hybrid model (Petoukhov et al. 2006).

An alternative approach, designed for complex particles consisting of multiple chains where one or more components have missing fragments of noticeable lengths, utilizes a pre-generated library of self-avoiding random DR linkers (Fig. 6.7). In the course of SA-based modelling, translations and rotations applied to the atomic models of the individual domains belonging to multiple components of the complex are not fully random. Only those rearrangements are accepted where the gap between N- and C-terminal portions of the subsequent domains belonging to one chain can be connected by a loop from the library with the appropriate end-to-end distance. If the query is successful, the loop is inserted as a placeholder of the missing linker and it contributes to the

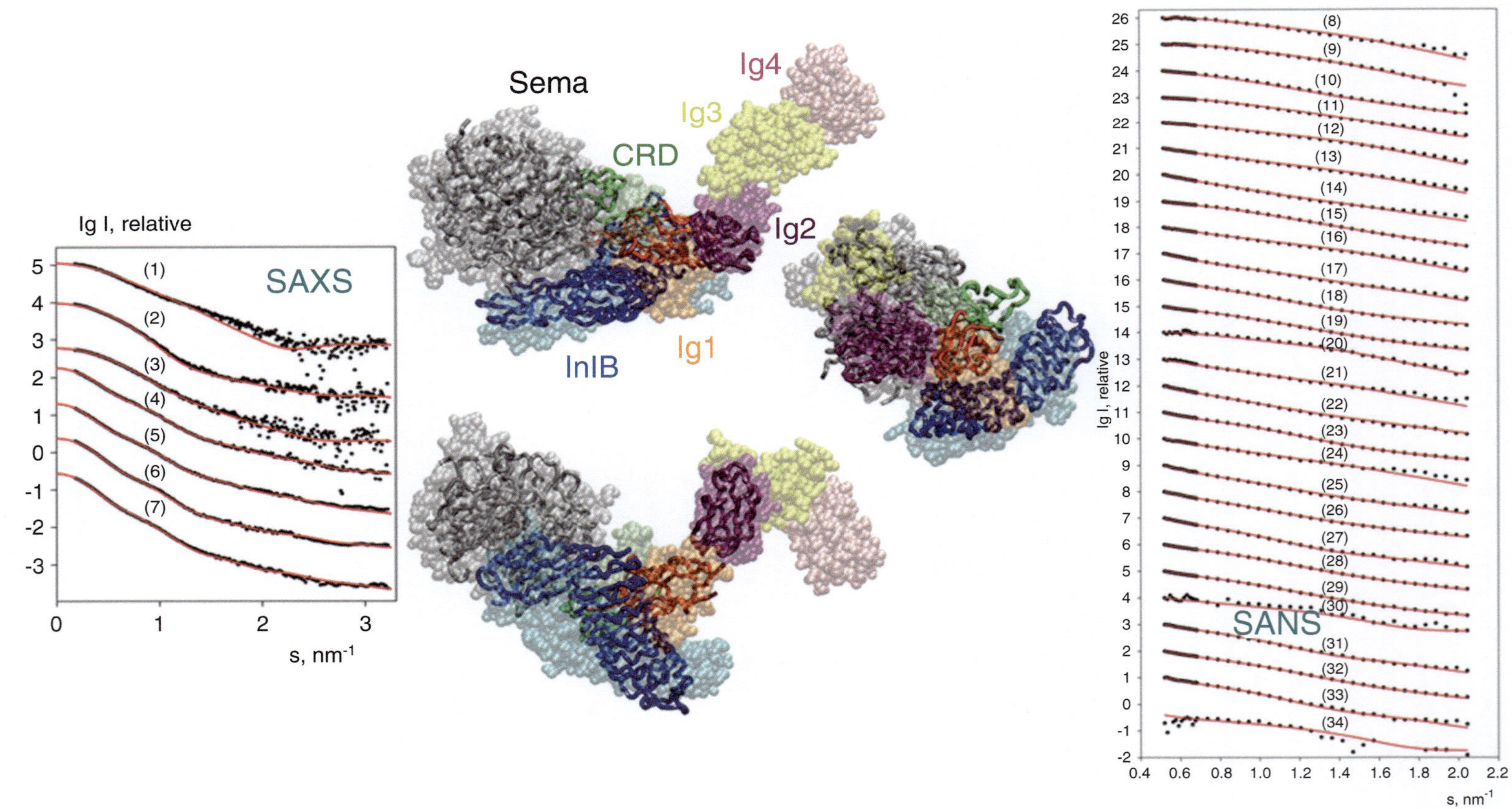


**Fig. 6.8** Rigid body modeling of the complex between multidomain Met protein with its InlB ligand. Simultaneous fitting of multiple SAXS and contrast variation SANS data (with specific deuteration of InlB) from the full-length complex and various deletion mutants of Met (lacking various portions of the C-terminal Ig-like domains) yielded rigid body model (shown as beads) consistent with independently obtained crystallographic model of a shorter Met construct in complex with InlB

computation of the scattering intensity of the system as well as to the other terms in the target function (e.g. overlaps, contact restraints etc.).

## 6.11 Quality Assessment of Rigid-Body Models and Cross-Validation with Other Methods

Similar to *ab initio* approaches, rigid-body modelling process is typically repeated multiple times in order to estimate the variability of the solution and to find the most typical model. The ambiguity of model reconstruction can be evaluated by hierarchical clustering on the basis of $C_\alpha$ atom RMSD values. Structural clustering provides an estimate of the stability of solutions and reveals common features in the models. However, high occurrence of a certain binding orientation between the subunits in the model ensemble does not directly mean that the model represents the most native-like binding. This ambiguity problem can be tackled using complementary experimental data. The reliability of rigid-body modelling improves significantly when it is done in combination with data obtained using other techniques such as dynamic light scattering (DLS), analytical ultracentrifugation (AUC) or other structural methods. Rigid-body models based on SAS-data can be compared and cross-validated against structural models of the same system obtained by EM, NMR or crystallography. On the other hand, non-structural methods such as AUC and DLS provide information about the overall dimensions of the particle giving further indications about the binding orientation and oligomeric state of the system. Site-directed mutagenesis studies of the amino acid residues on the predicted binding interfaces can either confirm or exclude rigid-body models. SAS analysis can be further improved by scoring rigid-body model ensembles in a post-processing step with the help of solely computational approaches. For rigid body models with atomic details, scoring functions similar to the molecular docking solution ranking using charge and surface complementarity and other bioinformatic tools can be employed (Andreani et al. 2013; Khashan et al. 2012; Hwang et al. 2010).

An example of using independent experimental data and computational analysis to validate rigid-body models is the study of the interaction between transglutaminase 2 (TG2) and an anti-TG2 antibody derived from a single gut IgA plasma cell of a celiac disease patient (Chen et al. 2015). The program SASREF was used to generate TG2 – antibody complex models to determine possible binding orientations using the crystal structures of the Fab fragment and TG2-GTP. The 17 different rigid-body models generated without any distance constraints could be classified into six different groups based on their binding interfaces (Fig. 6.9a–f). Additional information from complementary methods were then employed to overcome the ambiguity in the rigid-body results. Epitope mapping information obtained from hydrogen/deuterium exchange experiments and previous biological knowledge on the Fab fragment interaction motif indicated that the four models of group *f* most likely represent the correct binding orientation (Fig. 6.9b). To further optimize and to study the interaction interface at atomic level, nanosecond time-scale molecular dynamics (MD) simulations were performed using a representative model from the group *f*. The interaction amino acid residue predictions from the MD simulations fit well with results of hydrogen/deuterium exchange experiments and mutagenesis studies (Fig. 6.9d). The contributions of individual amino acid residues in the interaction were studied in detail by extracting the mean non-bonded interaction energies (electrostatic and van der Waals) from MD simulations on the native complex as well as several *in silico* mutants. MD simulations also revealed a tightly packed water network between residues of TG2 and the Fab fragment essential for their binding. Taken together, the combination of rigid-body modelling and advanced computational methods in this project enabled extraction of very detailed information about the binding interface.

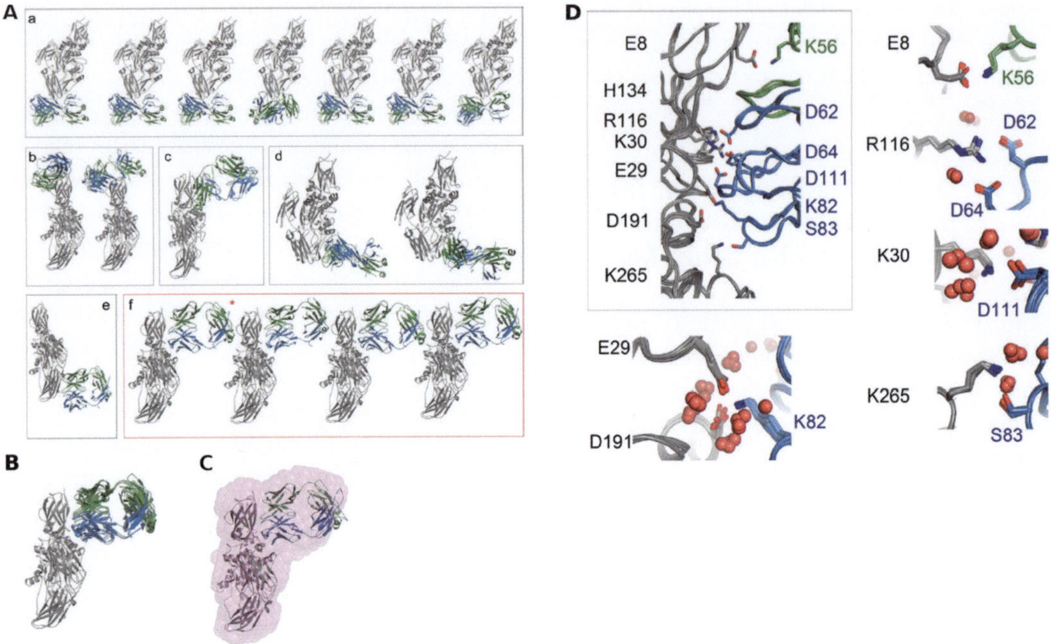


**Fig. 6.9** (**a**) Rigid body models of TG2-GTP – Fab fragment complex obtained by SASREF clustered into six groups (*a–f*). (**b**) Structural alignment of all models in the group *f*. (**c**) Structural alignment of the representative rigid body model of group *f* with the *ab initio* model of the same complex. (**d**) Structural models derived from MD simulations of the interaction between TG2 and the Fab fragment. The residues involved in the binding interface between the Fab and TG2 are shown as *sticks* and *red spheres* show water molecules. TG2 is represented in *gray* and the light chain and heavy chain of the Fab fragment are colored in *green* and *blue*, respectively

## 6.12 3D Reconstructions Against Polydisperse Data (Transient Complexes and Week Oligomers)

The requirement of monodispersity is an important prerequisite for reliable three-dimensional reconstructions from SAXS data. For some systems (like weak oligomers and transient complexes) polydispersity is however unavoidable being an inherent property of the sample. Typical indication of polydispersity can be concentration dependence of the SAXS profiles of the system and the disagreement of experimentally defined values of molecular weight and excluded volume of the solute with the expected ones. In case of oligomeric equilibrium where symmetric assembly partially dissociates into monomers, a specifically modified DR approach can be applied for the reconstruction of the quaternary structure of the entire oligomer (Petoukhov et al. 2013). The DR model generation of a symmetric homo-oligomer simply takes into account the presence of the dissociated state (Fig. 6.10), given the fact that the monomer represents the asymmetric part of the intact oligomer. Therefore, as compared to the monodisperse scenario, there is just a single additional fitting parameter, i.e. the volume fraction of the oligomeric particles in the mixture.

Further, for complex particles (including also symmetric oligomers), polydisperse samples can be analyzed in terms of rigid body modelling, if the composition of the dissociation products is known (Petoukhov et al. 2013). Indeed, the configurations of the dissociation products are deduced from that of the entire assembly. This means that for a given model of the intact assembly, besides its own scattering curve all scattering profiles of the dissociation products are known as

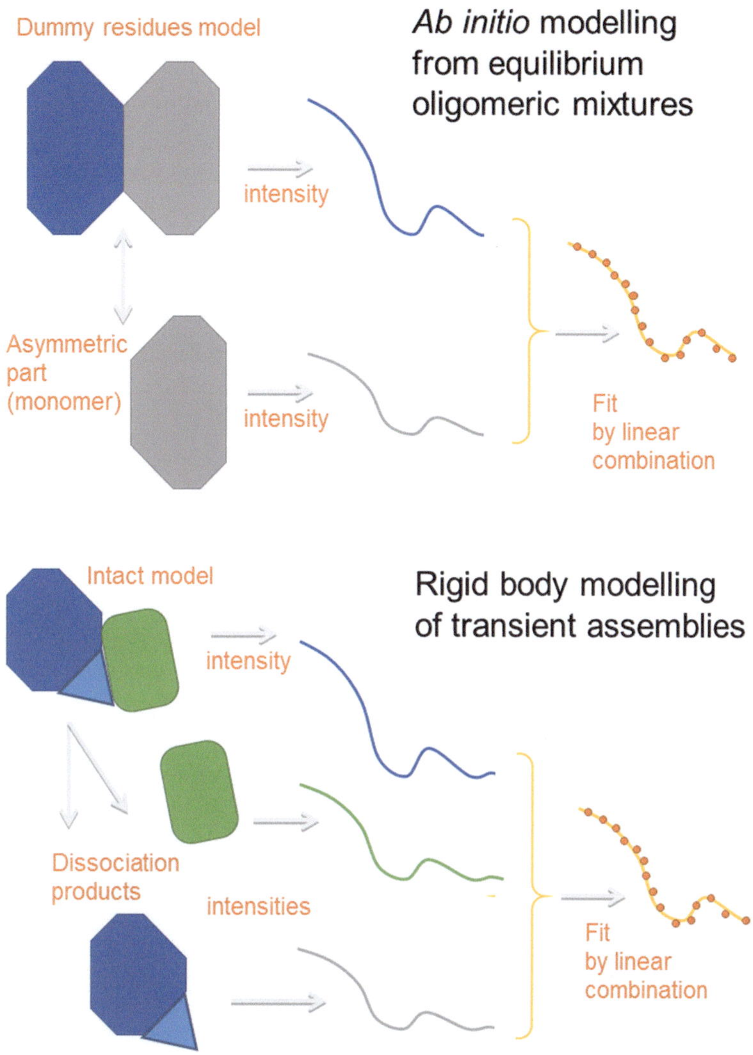


**Fig. 6.10** Principles of *ab initio* (*top*) and rigid body (*bottom*) modelling against polydisperse SAXS data. In case of oligomeric equilibrium, the intact oligomer is defined by the configuration of its monomer and the symmetry. In transient complexes, the organization of dissociation products can be deduced from the intact particle. In both scenarios, the experimental data from a mixture is fitted by a linear combination of the scattering profiles from its components (the complete particle and dissociated state(s))

well (Fig. 6.10). The experimental scattering data is then simply fitted by a linear combination of contributing individual profiles, whereby the volume fractions of the dissociation products in the mixture add just a few additional optimization parameters to the fitting as compared to the classic rigid body modelling against monodisperse data. Such an approach has been applied to the study of the association behavior and conformations of estrogen-related receptor ERRα in complex with inverse repeat IR3 DNA (Petoukhov et al. 2013). This system in solution was present as equilibrium between the protein hexamer in complex with three copies of DNA (6:3 assembly) and its dissociation product with a 2:1 stoichiometry (Fig. 6.11). Very often, the equilibrium between the full complex and the dissociation products is concentration-dependent and fitting multiple curves at different concentrations (typically, 3–4) allows to restrain the modelling. Also in this project, concentration series with different proportions of the intact particle and the dissociated state were fitted simultaneously.

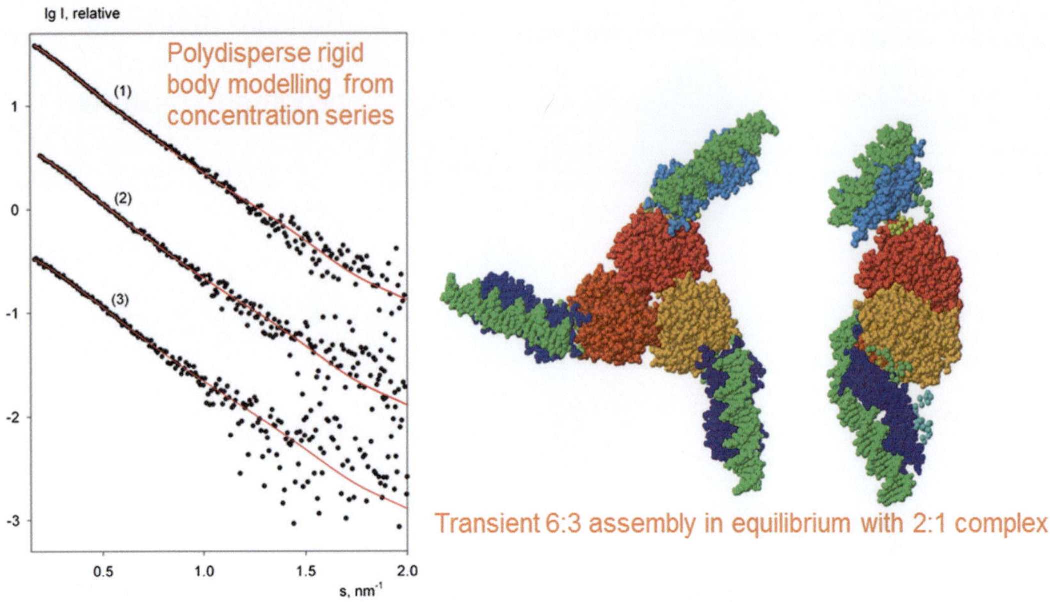


**Fig. 6.11** Rigid body modeling of ERRα – IR3 complex from polydisperse data. Scattering patterns at different concentrations are fitted simultaneously by adjustment of the volume fraction of the intact 6:3 assembly and the 2:1 dissociation product in the mixtures. Experimental data are shown by *black dots*, and the fits with *red solid lines*. Rigid body model consists of the hexameric core formed by the ligand binding domain (*orange, red* and *yellow*) and three copies of IR3 DNA (*green*) in complex with DNA-binding domain dimer (*blue shades*) attached to the core. The right view is rotated by 90 ° about the vertical axis

## References


Andreani J, Faure G, Guerois R (2013) InterEvScore: a novel coarse-grained interface scoring function using a multi-body statistical potential coupled to evolution. Bioinformatics 29:1742–1749

Chacon P, Moran F, Diaz JF, Pantos E, Andreu JM (1998) Low-resolution structures of proteins in solution retrieved from X-ray scattering with a genetic algorithm. Biophys J 74:2760–2775

Chen VB, Arendall WB, Headd JJ, Keedy DA, Immormino RM, Kapral GJ, Murray LW, Richardson JS, Richardson DC (2010) MolProbity: all-atom structure validation for macromolecular crystallography. Acta Crystallogr Sect D Biol Crystallogr 66:12–21

Chen X, Hnida K, Graewert MA, Andersen JT, Iversen R, Tuukkanen A, Svergun D, Sollid LM (2015) Structural basis for antigen recognition by transglutaminase 2-specific autoantibodies in celiac disease. J Biol Chem 290:21365–21375

Doreleijers JF, Vranken WF, Schulte C, Markley JL, Ulrich EL, Vriend G, Vuister GW (2012) NRG-CING: Integrated validation reports of remediated experimental biomolecular NMR data and coordinates in wwPDB. Nucleic Acids Res 40: D519–D524

Feigin LA, Svergun DI (1987) Structure analysis by small-angle X-ray and neutron scattering. Plenum Press, New York

Franke D, Svergun DI (2009) DAMMIF, a program for rapid ab-initio shape determination in small-angle scattering. J Appl Cryst 42:342–346

Gherardi E, Sandin S, Petoukhov MV, Finch J, Youles ME, Ofverstedt LG, Miguel RN, Blundell TL, Vande Woude GF, Skoglund U, Svergun DI (2006) Structural basis of hepatocyte growth factor/scatter factor and MET signalling. Proc Natl Acad Sci U S A 103:4046–4051

Glatter O, Kratky O (1982) Small angle X-ray scattering. Academic, London

Hwang H, Vreven T, Janin J, Weng Z (2010) Protein-protein docking benchmark version 4.0. Proteins Struct Funct Bioinf 78:3111–3114

Khashan R, Zheng W, Tropsha A (2012) Scoring protein interaction decoys using exposed residues (SPIDER): a novel multibody interaction scoring function based on frequent geometric patterns of interfacial residues. Proteins Struct Funct Bioinf 80:2207–2217

Konarev PV, Petoukhov MV, Svergun DI (2001) MASSHA – a graphic system for rigid body modelling of macromolecular complexes against solution scattering data. J Appl Crystallogr 34:527–532

Kozin MB, Svergun DI (2001) Automated matching of high- and low-resolution structural models. J Appl Crystallogr 34:33–41
Marino M, Zou P, Svergun D, Garcia P, Edlich C, Simon B, Wilmanns M, Muhle-Goll C, Mayans O (2006) The Ig doublet Z1Z2: a model system for the hybrid analysis of conformational dynamics in Ig tandems from titin. Structure 14:1437–1447
Marrott NL, Marshall JJ, Svergun DI, Crennell SJ, Hough DW, Danson MJ, van den Elsen JM (2012) The catalytic core of an archaeal 2-oxoacid dehydrogenase multienzyme complex is a 42-mer protein assembly. FEBS J 279:713–723
Nicastro G, Habeck M, Masino L, Svergun DI, Pastore A (2006) Structure validation of the Josephin domain of ataxin-3: conclusive evidence for an open conformation. J Biomol NMR 36:267–277
Niemann HH, Petoukhov MV, Hartlein M, Moulin M, Gherardi E, Timmins P, Heinz DW, Svergun DI (2008) X-ray and neutron small-angle scattering analysis of the complex formed by the met receptor and the listeria monocytogenes invasion protein InlB. J Mol Biol 377:489–500
Penczek PA (2010) Resolution measures in molecular electron microscopy. Methods Enzymol 482:73–100
Petoukhov MV, Svergun DI (2005) Global rigid body modeling of macromolecular complexes against small-angle scattering data. Biophys J 89:1237–1250
Petoukhov MV, Svergun DI (2006) Joint use of small-angle X-ray and neutron scattering to study biological macromolecules in solution. Eur Biophys J 35:567–576
Petoukhov MV, Svergun DI (2015) Ambiguity assessment of small-angle scattering curves from monodisperse systems. Acta Crystallogr D 71:1051–1058
Petoukhov MV, Monie TP, Allain FH, Matthews S, Curry S, Svergun DI (2006) Conformation of polypyrimidine tract binding protein in solution. Structure 14:1021–1027
Petoukhov MV, Franke D, Shkumatov AV, Tria G, Kikhney AG, Gajda M, Gorba C, Mertens HDT, Konarev PV, Svergun DI (2012) New developments in the ATSAS program package for small-angle scattering data analysis. J Appl Crystallogr 45:342–350
Petoukhov MV, Billas IM, Takacs M, Graewert MA, Moras D, Svergun DI (2013) Reconstruction of quaternary structure from X-ray scattering by equilibrium mixtures of biological macromolecules. Biochemistry 52:6844–6855
Stuhrmann HB (1970) Ein neues Verfahren zur Bestimmung der Oberflaechenform und der inneren Struktur von geloesten globularen Proteinen aus Roentgenkleinwinkelmessungen. Zeitschr Physik Chem Neue Folge 72:177–198
Svergun DI (1991) Mathematical methods in small-angle scattering data analysis. J Appl Crystallogr 24:485–492
Svergun DI (1992) Determination of the regularization parameter in indirect-transform methods using perceptual criteria. J Appl Crystallogr 25:495–503
Svergun DI (1999) Restoring low resolution structure of biological macromolecules from solution scattering using simulated annealing. Biophys J 76:2879–2886
Svergun D, I. (1993) A direct indirect method of small-angle scattering data treatment. J Appl Crystallogr 26:258–267
Svergun DI, Nierhaus KH (2000) A map of protein-rRNA distribution in the 70 S Escherichia coli ribosome. J Biol Chem 275:14432–14439
Svergun DI, Barberato C, Koch MHJ (1995) CRYSOL – a program to evaluate X-ray solution scattering of biological macromolecules from atomic coordinates. J Appl Crystallogr 28:768–773
Svergun DI, Volkov VV, Kozin MB, Stuhrmann HB (1996) New developments in direct shape determination from small-angle scattering 2. uniqueness. Acta Crystallogr A52:419–426
Svergun DI, Richard S, Koch MH, Sayers Z, Kuprin S, Zaccai G (1998) Protein hydration in solution: experimental observation by X-ray and neutron scattering. Proc Natl Acad Sci U S A 95:2267–2272
Svergun DI, Petoukhov MV, Koch MHJ (2001) Determination of domain structure of proteins from X-ray solution scattering. Biophys J 80:2946–2953
Tuukkanen AT, Kleywegt GJ, Svergun DI (2016) Resolution of *ab initio* shapes determined from small-angle scattering. IUCrJ 3:440–447
Volkov VV, Svergun DI (2003) Uniqueness of ab initio shape determination in small angle scattering. J Appl Crystallogr 36:860–864
Xu XF, Reinle WG, Hannemann F, Konarev PV, Svergun DI, Bernhardt R, Ubbink M (2008) Dynamics in a pure encounter complex of two proteins studied by solution scattering and paramagnetic NMR spectroscopy. J Am Chem Soc 130:6395–6403

# Structural Characterization of Highly Flexible Proteins by Small-Angle Scattering

7

Tiago N. Cordeiro, Fátima Herranz-Trillo, Annika Urbanek, Alejandro Estaña, Juan Cortés, Nathalie Sibille, and Pau Bernadó

**Abstract**

Intrinsically Disordered Proteins (IDPs) are fundamental actors of biological processes. Their inherent plasticity facilitates very specialized tasks in cell regulation and signalling, and their malfunction is linked to severe pathologies. Understanding the functional role of disorder requires the structural characterization of IDPs and the complexes they form. Small-angle Scattering of X-rays (SAXS) and Neutrons (SANS) have notably contributed to this structural understanding. In this review we summarize the most relevant developments in the field of SAS studies of disordered proteins. Emphasis is given to ensemble methods and how SAS data can be combined with computational approaches or other biophysical information such as NMR. The unique capabilities of SAS enable its application to extremely challenging disordered systems such as low-complexity regions, amyloidogenic proteins and transient biomolecular complexes. This reinforces the fundamental role of SAS in the structural and dynamic characterization of this elusive family of proteins.

T.N. Cordeiro • A. Urbanek • N. Sibille • P. Bernadó (✉)
Centre de Biochimie Structurale. INSERM, CNRS, Université de Montpellier.29, rue de Navacelles, 34090 Montpellier, France
e-mail: pau.bernado@cbs.cnrs.fr

F. Herranz-Trillo
Centre de Biochimie Structurale. INSERM, CNRS, Université de Montpellier.29, rue de Navacelles, 34090 Montpellier, France

Department of Pharmacy and Department of Drug Design and Pharmacology, University of Copenhagen, Universitetsparken 2, 2100 Copenhagen, Denmark

A. Estaña
Centre de Biochimie Structurale. INSERM, CNRS, Université de Montpellier.29, rue de Navacelles, 34090 Montpellier, France

LAAS-CNRS, Université de Toulouse, CNRS, Toulouse, France

J. Cortés
LAAS-CNRS, Université de Toulouse, CNRS, Toulouse, France



B. Chaudhuri et al. (eds.), *Biological Small Angle Scattering: Techniques, Strategies and Tips*, Advances in Experimental Medicine and Biology 1009, DOI 10.1007/978-981-10-6038-0_7

**Keywords**
Intrinsically disordered proteins • Small-angle x-ray scattering • Small-angle neutron scattering • Ensemble methods • Protein-protein interactions • Nuclear magnetic resonance • Low-complexity regions • Random coil models • Amyloids

## 7.1 Introduction

Intrinsically disordered Proteins or Regions (IDPs/IDRs) have emerged as key actors for a large variety of biological functions such as cell signalling and regulation (Kriwacki et al. 1996; Wright and Dyson 1999, 2015; Dunker et al. 2002). The main feature of IDPs and IDRs is their lack of permanent secondary or tertiary structure that provides them with an inherent malleability enabling highly specialized biological functions (Dunker et al. 2002). Eukaryotic genomes are highly enriched in genes coding for disordered proteins, and this observation has been linked to the major complexity of these organisms. The capacity of IDPs to adapt their conformation to specifically recognize one or several partners, and the low to moderate affinity for partners make IDPs ideal for protein-protein interactions (Tompa et al. 2015). In fact, it has been shown that interactome hubs are enriched in this family of proteins (Dunker et al. 2005; Kim et al. 2008). Partner recognition is normally performed through conserved and partially structured motifs of the protein, and their individual properties can be modulated by post-translational modifications (PTM) or alternative splicing. IDRs are highly flexible regions connecting well-folded globular proteins forming the so-called multi-domain proteins. Multi-domain protein topology, which is highly prevalent in eukaryotes, enables the presence of multiple biological activities performed by the globular domains in close proximity (Hawkins and Lamb 1995; Levitt 2009). In many of these cases, IDRs behave as entropic linkers with an inherent plasticity that can be tuned depending on the length and the specific amino acid sequence of the region.

The biological relevance of IDPs has fostered their structural characterization (Eliezer 2009). Identification of the conformational preferences of binding motifs, the detection of transient long-range contacts within the chain, the structural perturbations exerted by PTM, the shape of biomolecular complexes with disordered partners, and the spatial distribution of globular domains in multi-domain protein are structural features that must be characterized to understand the molecular bases of biological function. This characterization is far from being trivial as the inherent disorder of IDPs/IDRs precludes their crystallization. Nuclear Magnetic Resonance (NMR) has become the only technique that can provide atomic-resolution information on IDPs (Dyson and Wright 2004). Novel NMR experiments and modelling strategies have been developed to interpret experimental parameters in terms of structure (Jensen et al. 2009, 2013, 2014; Wright and Dyson 2015). However, some structural aspects remain elusive by NMR such as the overall shape and size of disordered proteins or the distribution of interdomain position in multi-domain proteins.

Small-Angle Scattering of X-rays or Neutrons have emerged as powerful techniques to probe the structure and dynamics of biomolecules in solution at low resolution (Feigin and Svergun 1987; Svergun and Koch 2003; Koch et al. 2003; Putnam et al. 2007; Jacques and Trewhella 2010). SAS provides unique information about the overall size and shape of individual macromolecules or their complexes in a rapid manner. Additionally, structural changes upon environmental perturbations, such as interactions with other molecules, can be addressed straightforwardly. Major advances in instrumentation and computational methods in the last decade have led to a tremendous increase in the applications of SAS in structural biology (Petoukhov and Svergun 2007; Mertens and Svergun 2010; Pérez and Nishino 2012; Rambo and Tainer 2013; Graewert and Svergun 2013). One of the major advances of SAXS in the last decade has been its extension to address biomolecular dynamics (Doniach 2001; Bernadó and Svergun 2012a, b; Receveur-Brechot and Durand 2012; Kikhney and Svergun 2015; Kachala et al. 2015). Although used in the past to study protein flexibility (Aslam et al. 2003), the availability of robust protocols to interpret SAS data in terms of ensembles of conformations have generalized these studies and, therefore, have enriched the spectrum of applications of the technique (Bernadó and Blackledge 2010).

The aim of this chapter is to provide an overview of the recent developments of SAS to study highly disordered proteins. A special emphasis is put on biological scenarios involving highly flexible proteins for which SAS is a crucial tool to tackle the structural/dynamic bases of biological function.

## 7.2 Scattering Properties of IDPs

The fact that IDPs sample an astronomical number of conformations has a strong impact on the scattering profiles measured and their comprehensive analysis in terms of structure. The experimental SAXS profile of an IDP corresponds to the average of all the conformations that the protein adopts in solution, inducing special features to the curves. Figure 7.1a displays the synthetic SAXS curves for seven conformations of $p15^{PAF}$, a 111 residue-long IDP, selected from a large pool of 5,000 (De Biasio et al. 2014). The individual conformations display several features along the complete momentum transfer range simulated. The initial part of the simulated curves, containing the lowest resolution structural information, presents distinct slopes indicating a large variety of possible sizes and shapes that an unstructured chain can adopt. The SAXS profile, obtained after averaging curves for the 5,000 conformations, presents a smoother behavior with essentially no features (Fig. 7.1b).

Traditionally, Kratky plots ($I(s)\cdot s^2$ as a function of *s*, where *I(s)* is the scattering intensity and the momentum transfer *s* is defined as $s = 4\pi\sin(\theta)\,\lambda^{-1}$, $2\theta$ is the scattering angle; and $\lambda$ is the X-ray wavelength) have been used to qualitatively identify disordered states and distinguish them from globular particles. The scattering intensity of a globular protein behaves approximately as $1/s^4$ for *s* significantly greater than $1/R_g$, conferring a bell-shaped Kratky plot with a well-defined maximum. Conversely, an ideal Gaussian chain has a $1/s^2$ dependence of $I(s)$ and therefore presents a plateau at large *s* values. In the case of a chain with negligible thickness, $t << 1/s$, the Kratky plot presents a plateau over a specific range of *s*, which is followed by a monotonic increase. This last behavior is normally observed experimentally in unfolded proteins. The Kratky representation has the capacity to enhance particular features of scattering profiles that allows an easier identification of different degrees of compactness (Doniach 2001). This is shown in Fig. 7.1c where different degrees of compactness for the conformations are observed. Multi-domain proteins, which are composed by two or more globular domains connected by intrinsically disordered linkers, present a dual (folded/disordered) behavior, and consequently SAXS profiles and Kratky plots display contributions from both structurally distinct regions. Pair-wise distance distributions, *p(r)*, derived from disordered proteins also present specific properties (Fig. 7.1d). The most characteristic feature is the smooth decrease towards large intramolecular distances. Maximum intramolecular distances, $D_{max}$, which represent the maximum distance within one of the accessible conformations of the protein, are very large in disordered proteins. It is worth noting that due to the low population of highly extended conformations in the ensembles, experimental $D_{max}$ values are systematically underestimated (Bernadó 2010).

Unstructured proteins, due to the presence of extended conformations, are characterized by large average sizes compared to globular

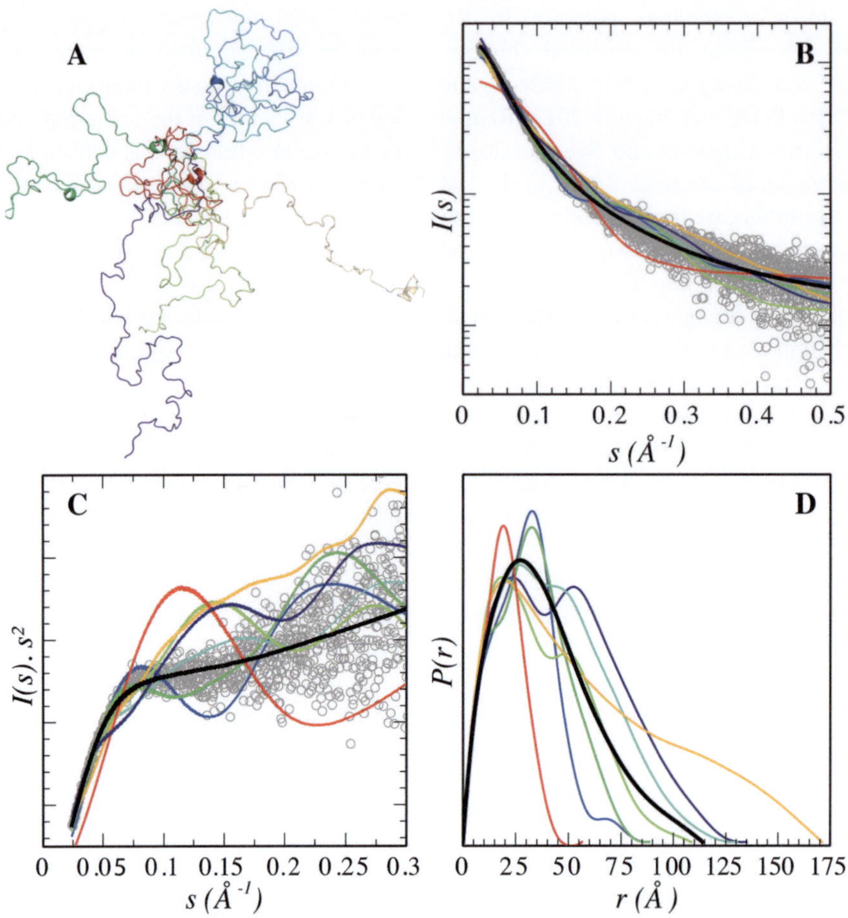


**Fig. 7.1** (**a**) Seven representative conformers randomly selected from an ensemble of 5000 explicit all-atoms models generate for p15[PAF] (De Biasio et al. 2014). *Solid lines* correspond to their computed SAXS *curves* (**b**) and Kratky plots (**c**) and are *colored* as in planel *A. T.* The average over the ensemble of 5,000 conformations yields a featureless curve that is in very good agreement with the experimental data (*gray circles*). (**d**) p (r) functions computed for the seven conformers and the complete ensembles in the same color code as in panels (**a–c**)

proteins. The radius of gyration, $R_g$, which can be directly obtained from a SAXS curve using a classical Guinier approximation, is the most common descriptor to quantify the overall size of molecules in solution (Guinier 1939). In the Guinier approximation, it is assumed that at very small angles ($s < 1.3/R_g$) the scattering intensity can be represented as $I(s) = I(0)\exp(-(sR_g)^{2/3})$, and the $R_g$ is obtained by a simple linear fit in logarithmic scale. Debye's equation (Eq. 1), which describes the scattering from an ensemble of monodisperse random coils, can be more precise than Guinier's approximation to derive $R_g$ values as its validity extends to larger momentum transfer ranges (Calmettes et al. 1994).

$$\frac{I(s)}{I(0)} = \frac{2}{x^2}(x - 1 + e^{-x}); x = s^2\mathrm{Rg}^2 \qquad (7.1)$$

Alternatively, the *p(r)* function calculated from the complete scattering profile using a Fourier transformation also yields precise $R_g$ values for disordered proteins.

The experimental $R_g$ is a single value representation of the size of the molecule, which for disordered states represents a $z$-average over all accessible conformations in solution (Feigin and Svergun 1987). The most common quantitative interpretation of $R_g$ for unfolded proteins, which is based on Flory's studies in polymer science, relates this parameter to the length of the protein chain through a power law (Flory 1953),

$$R_g = R_0 \cdot N^{\nu} \tag{7.2}$$

where N is the number of residues in the polymer chain, $R_0$ is a constant that depends on several factors, in particular, on the persistence length, and $\nu$ is an scaling exponent. For an excluded-volume polymer, Flory estimated $\nu$ to be $\approx 0.6$, and more accurate theoretical estimates established a value of 0.588 (LeGuillou and Zinn-Justin 1977). A recent compilation of $R_g$ values measured for 26 chemically denatured proteins sampling broad range of chain lengths found a $\nu$ value of $0.598 \pm 0.028$, and a $R_0$ value of $1.927 \pm 0.27$ (Kohn et al. 2004). The agreement between the $\nu$ value obtained experimentally and the theoretical models suggest the random coil nature of the chemically denatured proteins. However, the question whether the conformational sampling in the chemically denatured state is equivalent to that found for IDPs in native conditions must be clarified (Stumpe and Grubmüller 2007 and references therein). Using atomistic ensemble models of several disordered proteins, Flory's equation has been parametrized as a function of the number of residues of the IDP, $N$ (Bernadó and Blackledge 2009):

$$R_g = (2.54 \pm 0.01) \cdot N^{(0.522 \pm 0.01)} \tag{7.3}$$

The exponential value obtained from the parametrization, $\nu = 0.522 \pm 0.01$, is notably smaller than that derived from the dataset of denatured proteins, $\nu = 0.598 \pm 0.028$, indicating that IDPs are more compact than chemically denatured proteins. This observation is in line with NMR studies that indicated that urea denatured proteins have an enhanced sampling (around 15%) of extended conformations compared with IDPs (Meier et al. 2007).

Table 7.1 compiles $R_g$ data from 74 IDPs from the literature, which are plotted as a function of the chain length in Fig. 7.2. As expected, the $R_g$s collected display a correlation with the number of residues of the chain. This linear relationship is closer to the above-mentioned parametrization of IDPs than to that established for chemically denatured proteins. As some IDPs are expected to have certain populations of secondary or tertiary structure, this relationship can be used as an interpretative tool. Thus, deviations from the average IDP random coil model indicate enhanced degrees of compactness or extendedness within the protein.

## 7.3 Molecular Modelling of Intrinsically Disordered Proteins

The interpretation of the SAXS parameters such as $R_g$, $p(r)$ and $D_{max}$ from disordered proteins in terms of structure is limited to overall molecular information. In order to fully exploit the structural and dynamic information encoded in SAXS data, the use of realistic three-dimensional models is necessary. However, the generation of conformational ensembles of disordered proteins is extremely challenging (Zhou 2004). IDPs present a relatively flat (non-funnelled) energy landscape, with an extremely large number of local minima separated by low-energy barriers. This, combined with their large size, makes the analysis of their energy landscape a challenging problem for computational methods. Most of the available computational methods aim at collecting an ensemble representation of IDPs (Bernadó et al. 2007; Jensen et al. 2014; Wright and Dyson 2015). This requires an extensive and statistically correct exploration of the conformational space to obtain a representative set of states. Three main families of approaches have been proposed to generate conformational ensembles: molecular dynamics (MD) simulations, Monte Carlo (MC) methods, and experimentally parametrized statistical approaches. These three families of methods are succinctly explained next.

**Table 7.1** SAXS studies performed on IDPs

| Protein | #Residues[a] | $R_g^{exp}$ (Å) | $R_g^{RC}$ (Å)[b] | References |
|---|---|---|---|---|
| CyaA toxin (1006–1707) | 702 | 84 | 77.7 | O'Brien et al. (2015) |
| MeCP2 | 486 | 62.5 ± 4.5 | 64.2 | Yang et al. (2011) |
| Map2C | 467 | 83 ± 1[d] | 62.8 | Borysik et al. (2015) |
| Rec1-resilin | 310 | 43.4 ± 0.8 | 50.7 | Balu et al. (2015) |
| Msh6 N-term | 304 | 56 ± 2 | 50.2 | Shell et al. (2007) |
| Ki-1/57 | 292 | 47.5 ± 1 | 49.2 | Bressan et al. (2008) |
| Juxtanodin (1–282) | 282 | 55.9 | 48.3 | Ruskamo et al. (2012) |
| Dehydrin ERD10 (2–260) | 259 | 60 ± 1.0[d] | 46.2 | Borysik et al. (2015) |
| MeCP2 (78–305) | 228 | 37.0 ± 0.9 | 43.3 | Yang et al. (2011) |
| PGC-1α 220 (2–220) | 219 | 61.3 | 42.3 | Devarakonda et al. (2011) |
| Synthetic Resilin | 185 | 50 ± 5 | 38.8 | Nairn et al. (2008) |
| CFTR (654–838) | 185 | 32.5 ± 1.8 | 38.8 | Marasini et al. (2013) |
| pCFTR (654–838) | 185 | 29.1 ± 1.8 | 38.8 | Marasini et al. (2013) |
| Dehydrin ERD14 (2–185) | 184 | 60 ± 1.0[d] | 38.6 | Borysik et al. (2015) |
| Juxtanodin (1–170) | 170 | 42.9 | 37.1 | Ruskamo et al. (2012) |
| Myelin basic protein | 170 | 33 | 37.1 | Stadler et al. (2014) |
| AavLEA1 | 163 | 41 ± 1[d] | 36.3 | Borysik et al. (2015) |
| SRC-1 (617–769) | 153 | 33.96 | 35.1 | Pavlin et al. (2014) |
| Osteopontin | 150 | 37.90 ± 0.08 | 34.7 | Lenton et al. (2015) |
| Pig Calpastatin domain I | 148 | 35.4 | 34.5 | Konno et al. (1997) |
| HrpO | 147 | 35.0 | 34.4 | Gazi et al. (2008) |
| Ii-1 | 141[c] | 41.0 ± 1 | 33.6 | Boze et al. (2010) |
| α-Synuclein, pH 7.5 | 140 | 40 ± 1 | 33.5 | Li et al. (2002) |
| α-Synuclein, pH 3.0 | 140 | 30 ± 1 | 33.5 | Li et al. (2002) |
| N-tail nucleoprotein MV | 139 | 27.2 ± 0.5 | 33.4 | Longhi et al. (2003) |
| β-synuclein | 137 | 49 ± 1 | 33.1 | Uversky et al. (2002) |
| *Human* NHE1 cdt (5 °C) | 131 | 37.1[d] | 32.4 | Nørholm et al. (2011) |
| *Human* NHE1 cdt (45 °C) | 131 | 35.3[d] | 32.4 | Kjaergaard et al. (2010) |
| ERM transactivation domain | 130 | 39.6 ± 0.7 | 32.2 | Lens et al. (2010) |
| *Human* NL3 (731–848) | 118 | 31.5 ± 1.0 | 30.6 | Paz et al. (2008) |
| elF4E binding protein (4E–BP) | 117 | 48.8 ± 0.2 | 30.5 | Gosselin et al. (2011) |
| Juxtanodin (172–282) | 111 | 31.5 | 29.7 | Ruskamo et al. (2012) |
| $p15^{PAF}$(2–111) | 110 | 28.1 ± 0.3 | 29.5 | De Biasio et al. (2014) |
| Prothymosin α, pH 7.5 | 109 | 37.8 ± 0.9 | 29.4 | Uversky et al. (1999) |
| Prothymosin α, pH 2.5 | 109 | 27.6 ± 0.9 | 29.4 | Uversky et al. (1999) |
| *pa*NHE1 cdt (5 °C) | 107 | 32.8[d] | 29.1 | Nørholm et al. (2011) |
| *pa*NHE1 cdt (45 °C) | 107 | 32.9[d] | 29.1 | Kjaergaard et al. (2010) |
| N-protein of bacteriophage λ | 107 | 33 ± 2[e] | 29.1 | Johansen et al. (2011a) |
| N-protein of bacteriophage λ | 107 | 38 ± 3.5 | 29.1 | Johansen et al. (2011b) |
| *Human* NCBD domain | 105 | 33 ± 1 | 28.8 | Borysik et al. (2015) |
| FEZ1 monomer | 103 | 36 ± 1 | 28.5 | Alborghetti et al. (2010) |
| HIV-1 $Tat_{133}$ | 101 | 33.0 ± 1.5 | 28.3 | Foucault et al. (2010) |
| *Human* Calpastatin (137–237) | 100 | 39.0 ± 1.5 | 28.1 | Borysik et al. (2015) |
| p53 (1–93) | 93 | 28.7 ± 0.3 | 27.1 | Wells et al. (2008) |
| Sic1 | 92 | 34.7 | 26.9 | Mittag et al. (2010) |
| pSic1 (hexaphosphorylated) | 92 | 34.0 | 26.9 | Mittag et al. (2010) |
| Juxtanodium (103–282) | 79 | 37.4 | 38.1 | Ruskamo et al. (2012) |
| PIR domain | 75 | 26.5 ± 0.5 | 24.2 | Moncoq et al. (2004) |
| N-term NRG1 type III | 75 | 26.8[d] | 24.2 | Chukhlieb et al. (2015) |

(continued)

**Table 7.1** (continued)

| Protein | #Residues[a] | $R_g^{exp}$ (Å) | $R_g^{RC}$ (Å)[b] | References |
|---|---|---|---|---|
| IB5 | 73[b] | 27.9 ± 1.0 | 23.8 | Boze et al. (2010) |
| ACTR (5 °C) | 71 | 25.8[d] | 23.5 | Kjaergaard et al. (2010) |
| ACTR (45 °C) | 71 | 23.8[d] | 23.5 | Kjaergaard et al. (2010) |
| PaaA2 (1–63) | 70 [a] | 22.15 ± 0.87[d] | 23.3 | Sterckx et al. (2014) |
| N-term VS virus phosphoprotein | 68 | 26 ± 1[f] | 23.0 | Leyrat et al. (2011) |
| E3 ubiquitin ligase RNF4 (32–82) | 57 | 25.8 | 21.0 | Kung et al. (2014) |
| Histatin 5 | 24 | 13.3 | 13.3 | Cragnell et al. (2016) |
| R/S peptide | 24 | 12.6 ± 0.1 | 13.3 | Rauscher et al. (2015) |
| Constructions of Tau protein | | | | |
| Tau ht40 | 441 | 65 ± 3 | 61.0 | Mylonas et al. (2008) |
| Tau K32 | 202 | 42 ± 3 | 40.6 | Mylonas et al. (2008) |
| Tau K16 | 174 | 39 ± 3 | 37.5 | Mylonas et al. (2008) |
| Tau K18 | 130 | 38 ± 3 | 32.2 | Mylonas et al. (2008) |
| Tau ht23 | 352 | 53 ± 3 | 54.2 | Mylonas et al. (2008) |
| Tau K27 | 171 | 37 ± 2 | 37.2 | Mylonas et al. (2008) |
| Tau K17 | 143 | 36 ± 2 | 33.9 | Mylonas et al. (2008) |
| Tau K19 | 99 | 35 ± 1 | 28.0 | Mylonas et al. (2008) |
| Tau K44 | 283 | 52 ± 2 | 48.4 | Mylonas et al. (2008) |
| Tau K10 | 167 | 40 ± 1 | 36.7 | Mylonas et al. (2008) |
| Tau K25 | 185 | 41 ± 2 | 38.7 | Mylonas et al. (2008) |
| Tau K23 | 254 | 49 ± 2 | 45.7 | Mylonas et al. (2008) |
| Tau K32 AT8 AT100 | 202 | 41 ± 3 | 40.6 | Mylonas et al. (2008) |
| Tau ht23 S214E | 352 | 54 ± 3 | 54.2 | Mylonas et al. (2008) |
| Tau ht23 AT8 AT100 | 352 | 52 ± 3 | 54.2 | Mylonas et al. (2008) |
| Tau K18 P301L | 130 | 35 ± 2 | 32.2 | Mylonas et al. (2008) |
| Tau ht40 AT8 AT100 PHF1 (10 °C) | 441 | 66 ± 3 | 61.0 | Shkumatov et al. (2011) |
| Tau ht40 AT8 AT100 PHF1 (50 °C) | 441 | 67 ± 3 | 61.0 | Shkumatov et al. (2011) |

[a]When present, purification tags or extra terminal residues resulting from cloning were considered as part of the protein
[b]Threshold $R_g$ value obtained from the parametrization of Flory's relationship with the coil database
[c]Length of the most populated isoform of the samples was used
[d]$R_g$ derived from averaging conformations selected with EOM
[e]Data measured by SANS in highly crowded conditions (130 mg/ml of BPTI)
[f]Data derived from the 10 mM Arg/Glu buffer

MD simulations analyze the evolution of the system under study by solving Newton's equations of motion (Karplus and McCammon 2002; Piana et al. 2014). Theoretically, MD is a suitable method to correctly sample the conformational space of IDPs. Nevertheless, in practice, the high-dimensionality and the wideness of the energy landscape hampers its exhaustive exploration. Several approaches have been proposed to enhance conformational exploration with MD methods. A particularly effective one is Replica Exchange MD (REMD) that runs multiple simulations in parallel with different settings (usually different temperatures) and exchanges states between these processes (Trakawa and Takada 2011; Zerze et al. 2015; Chebaro et al. 2015). Going further in this direction, a very recent method called Multiscale Enhanced Sampling (MSES) couples temperature replica exchange and Hamiltonian replica exchange, using a coarse-grained model to guide atomistic conformational sampling (Lee and Chen 2016). The performance of MD-based methods can also be improved by the integration of experimental data to restrain the exploration of the most relevant regions of the conformational space (Lindorff-Larsen et al. 2004; Dedmon et al. 2005; Wu et al. 2009).

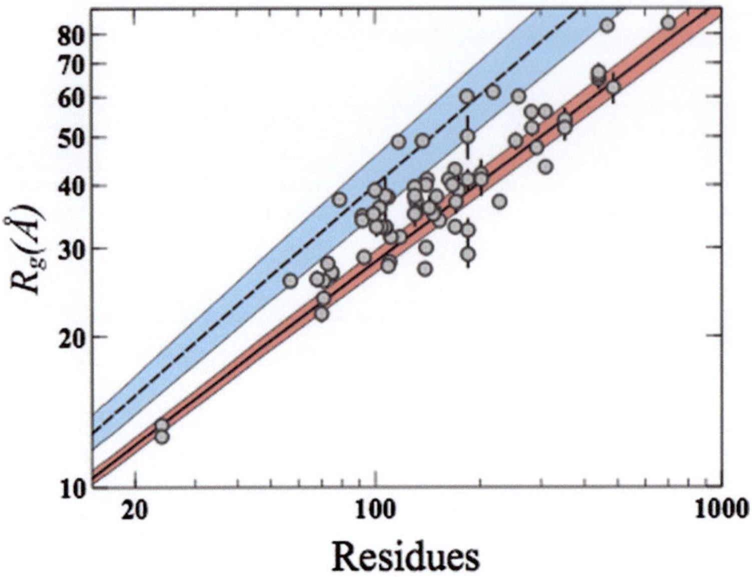


**Fig. 7.2** $R_g$ values from Table 7.1 (*gray dots*) as a function of the number of residues are plotted in Log-Log scale. *Straight lines* correspond to Flory's relationships parametrized for denatured proteins (*blue-dashed*) (Kohn et al. 2004) and IDPs (*red-solid*) (Bernadó and Blackledge 2009). *Colored bands* correspond to uncertainty of the parametrization for both models. Some IDPs are not fully disordered and are globally more extended or more compact than expected for a random coil. These structural features even if transient affect the experimental $R_g$

MC methods are a classical alternative to MD, the Markov chain Metropolis scheme (Metropolis et al. 1953) being the most widely used sampling technique (Vitalis and Pappu 2009a). The system is randomly perturbed and the new conformation is accepted with a probability that depends on the energy change between the new conformation and the previous one. Particular mention deserves a recently proposed variant called Hamiltonian Switch Metropolis Monte Carlo (HS-MMC), which has been specially conceived to study IDRs tethered to globular domains. Proteins including IDRs present energy minima due to the contact of the disordered and ordered regions. To avoid being trapped in such minima, the HS-MMC switches between an all-atom Hamiltonian to an excluded volume Hamiltonian to push the IDR away from the ordered domain.

Both, MD-based and MC-based approaches may suffer from inaccuracies of current energy models, which are better suited to globular proteins, and tend to provide structurally biased ensembles that do not properly reflect the conformational behaviour in solution of unstructured proteins (Best et al. 2014; Henriques et al. 2015). The development of more suitable force-fields and solvation models for IDPs are key issues for a correct performance of computational methods (Vitalis and Pappu 2009b; Emperador et al. 2015).

Knowledge-based statistical approaches are an alternative to physics-based energy functions. The most representative knowledge-based method for the generation of atomistic models of disordered proteins is Flexible-Meccano (FM) (Bernadó et al. 2005; Ozenne et al. 2012), although other similar methods have been described (Jha et al. 2005). The FM algorithm uses an amino acid-specific statistical coil derived from crystallographic structures. In FM, each conformation is built by assembling peptide plane units in a consecutive manner using a residue-specific coil library derived from crystallographic structures. To avoid the collapse of the chain, a coarse-grained description of side chains is also used. Based on this set of conformations, experimentally measurable NMR parameters and SAXS curves can be estimated, which has permitted the validation of the resulting models (see below).

Despite the efforts that have been made to precisely describe the conformational states of disordered proteins, there are still many technical and conceptual issues that must be addressed to correctly describe their energy landscape and, as a consequence, their associated experimental observables.

## 7.4 Ensemble Approaches

IDPs sample a large number of conformations. Therefore, ensembles of conformations are the most appropriate framework to structurally represent this family of proteins. In recent years, structural biologists have addressed the challenge of describing dynamic systems in terms of ensembles of reliable conformations guided by experimental data that represents average values for the complete ensemble of conformations (Bernadó and Blackledge 2010). SAXS has not been exempt from this tendency and several approaches have been developed to characterize protein mobility: Ensemble Optimization Method (EOM) (Bernadó et al. 2007; Tria et al. 2015), Minimal Ensemble Search (MES) (Pelikan et al. 2009), Basis-Set Supported SAXS (BSS-SAXS) (Yang et al. 2010), Maximum Occurrence (MAX-Occ) (Bertini et al. 2010), Ensemble Refinement of SAXS (EROS) (Rozycki et al. 2011), Broad Ensemble Generator with Re-weighting (BEGR) (Daughdrill et al. 2012), and Bayesian Ensemble SAXS (BE-SAXS) (Antonov et al. 2016). These methods are based on a common strategy that consists of three consecutive steps: (i) computational generation of a large ensemble describing the conformational landscape available to the protein, (ii) computation of the theoretical SAXS curves from the individual conformations, and (iii) selection of a subensemble of conformations that collectively describes the experimental profile using multiparametric optimization methods. Despite the common philosophy, these programs present distinct features in the three steps. Readers are referred to the original articles for a detailed description of the approaches.

The availability of ensemble methods has revolutionized the study of flexible proteins by SAS. Ensemble methods provide a description in terms of statistical distributions of structural parameters or conformations that represents a crucial step forward with respect to traditional analysis based on averaged parameters extracted from raw data such as $R_g$ or $D_{max}$. In that context, conformational perturbations exerted by temperature (Shkumatov et al. 2011; Kjaergaard et al. 2010), buffer composition (Leyrat et al. 2011), or mutations (Stott et al. 2010) can be monitored in terms of ensembles of accessible conformations.

The main approximation of ensemble methods is the discrete description of entities that probe an astronomical number of conformations. It is therefore reasonable to argue about the real meaning of the SAXS-derived ensembles. An additional problem is the statistically significant size of the derived ensembles based on experimental data with a very limited amount of information (Hammel 2012; Yang 2014). The described strategies use distinct philosophies to address these issues. In some cases such as EOM 2.0 and MES, programs search for the minimal number of conformations required to describe the data to limit or abolish over-fitting. BSS-SAXS and BE-SAXS use Bayesian statistics to address the confidence of the derived populations. In other cases, such as EROS and BEGR, populations of the conformers of an initially built ensemble are slightly modified (re-weighted) in order to describe the data.

SAXS-derived ensembles are representations of the conformational landscape sampled by proteins in solution, but not necessarily the exact states. Ensemble approaches are inherently ill-defined problems, and this is especially severe in SAXS that codes for a very limited amount of structural information. In that sense, it is more adequate to represent highly flexible proteins as distributions of accessible structural parameters such as $R_g$ or $D_{max}$. These representations are less prone to over-fitting artifacts (Bernadó et al. 2007).

In disordered proteins, SAS reports on the overall size and shape of the protein in solution. The presence of extendedness or compactness can be probed by SAS, but regions causing these structural biases can not be identified unambiguously due to the low-resolution nature of the data. An interesting way to enrich the structural content of SAXS data to more precisely localize partially structured regions in IDPs has been proposed. In this strategy, SAXS curves are measured for multiple deletion mutants of the disordered chain, and simultaneously fitted in terms of a common ensemble. Note that this strategy is only valid if the structural elements of the full-length protein remain intact in the deletion mutants. This approach is described in detail in the original article of EOM where it was applied to a flexible multi-domain kinase (Bernadó et al. 2007). The most notable example of this approach was the study of two isoforms of Alzheimer-related tau, ht23 and ht40 (Mylonas et al. 2008). SAXS data for full-length ht23 and ht40 and for five and three deletion mutants for each isoform were measured, respectively. The simultaneous fit of all curves with EOM unambiguously identified the so-called repeat region as the source of residual secondary structure in tau. For the ht23 isoform, with three repeats, the maximum separation is found within the repeat domain itself. Conversely, the ht40 isoform, with four repeats, presented an enhanced separation between the repeat domain and the preceding region. These results suggest that the different number of turns (one per repeat) may lead to different global arrangements of the chain in that region.

In addition to the previously described strategy, the information content of SAXS curves measured for IDPs can be enriched with data from other techniques. NMR is by far the most common technique used synergistically with SAXS (Sibille and Bernadó 2012), and a special section of this chapter is devoted to it. Additionally, experimental data coming from Electron Paramagnetic Resonance (EPR) and single molecule Fluorescence Resonance Energy Transfer (smFRET) have been synergistically used with SAXS to characterize disordered proteins (Boura et al. 2012). In that sense, the development of robust ways to integrate other biophysical measurements in ensemble approaches is an unavoidable future direction.

## 7.5 Application of SANS to Study IDPs

The physical bases and the structural information that can be derived form SAXS and SANS are equivalent and, in principle, both techniques can be used to characterize biomolecules. The more general use of SAXS is based on the higher beam intensity and more widespread availability of X-ray beam-lines. SANS however offers some advantages with respect to SAXS. The first one is the absence of radiation damage that makes it a non-destructive technique. The second one is based on the possibility to perform contrast-match experiments, where some components of the sample are made invisible by finely tuning the degree of deuteration of sample components and the ratio of $H_2O/D_2O$ of the buffer. Contrast matching has been widely used in structural biology and the reader is refereed to the excellent reviews available on the subject (Jacrot 1976; Heller 2010; Gabel 2012).

Although SANS has been used to study IDPs or IDRs (Krueger et al. 2011) or radiation sensitive systems (Greving et al. 2010; Stanley et al. 2011; Perevozchikova et al. 2014), its main application is in contrast-match experiments. An example of this family of experiments is the study of the interaction of β-amyloid (1–40) with the detergent SDS (Jeng et al. 2006). In this study the use of deuterated SDS in SANS experiments enabled the characterization of the peptide in the presence of SDS and showed that β-amyloid adopts a short rod-like shape. Interestingly the authors observed that SDS suppress fibrils by forming complexes with a 30:1 molar ratio between detergent and protein.

IDPs may be particularly sensitive to the effects of molecular crowding, and contrast-matching SANS experiments is a powerful tool to study these effects. By choosing appropriate levels of deuteration of the protein of interest and the crowding agent, the scattering contribution of the latter can be made negligible, enabling the study of the structural perturbations exerted by macromolecular crowding on disordered proteins. This strategy was used to study the disordered N-protein of bacteriophage λ in the presence of high concentrations of bovine pancreatic trypsin inhibitor (BPTI), a small globular protein (Johansen et al. 2011a; Goldenberg and Argyle 2014). In 46% $D_2O$, high concentrations of non-deuterated BPTI were contrast matched, and only the signal of the 85% deuterated N-protein was visible. The study demonstrated that molecular crowding exerted a compaction effect on N-protein. However, this effect was non-linear and the effects observed at 65 mg/mL of BPTI were equivalent to those at 130 mg/mL.

## 7.6 SAXS Studies of IDPs Within Macromolecular Assemblies

Given the structural plasticity of disordered proteins, they are regarded as interacting specialists and have a special place in cell signalling using short motifs or domain-sized disordered segments for partner recognition (Wright and Dyson 2015; Tompa et al. 2015). Providing a comprehensive description of the biomolecular recognition processes for IDPs is thus of great importance for the understanding of key biological functions that are orchestrated by this family of proteins. In this regard, SAXS is emerging as an extremely valuable tool for characterizing biomolecular interactions involving highly flexible proteins, which are highly challenging for crystallography. The application of solution NMR also encounters severe limitations when characterizing large biomolecular complexes. Conversely, SAXS, which is not limited by size, offers a source of structural and dynamic information that by itself or combined with NMR and/or X-ray crystallography is a promising alternative for the structural characterization of highly dynamic proteins and complexes in solution.

The binding of an IDP to its target produces specific SAXS signatures that enable the detection of the interaction. Typically, the mixing of two interacting partners will result in a rise of $R_g$ values, otherwise, if the proteins do not form a complex the $R_g$ value will follow an population-weighted average of the two. Insightful information can be obtained by inspecting the $p(r)$ function in the absence and presence of the disordered interaction partner. The scattering of globular proteins generally give a symmetrical bell-shaped $p(r)$ function, while the interaction with a sufficiently large IDP results in tailing of the peak shape to higher values of $r$, culminating at a large $D_{max}$. The Kratky and Porod-Debye analyses are also powerful indicators of flexibility within macromolecular assemblies (Rambo and Tainer 2011).

It is possible to define 3D molecular envelopes describing the low-resolution shapes of flexible macromolecular assemblies involving IDPs (Longhi et al. 2003; Marsh et al. 2010; Gosselin et al. 2011; Devarakonda et al. 2011). However, the resulting fixed low-resolution structure is not the most appropriate structural description of a disordered protein. Ensemble analysis of explicit or coarse-grained models should provide a more accurate characterization of flexible macromolecular assemblies, particularly if combined with available structure coordinates of folded-domains and data affordable by NMR. Moreover, weak or moderate affinities are a hallmark of many IDP molecular recognition events, which entails the presence of multiple species in solution at standard experimental concentrations. In these cases, any modeling strategy envisioned should account for this species polydispersity to reliably describe SAS data.

Several SAXS studies have been devoted to the interactions of different IDPs with other proteins or DNA, as the only source of structural information or in combination with high-resolution methods. Some examples will be presented.

The complex of Msh2 and Msh6 recognizes mismatched bases in DNA during mismatch repair. The N-terminal region of Msh6, a 304 residue long IDR, recognizes PCNA, a protein that controls processivity of DNA polymerases. Shell and co-workers have demonstrated this direct interaction with SAXS (Shell et al. 2007). A comparison of the $R_g$, Kratky plots and $p(r)$ functions of the isolated partners and the complex showed that PCNA does not induce substantial structure to the N-terminal region of Msh6, which remains mainly disordered and proteolytically accessible upon binding. The interaction of the Msh2-Msh6 complex with PCNA was also addressed by SAXS. The interaction was shown to produce a complex that could be considered as a highly flexible dumbbell where both globular domains are tethered by the N-terminal Msh6 fragment that acts as a molecular leash. These observations were further confirmed in an additional experiment with a biologically active deletion mutant of Msh6 containing a notably shorter N-terminal tail. Under these conditions the important size changes upon binding were easily monitored using the $p(r)$ and $D_{max}$.

The tumor suppressor p53 is a multifunctional protein that plays a crucial role in processes like apoptosis control and DNA repair. P53 is a homotetramer with two folded domains that are tethered and flanked by unstructured regions. Rigid-body modeling of SAXS data measured for p53 suggests that the protein is a rather open cross-like tetrameric assembly, which collapses to tightly embrace DNA (Tidow et al. 2007). This is how the flexibility of IDPs helps the protein to fulfill its function.

Nuclear receptors (NR) are engaged in gene transcription regulation in response to binding of specific ligands. Signal transduction from ligand binding to gene expression requires the recruitment of co-regulator proteins. Most NR co-regulators function as flexible scaffolds for chromatin modifiers and transcriptional machinery (Millard et al. 2013). Structural analyses of their interaction have been restricted to small peptides of the regulators and the nuclear receptor ligand-binding domain. More recently, several SAXS studies have provided new insights into the stoichiometry and binding mode of the complexes formed by NR and disordered co-regulators (Jin et al. 2008; Rochel et al. 2011; Devarakonda et al. 2011; Pavlin et al. 2014).

β-thymosin/WH2 (β-t/WH2) domains are widespread short disordered regions (25–50 residues) able to recognize G-actin and regulate actin-assembly. When bound to G-actin the N-terminal half of β-t/WH2 adopts a well-ordered amphipathic helix. SAXS analysis at different ionic strengths revealed that the C-terminal regions of different β-t/WH2 domains display distinct dynamics, which correlate with functional differences. At low ionic strength β-t/WH2 sequesters G-actin in a polymerization incompetent state, where the dynamic interactions of the C-terminal part are restrained to a single conformational state. This SAXS-driven observation prompted the study of the β-t/WH2:G-actin complex by NMR at different ionic strengths revealing that a single intermolecular salt-bridge controls the assembly (Didry et al. 2012).

## 7.7 Low-Complexity Regions

Low-complexity regions (LCRs) are unusually simple protein sequences with a strong amino acid composition bias, and include homo-repeats of a single amino acid, short period repeats or aperiodic mosaics of a few residues (Wootton and Federhen 1993). Protein sequences from all three kingdoms of life contain LCRs, but they are more common in eukaryotes, with *Plasmodium falciparum* being an extreme case as ~90% of its proteins host LCRs (Marcotte et al. 1999; Pizzi and Frontali 2001). Despite their high abundance and doubtless biological relevance, not many studies are focused on the structural characterization of LCRs mainly due to the technical challenges they pose. LCRs are often inserted in IDRs precluding their crystallization (Haerty and

Golding 2010), and the NMR sequence assignment is complicated (or impossible) by the similarity of nuclear chemical environments. In that context, SAS is a powerful technique to investigate the structure and dynamics of this elusive family of proteins.

Prothymosin α was the first and probably the most well characterized LCR-hosting protein (Gast et al. 1995). Roughly half of the 109 residues of Prothymosin α are acidic (Asp and Glu), leading to a net charge of −44 at neutral pH. A SAXS analysis of prothymosin α at near neutral (7.5) and acidic (2.5) pH showed that while it was unstructured under both conditions, a dramatic reduction in $R_g$ (from 37.8 ± 0.9 Å to 27.6 ± 0.8 Å) could be observed (Uversky et al. 1999). The pH-induced reduction in protein size can be explained by the neutralization of the negatively charged acidic residues due to the decrease in pH. A similar reduction of size was observed after the addition of 15 mM $Zn^{2+}$ at neutral pH ($R_g$ = 28.1 ± 0.8 Å) suggesting the electrostatic screening of the negative charges by cations (Uversky et al. 2000). A similar situation was observed for the basic proline-rich salivary proteins IB5 and Il-1. Proline-rich salivary proteins bind polyphenolic plant compounds (e.g. tannins) and form aggregates upon binding high concentrations of these compounds. At sequence level IB5 and Il-1 contain repetitions of Pro, Gly and Gln or Glu residues, and they are predicted to be disordered. Boze et al. used SAXS to study the conformations of IB5 and Il-1, and to see if there are any functional advantages linked to the respective conformations (Boze et al. 2010). Both IB5 and Il-1 showed an $R_g$ larger than the theoretical one for IDPs of their lengths, $R_g$ = 27.9 Å ± 1.0 Å and 41.0 Å ± 1.0 Å, respectively. In addition, the experimental $D_{max}$ (110 ± 10 Å and 155 Å ± 10 Å, respectively) also indicated the presence of highly extended conformations that could contain poly-proline II helices.

Another example for a protein with low sequence complexity is the crosslinked elastomeric protein resilin. Resilin is rich in Gly, Ser and Pro residues and is present in most insects where it is critical for flight and jumping. Due to its low stiffness, high fatigue lifetime and high resilience, resilin has been of great interest for the production of biomaterials for biomedical applications. SAXS measurements on a recombinant resilin (rec1-resilin) comprising 18 copies of the N-terminal repeat motif (GGRPSDSYGAPGGGN), yielded an $R_g$ of 43.4 ± 0.8 Å and a $D_{max}$ of 200 Å (Balu et al. 2015). Since the expected $R_g$ of a 310 residue-long protein with a compact structure would be ~19.6 Å and that of an IDP would be ~59.5 Å, these data suggest that rec1-resilin is largely unfolded but not completely disordered.

Homo-repeat proteins represent extreme cases for the structural characterization, and huntingtin protein (htt) is the prototypical example of this family of proteins. Htt exon 1 contains two homo-repeat regions consisting of poly-Gln and poly-Pro, respectively. SAXS has been used for the characterization of the overall properties of this protein. Kratky plots of htt exon1 constructs fused to thioredoxin showed broad peaks with less decrease at higher scattering angles, consistent with flexible or unfolded proteins. Interestingly, the observed radii of gyration for constructs with 16 Gln or 39 Gln were very similar, $R_g$ = 49 Å or 52 Å, respectively (Owens et al. 2015). These data were consistent with previous experimental observations and MD simulations indicating that poly-Gln tracts form disordered, collapsed globular structures in solution (Vitalis et al. 2008; Dougan et al. 2009).

An example for a protein with low sequence complexity that forms higher order structures is silk fibroin. The heavy chain of silk fibroin from *Bombyx mori* is 5,263 amino acids long and dominated (94%) by the repetition of Gly-X repeats where X is mainly Ala (65%), Ser (23%), or Tyr (9%) (Zhou et al. 2001). In a pioneering study, Greving et al. characterized native (SF) and reconstituted silk fibroin (RSF) by SANS (Greving et al. 2010). Their study identified significant differences between the molecular weights and $R_g$s of native and reconstituted silk fibroin, as well as for RSF samples prepared under different conditions.

## 7.8 Application of SAS to Aggregating IDPs

Through a complex self-assembly process, some IDPs form amyloid fibrils that are the hallmark of disorders such as Alzheimer's, Parkinson's or diabetes (Chiti and Dobson 2006). Fibrillation is a very complex process that involves multiple oligomeric species that are transformed following intricate pathways. Interestingly, there are indications that soluble oligomers, often precursors of fibrils, are the main cause of cytotoxicity and neuronal damage. The structural analysis of the distinct species involved in fibrillation is a major challenge due to their instability, low relative concentration, difficulties of isolation, and the equilibrium between species of very different sizes, present at any time point during the fibrillation process.

The aggregation process of huntingtin exon 1 (htt, see above) has been followed by SANS (Stanley et al. 2011). For this study, the aggregation of a pathological truncated version of the protein encompassing 42 consecutive Gln residues ($NtQ_{42}P_{10}$) was followed in a time-dependent manner. The study indicated that the aggregation begins with dimeric and trimeric forms of $NtQ_{42}P_{10}$. In a very fast process, large oligomeric species, estimated to be 14-mers, are formed, and after that point the protein becomes predominately fibrillar. The same group addressed the effects of the Gln tract length on the fibrillation mechanism by investigating a non-pathological htt version ($NtQ_{22}P_{10}$) also using SANS (Perevozchikova et al. 2014). Their results indicate that the length of the homo-repeat dictates the size of the initial species and the aggregation pathway followed by htt. In these studies, however, the complex nature of the samples, which are composed by multiple oligomeric species, is not taken into account, and the average values of the derived parameters are analysed. The quantitative interpretation of SAS data of fibrillating proteins requires their decomposition into species-pure SAS profiles, reporting on the structure of each species present, and their relative concentration, reporting on the kinetics of the process. In a pioneering study this challenging SAXS data decomposition was successfully achieved (Giehm et al. 2011). In this study, the fibrillation of α-synuclein was monitored using SAXS by measuring profiles every 30 min during 24 h. Measured profiles contain contributions from all species present at each time-point. Singular Value Decomposition (SVD) analysis indicated that three species are enough to describe the complete SAXS dataset, which were assigned to monomer/dimer, mature fibril, and a third species of unknown nature. Using SAXS profiles measured separately for the two extreme components (monomer/dimer and mature fibril) as fixed contributions, the SAXS curve for the third component was derived (Vestergaard et al. 2007). This decomposed species turned out to be a building block for fibril formation. From the analysis of the time-dependent relative populations, the kinetics of the aggregation was also derived. Despite the novelty of the study, it seems clear that more robust and objective approaches to decompose complex SAXS data from aggregating proteins are necessary.

## 7.9 Combined Use of SAXS and Nuclear Magnetic Resonance

NMR is the only technique that can provide atomic-resolution information of IDPs (Dyson and Wright 2004). The first step to study an IDP by NMR is to assign a resonance frequency to all magnetically active nuclei ($^{1}H$, $^{15}N$, and $^{13}C$) of the protein. Due to very low amide proton dispersion, assignment of IDP spectra is challenging. However, the use of high magnetic field spectrometers and several methodological developments allow to routinely assign NMR frequencies (Narayanan et al. 2010). NMR is a highly versatile technique and multiple observables reporting on protein structure and dynamics can be measured. In the following paragraphs the most relevant ones will be succinctly described.

- **Chemical shifts (CSs):** CSs correspond to the resonance frequencies of the nuclei, and are the primary information that can be derived from a NMR experiment. CSs are very sensitive to the chemical environment of nuclei and reveal the presence of secondary structural elements. A chemical shift index (CSI) has been established to highlight regions that deviate from purely random coil to form secondary structural elements (Wishart and Sykes 1994; Wishart et al. 1995). With the discovery of IDPs, the interest in using CSs to detect partially structured elements has been renewed, and several databases have appeared based on small synthetic peptides (Schwarzinger et al. 2000; Kjaergaard et al. 2011) or IDPs (Tamiola et al. 2010) to identify these regions.
- **Scalar $^3J_{HNHA}$Couplings:** The scalar couplings between $H^N$ and $H^\alpha$ are sensitive to backbone conformations and can be converted into the $\Phi$ torsion angle using Karplus-like relationships (Karplus 1959). Secondary structures can be unveiled using $^3J_{HNHA}$. Values below 5 Hz suggest the presence of α-helices, whereas values above 8 Hz suggest extended regions. Values lying in between correspond to random coil conformations. In combination with other NMR observables, $^3J_{HNHA}$ can be useful to discriminate between extended structures and poly-proline-II helices (Oh et al. 2012).
- **Residual Dipolar Couplings (RDCs):** RDCs, measured in magnetically aligned samples, are the most sensitive experimental measurement to probe conformational sampling in IDPs (Jensen et al. 2009). Negative NH RDC values are observed in random coils (Louhivuori et al. 2003). Interestingly, more positive and more negative RDCs than expected for a random coil are associated to α-helices and extended conformations, respectively (Mohana-Borges et al. 2004). This is an excellent indication to qualitatively assess the presence of distinct types of secondary structural elements. More quantitative interpretation of RDCs can be derived when applying atomistic models of disordered chains (Jha et al. 2005; Bernadó et al. 2005; Marsh et al. 2008). The measurement of multiple backbone RDCs enriches the description of residue-specific structural preferences (Jensen et al. 2008).
- **Paramagnetic Relaxation Enhancement (PRE) experiments:** PREs are measured as enhanced relaxation rates in residues that are close (within 15–30 Å) to a paramagnetic tag engineered in a specific position of the chain (Gillespie and Shortle 1997). Although nitroxyl spin labels are normally used, paramagnetic cations can also be attached providing, in addition to PREs, other structural observables such as pseudocontact shifts (PCS) (Otting 2008). Therefore, PREs are suitable observables to probe transient long-range interactions in IDPs.

The complementarity between NMR and SAS is based on the distinct resolution of the information provided. Whereas SAS probes the overall properties of molecules, NMR information reports on atomic or residue-specific information. Therefore, the simultaneous description of both observables strongly suggests the appropriateness of the derived model. In that context, SAXS can be used to validate structural models of IDPs refined with NMR data. In this approach, the residue-specific conformational preferences of an IDP are refined using RDCs and CSs using Flexible-Meccano (Bernadó et al. 2005; Ozenne et al. 2012). The final model contains percentages of secondary structural elements in localized regions that have been imposed to properly describe the NMR data. The resulting ensemble can be validated by simply comparing the average SAXS curve computed from the ensemble with that experimentally measured. This strategy has been applied to the partially folded Sendai virus PX (Bernadó et al. 2005), the transactivation domain of p53 (Wells et al. 2008), the K18 construct of Tau protein (Mylonas et al. 2008; Mukrasch et al. 2007), and the oncogene $p15^{PAF}$ (De Biasio et al. 2014). A similar approach has been performed to study PaaA2 antitoxin (Sterckx et al. 2014). In this last study the NMR derived ensemble was used as starting pool for a SAXS EOM refinement demonstrating that the protein exists in solution as two preformed helices, connected by a flexible linker.

The best manner to exploit the complementarity of both techniques is to integrate the experimental data into the same refinement protocol. Some of these integrative approaches have been applied to IDPs. One of these is ENSEMBLE, a program that derives ensembles of disordered proteins by collectively describing SAXS curves in addition to several NMR observables: CS, J-couplings, RDCs, PREs, Nuclear Overhauser effects, hydrodynamic radius, solvent accessibility restraints, hydrogen-exchange protection factors, and $^{15}N\ R_2$ relaxation rates (Marsh et al. 2007; Krzeminski et al. 2013). A large number of random structures are computed with FOLDTRAJ or TRADES (Feldman and Hogue 2000, 2002), and a Monte Carlo algorithm is used to select a subset of these structures that are collectively consistent with the experimental restraints. This subset is used as a basis for the generation of new structures, and the process is repeated until a final ensemble consistent with all of the experimental measurements is obtained. This approach addresses the intrinsic problem of under-restraining and consequent over-fitting by finding the smallest ensemble that is consistent with all experimental restraints imposed. ENSEMBLE has been recently applied to characterize the protein Sic1 and its hexaphosphorylated version pSic1 by combining SAXS data with several NMR parameters, including CS, PREs, RDCs, and $^{15}N\ R_2$ (Mittag et al. 2010). Moreover, a structural model of the complex between pSic1, which contains several binding regions, and its partner Cdc4 was generated by combining restraints of the free form of pSic1 with sparse NMR data of the complex suggesting a fuzzy interaction. ASTEROIDS is another program that allows the synergistic interpretation of NMR and SAXS data (Jensen et al. 2008). The power of ASTEROIDS is illustrated in a recent study of tau and α-synuclein using NMR (CSs, RDCs, PREs) and SAXS data (Schwalbe et al. 2014). Using extensive cross-validation, the authors showed that five different types of independent experimental parameters are predicted more accurately by selected ensembles than by statistical coil descriptions. With this method, they could highlight that tau and α-synuclein sample poly-proline II region in the aggregation-nucleation sites.

## 7.10 Final Remarks

Disordered systems represent a challenge for structural biology. The presence of multiple conformational states hampers the application of traditional approaches and has fostered the development of strategies with the capacity to capture their inherent plasticity. Despite the limited amount of information coded, SAS has played (and will play) a crucial role in the structural biology of disordered proteins. The appearance of several ensemble approaches to describe SAS data in terms of multiple conformations has been a revolution in the field as they have opened the characterization of the size and shape properties of highly flexible proteins in solution. These methods, however, are model-dependent as their results can be biased depending on the approach used to compute atomic structures of disordered states. New strategies will have to be developed to accurately explore the energetically plausible conformational landscape of disordered proteins without dramatically increasing computational costs. A related aspect that remains poorly understood in IDPs is hydration. The accurate description of surrounding water molecules in disordered chains is necessary to describe their scattering properties.

Alternatively, the integration of experimental information derived from other biophysical methods, mainly NMR, enriches the resolution of the derived conformational ensembles. This is a research field that despite its continuous evolution will require additional efforts. Indeed, a database, pE-DB (http://pedb.vib.be), has been created to compile structural ensembles derived either from NMR, SAXS or their combination to foster integrative approaches and the development of validation strategies in disordered proteins (Varadi et al. 2014).

Disordered biomolecular complexes exemplify the challenges of the field and the need to properly combine experimental and computational approaches. Their characterization requires the detailed description of both the interacting surface and the disordered regions. In that context, SAS can be the key technique to merge in a unified picture the information derived from high-resolution techniques such as NMR and crystallography. Moreover, many of the involving IDPs have a low or moderate affinity causing species polydispersity in typical SAS experiments. This incorporates an additional level of complexity that has to be overcome.

In the last decade there has been a general realization that protein dynamics is fundamental to understand biomolecular function. SAS has a brilliant future in that emerging field. The exploitation of SAS information to address this recent field in structural biology will require further experimental, computational and conceptual developments.

**Acknowledgements** This work was supported by the European Research Council under the European Union's H2020 Framework Programme (2014-2020) / ERC Grant agreement n° [648030], the French Infrastructure for Integrated Structural Biology (FRISBI – ANR-10-INSB-05-01), and Labex EpiGenMed, an « Investissements d'avenir » program, (ANR-10-LABX-12-01) to PB. FHT is supported by INSERM and the Sapere Aude programme SAFIR of the University of Copenhagen. AU is supported by a grant from the Fondation pour Recherche Médicale.

## References

Alborghetti MR, Furlan AS, Silva JC, Paes Leme AF, Torriani ICL, Kobarg J (2010) Human FEZ1 protein forms a disulfide bond mediated dimer: implications for cargo transport. J Proteome Res 9:4595–4603

Antonov LD, Olsson S, Boomsma W, Hamelryck T (2016) Bayesian inference of protein ensembles from SAXS data. Phys Chem Chem Phys 18:5832–5838

Aslam M, Guthridge JM, Hack BK, Quigg RJ, Holers VM, Perkins SJ (2003) The extended multidomain solution structures of the complement protein Crry and its chimeric conjugate Crry-Ig by scattering, analytical ultracentrifugation and constrained modelling: implications for function and therapy. J Mol Biol 329:525–550

Balu R, Knott R, Cowieson NP, Elvin CM, Hill AJ, Choudhury NR, Dutta NK (2015) Structural ensembles reveal intrinsic disorder for the multi-stimuli responsive bio-mimetic protein Rec1-resilin. Sci Rep 5:10896

Bernadó P (2010) Effect of interdomain dynamics on the structure determination of modular proteins by small-angle scattering. Eur Biophys J 39:769–780

Bernadó P, Blackledge M (2009) A self-consistent description of the conformational behavior of chemically denatured proteins from NMR and small angle scattering. Biophys J 97:2839–2845

Bernadó P, Blackledge M (2010) Structural biology: proteins in dynamic equilibrium. Nature 468:1046–1048

Bernadó P, Blanchard L, Timmins P, Marion D, Ruigrok RW, Blackledge M (2005) A structural model for unfolded proteins from residual dipolar couplings and small-angle x-ray scattering. Proc Natl Acad Sci U S A 102:17002–17007

Bernadó P, Mylonas E, Petoukhov MV, Blackledge M, Svergun DI (2007) Structural characterization of flexible proteins using small-angle X-ray scattering. J Am Chem Soc 129:5656–5664

Bernadó P, Svergun DI (2012a) Structural analysis of intrinsically disordered proteins by small-angle X-ray scattering. Mol BioSyst 8:151–167

Bernadó P, Svergun DI (2012b) Analysis of intrinsically disordered proteins by small-angle X-ray scattering. Methods Mol Biol 896:107–122

Bertini I, Giachetti A, Luchinat C, Parigi G, Petoukhov MV, Pierattelli R, Ravera E, Svergun DI (2010) Conformational space of flexible biological macromolecules from average data. J Am Chem Soc 132:13553–13558

Best RB, Zheng W, Mittal J (2014) Balanced protein-water interactions improve properties of disordered proteins and non-specific protein associantion. J Chem Theory Comput 10:5113–5124

Borysik AJ, Kovacs D, Guharoy M, Tompa P (2015) Ensemble methods enable a new definition for the solution to gas-phase transfer of intrinsically disordered proteins. J Am Chem Soc 137:13807–13817

Boura E, Różycki B, Chung HS, Herrick DZ, Canagarajah B, Cafiso DS, Eaton WA, Hummer G, Hurley JH (2012) Solution structure of the ESCRT-I and -II supercomplex: implications for membrane budding and scission. Structure 20:874–886

Boze H, Marlin T, Durand D, Pérez J, Vernhet A, Canon F, Sarni-Manchado P, Cheynier V, Cabane B (2010) Proline-rich salivary proteins have extended conformations. Biophys J 99:656–665

Bressan GC, Silva JC, Borges JC, Dos Passos DO, Ramos CHI, Torriani IL, Kobarg J (2008) Human regulatory protein Ki-1/57 has characteristics of an intrinsically unstructured protein. J Proteome Res 7:4465–4474

Calmettes P, Durand D, Desmadril M, Minard P, Receveur V, Smith JC (1994) How random is a highly denatured protein? Biophys Chem 53:105–114

Chebaro Y, Ballard AJ, Chakraborty D, Wales DJ (2015) Intrinsically disordered energy landscapes. Sci Rep 5:10386

Chiti F, Dobson CM (2006) Protein misfolding, functional amyloid, and human disease. Annu Rev Biochem 75:333–366

Chukhlieb M, Raasakka A, Ruskamo S, Kursula P (2015) The N-terminal cytoplasmic domain of neuregulin 1 type III is intrinsically disordered. Amino Acids 47:1567–1577

Cragnell C, Durand D, Cabane B, Skepö M (2016) Coarse-grained modelling of the intrinsically disordered protein Histatin 5 in solution. Monte Carlo simulations in combination with SAXS. Proteins 84 (6)777–791

Daughdrill GW, Kashtanov S, Stancik A, Hill SE, Helms G, Muschol M, Receveur-Bréchot V, Ytreberg FM (2012) Understanding the structural ensembles of a highly extended disordered protein. Mol BioSyst 8:308–319

De Biasio A, Ibáñez de Opakua A, Cordeiro TN, Villate M, Merino N, Sibille N, Lelli M, Diercks T, Bernadó P, Blanco FJ (2014) p15PAF is an intrinsically disordered protein with nonrandom structural preferences at sites of interaction with other proteins. Biophys J 106:865–874

Dedmon M, Lindorff-Larsen K, Christodoulou J, Vendruscolo M, Dobson CM (2005) Mapping long-range interactions in alpha-synuclein using spin-label NMR and ensemble molecular dynamics simulations. J Am Chem Soc 127:4767

Devarakonda S, Gupta K, Chalmers MJ, Hunt JF, Griffin PR, Van Duyne GD, Spiegelman BM (2011) Disorder-to-order transition underlies the structural basis for the assembly of a transcriptionally active PGC-1α/ERRγ complex. Proc Natl Acad Sci U S A 108:18678–18683

Didry D, Cantrelle F-X, Husson C, Roblin P, Moorthy AME, Perez J, Le Clainche C, Hertzog M, Guittet E, Carlier M-F, van Heijenoort C, Renault L (2012) How a single residue in individual β-thymosin/WH2 domains controls their functions in actin assembly. EMBO J 31:1000–1013

Doniach S (2001) Changes in biomolecular conformation seen by small angle X-ray scattering. Chem Rev 101:1763–1778

Dougan L, Li J, Badilla CL, Berne BJ, Fernandez JM (2009) Single homopolypeptide chains collapse into mechanically rigid conformations. Proc Natl Acad Sci U S A 106:12605–12610

Dunker AK, Brown CJ, Lawson JD, Iakoucheva LM, Obradovic Z (2002) Intrinsic disorder and protein function. Biochemistry 41:6573–6582

Dunker AK, Cortese MS, Romero P, Iakoucheva LM, Uversky VN (2005) Flexible nets. The roles of intrinsic disorder in protein interaction networks. FEBS J 272:5129–5148

Dyson HJ, Wright PE (2004) Unfolded proteins and protein folding studied by NMR. Chem Rev 104:3607–3622

Eliezer D (2009) Biophysical characterization of intrinsically disordered proteins. Curr Opin Struct Biol 19:23–30

Emperador A, Sfriso P, Villarreal MA, Gelpí JL, Orozco M (2015) PACSAB: coarse-grained force field for the study of protein-protein interactions and conformational sampling in multiprotein systems. J Chem Theory Comput 11:5929–5938

Feigin LA, Svergun DI (1987) Structure analysis by small-angle X-ray and neutron scattering. Plenum Press, New York

Feldman HJ, Hogue CW (2000) A fast method to sample real protein conformational space. Proteins 39:112–131

Feldman HJ, Hogue CW (2002) Probabilistic sampling of protein conformations: new hope for brute force? Proteins 46:8–23

Flory PJ (1953) Principles of polymer chemistry. Cornell University Press, Ithaca

Foucault M, Mayol K, Receveur-Bréchot V, Bussat M-C, Klinguer-Hamour C, Verrier B, Beck A, Haser R, Gouet P, Guillon C (2010) UV and X-ray structural studies of a 101-residue long tat protein from a HIV-1 primary isolate and of its mutated, detoxified, vaccine candidate. Proteins 78:1441–1456

Gabel F (2012) Small angle neutron scattering for the structural study of intrinsically disordered proteins in solution: a practical guide. Methods Mol Biol 896:123–135

Gast K, Damaschun H, Eckert K, Schulze-Forster K, Maurer HR, Müller-Frohne M, Zirwer D, Czarnecki J, Damaschun G (1995) Prothymosin alpha: a biologically active protein with random coil conformation. Biochemistry 34:13211–13218

Gazi AD, Bastaki M, Charova SN, Gkougkoulia EA, Kapellios EA, Panopoulos NJ, Kokkinidis M (2008) Evidence for a coiled-coil interaction mode of disordered proteins from bacterial type III secretion systems. J Biol Chem 283:34062–14068

Giehm L, Svergun DI, Otzen DE, Vestergaard B (2011) Low-resolution structure of a vesicle disrupting alpha-synuclein oligomer that accumulates during fibrillation. Proc Natl Acad Sci U S A 108:3246–3251

Gillespie JR, Shortle D (1997) Characterization of long-range structure in the denatured state of staphylococcal nuclease. I. Paramagnetic relaxation enhancement by nitroxide spin labels. J Mol Biol 268:158–169

Goldenberg DP, Argyle B (2014) Minimal effects of macromolecular crowding on an intrinsically disordered protein: a small-angle neutron scattering study. Biophys J 106:905–914

Gosselin P, Oulhen N, Jam M, Ronzca J, Cormier P, Czjzek M, Cosson B (2011) The translational repressor 4E-BP called to order by eIF4E: new structural insights by SAXS. Nucleic Acids Res 39:3496–3503

Graewert MA, Svergun DI (2013) Impact and progress in small and wide angle X-ray scattering (SAXS and WAXS). Curr Opin Struct Biol 23:748–754

Greving I, Dicko C, Terry A, Callow P, Vollrath F (2010) Small angle neutron scattering of native and reconstituted silk fibroin. Soft Matter 6:4389

Guinier A (1939) La diffraction des rayons X aux tres petits angles; application a l'etude de phenomenes ultramicroscopiques. Ann Phys (Paris) 12:161–237

Haerty W, Golding GB (2010) Low-complexity sequences and single amino acid repeats: not just "junk" peptide sequences. Genome 53:753–762

Hammel M (2012) Validation of macromolecular flexibility in solution by small-angle X-ray scattering (SAXS). Eur Biophys J 41:789–799

Hawkins AR, Lamb HK (1995) The molecular biology of multidomain proteins. Eur J Biochem 232:7–18

Heller WT (2010) Small-angle neutron scattering and contrast variation: a powerful combination for studying biological structures. Acta Crystallogr D Biol Crystallogr 66:1213–1217

Henriques J, Cragnell C, Skepö M (2015) Molecular dynamics simulations of intrinsically disordered proteins: force field evaluation and comparison with experiment. J Chem Theory Comput 11:3420–3431

Jacques DA, Trewhella J (2010) Small-angle scattering for structural for structural biology-expanding the frontier while avoiding the pitfalls. Protein Sci 19:642–657

Jacrot B (1976) The study of biological structures by neutron scattering from solution. Rep Prog Phys 39:911–953

Jeng U-S, Lin T-L, Lin JM, Ho DL (2006) Contrast variation SANS for the solution structure of the β-amyloid peptide 1–40 influenced by SDS surfactants. Phys B Condens Matter 385-386:865–867

Jensen MR, Houben K, Lescop E, Blanchard L, Ruigrok RW, Blackledge M (2008) Quantitative conformational analysis of partially folded proteins from residual dipolar couplings: application to the molecular recognition element of Sendai virus nucleoprotein. J Am Chem Soc 130:8055–8061

Jensen MR, Markwick PR, Meier S, Griesinger C, Zweckstetter M, Grzesiek S, Bernadó P, Blackledge M (2009) Quantitative determination of the conformational properties of partially folded and intrinsically disordered proteins using NMR dipolar couplings. Structure 17:952–960

Jensen MR, Ruigrok RWH, Blackledge M (2013) Describing intrinsically disordered proteins at atomic resolution by NMR. Curr Opin Struct Biol 23:426–435

Jensen MR, Zweckstetter M, Huang JR, Blackledge M (2014) Exploring free-energy landscapes of intrinsically disordered proteins at atomic resolution using NMR spectroscopy. Chem Rev 114:6632–6660

Jha AK, Colubri A, Freed KF, Sosnick TR (2005) Statistical coil model of the unfolded state: resolving the reconciliation problem. Proc Natl Acad Sci U S A 102:13099–13104

Jin KS, Park JK, Yoon J, Rho Y, Kim J-H, Kim EE, Ree M (2008) Small-angle X-ray scattering studies on structures of an estrogen-related receptor alpha ligand binding domain and its complexes with ligands and coactivators. J Phys Chem B 112:9603–9612

Johansen D, Jeffries CMJ, Hammouda B, Trewhella J, Goldenberg DP (2011a) Effects of macromolecular crowding on an intrinsically disordered protein characterized by small-angle neutron scattering with contrast matching. Biophys J 100:1120–1128

Johansen D, Trewhella J, Goldenberg DP (2011b) Fractal dimension of an intrinsically disordered protein: small-angle X-ray scattering and computational study of the bacteriophage λ N protein. Protein Sci 20:1955–1970

Kachala M, Valentini E, Svergun DI (2015) Application of SAXS for the structural characterization of IDPs. Adv Exp Med Biol 870:261–289

Karplus M (1959) Contact electron-spin coupling of nuclear magnetic moments. J Chem Phys 30

Karplus M, McCammon JA (2002) Molecular dynamics simulations of biomolecules. Nat Struct Biol 9:646–652

Kikhney AG, Svergun DI (2015) A practical guide to small angle X-ray scattering (SAXS) of flexible and intrinsically disordered proteins. FEBS Lett 589:2570–2577

Kim PM, Sboner A, Xia Y, Gerstein M (2008) The role of disorder in interaction networks: a structural analysis. Mol Syst Biol 4:179

Kjaergaard M, Brander S, Poulsen FM (2011) Random coil chemical shift for intrinsically disordered proteins: effects of temperature and pH. J Biomol NMR 49:139–149

Kjaergaard M, Nørholm AB, Hendus-Altenburger R, Pedersen SF, Poulsen FM, Kragelund BB (2010) Temperature-dependent structural changes in intrinsically disordered proteins: formation of alpha-helices or loss of polyproline II? Protein Sci 19:1555–1564

Koch MHJ, Vachette P, Svergun DI (2003) Small-angle scattering: a view on the properties, structures and structural changes of biological macromolecules in solution. Q Rev Biophys 36:147–227

Kohn JE, Millett IS, Jacob J, Zagrovic B, Dillon TM, Cingel N, Dothager RS, Seifert S, Thiyagarajan P, Sosnick TR, Hasan MZ, Pande VS, Ruczinski I, Doniach S, Plaxco KW (2004) Random-coil behavior and the dimensions of chemically unfolded proteins. Proc Natl Acad Sci U S A 101:12491–12496

Konno T, Tanaka N, Kataoka M, Takano E, Maki M (1997) A circular dichroism study of preferential hydration and alcohol effects on a denatured protein, pig calpastatin domain I. Biochim Biophys Acta 1342:73–82

Kriwacki RW, Hengst L, Tennant L, Reed SI, Wright PE (1996) Structural studies of p21Waf1/Cip1/Sdi1 in the free and Cdk2-bound state: conformational disorder mediates binding diversity. Proc Natl Acad Sci U S A 93:11504–11509

Krueger S, Shin J-H, Raghunandan S, Curtis JE, Kelman Z (2011) Atomistic ensemble modeling and small-angle neutron scattering of intrinsically disordered protein complexes: applied to minichromosome maintenance protein. Biophys J 101:2999–3007

Krzeminski M, Marsh JA, Neale C, Choy WY, Forman-Kay JD (2013) Characterization of disordered proteins with ENSEMBLE. Bioinformatics 29:398–399

Kung CC-H, Naik MT, Wang S-H, Shih H-M, Chang C-C, Lin L-Y, Chen C-L, Ma C, Chang C-F, Huang T-H (2014) Structural analysis of poly-SUMO chain recognition by the RNF4-SIMs domain. Biochem J 462:53–65

Lee KH, Chen J (2016) Multiscale enhanced sampling of intrinsically disordered protein conformations. J Comput Chem 37:550–557

LeGuillou JC, Zinn-Justin J (1977) Critical exponents for the n-vector model in three dimensions from field theory. Phys Rev Lett 39:95–98

Lens Z, Dewitte F, Monté D, Baert J-L, Bompard C, Sénéchal M, Van Lint C, de Launoit Y, Villeret V, Verger A (2010) Solution structure of the N-terminal transactivation domain of ERM modified by SUMO-1. Biochem Biophys Res Commun 399:104–110

Lenton S, Seydel T, Nylander T, Holt C, Härtlein M, Teixeira S, Zaccai G (2015) Dynamic footprint of sequestration in the molecular fluctuations of osteopontin. J R Soc Interface 12:0506

Levitt M (2009) Nature of the protein universe. Proc Natl Acad Sci U S A 106:11079–11084

Leyrat C, Jensen MR, Ribeiro EA Jr, Gérard FC, Ruigrok RW, Blackledge M, Jamin M (2011) The N(0)-binding region of the vesicular stomatitis virus phosphoprotein is globally disordered but contains transient α-helices. Protein Sci 20:542–556

Li J, Uversky VN, Fink AL (2002) Conformational behavior of human alpha-synuclein is modulated by familial Parkinson's disease point mutations A30P and A53T. Neurotoxicology 23:553–567

Lindorff-Larsen K, Kristjansdottir S, Teilum K, Fieber W, Dobson C, Poulsen F, Vendruscolo M (2004) Determination of an ensemble of structures representing the denatured state of the bovine acyl-coenzyme a binding protein. J Am Chem Soc 126:3291

Longhi S, Receveur-Bréchot V, Karlin D, Johansson K, Darbon H, Bhella D, Yeo R, Finet S, Canard B (2003) The C-terminal domain of the measles virus nucleoprotein is intrinsically disordered and folds upon binding to the C-terminal moiety of the phosphoprotein. J Biol Chem 278:18638–18648

Louhivuori M, Pääkkönen K, Fredriksson K, Permi P, Lounila J, Annila A (2003) On the origin of residual dipolar couplings from denatured proteins. J Am Chem Soc 125:15647–15650

Marasini C, Galeno L, Moran O (2013) A SAXS-based ensemble model of the native and phosphorylated regulatory domain of the CFTR. Cell Mol Life Sci 70:923–933

Marcotte EM, Pellegrini M, Yeates TO, Eisenberg D (1999) A census of protein repeats. J Mol Biol 293:151–160

Marsh JA, Baker JM, Tollinger M, Forman-Kay JD (2008) Calculation of residual dipolar couplings from disordered state ensembles using local alignment. J Am Chem Soc 130:7804–7805

Marsh JA, Dancheck B, Ragusa MJ, Allaire M, Forman-Kay JD, Peti W (2010) Structural diversity in free and bound states of intrinsically disordered protein phosphatase 1 regulators. Structure 18:1094–1103

Marsh JA, Neale C, Jack FE, Choy WY, Lee AY, Crowhurst KA, Forman-Kay JD (2007) Improved structural characterizations of the drkN SH3 domain unfolded state suggest a compact ensemble with native-like and non-native structure. J Mol Biol 367:1494–1510

Meier S, Grzesiek S, Blackledge M (2007) Mapping the conformational landscape of urea-denatured ubiquitin using residual dipolar couplings. J Am Chem Soc 129:9799–9807

Mertens HD, Svergun DI (2010) Structural characterization of proteins and complexes using small-angle X-ray scattering. J Struct Biol 172:128–141

Metropolis N, Rosenbluth AW, Rosenbluth MN, Teller AH, Teller E (1953) Equation of state calculations by fast computing machines. J Chem Phys 21:1087–1092

Millard CJ, Watson PJ, Fairall L, Schwabe JWR (2013) An evolving understanding of nuclear receptor coregulator proteins. J Mol Endocrinol 51:T23–T36

Mittag T, Marsh J, Grishaev A, Orlicky S, Sicheri F, Tyers M, Forman-Kay JD (2010) Structure/function implications in a dynamic complex of the intrinsically disordered Sic1 with the Cdc4 subunit of an SCF ubiquitin ligase. Structure 18:494–506

Mittal A, Lyle N, Harmon TS, Pappu RV (2014) Hamiltonian switch metropolis Monte Carlo simulations for improved conformational sampling of intrinsically disordered regions tethered to ordered domains of proteins. J Chem Theory Comput 10:3550–3562

Mohana-Borges R, Goto NK, Kroon GJ, Dyson HJ, Wright PE (2004) Structural characterization of unfolded states of apomyoglobin using residual dipolar couplings. J Mol Biol 340:1131–1142

Moncoq K, Broutin I, Craescu CT, Vachette P, Ducruix A, Durand D (2004) SAXS study of the PIR domain from the Grb14 molecular adaptor: a natively unfolded protein with a transient structure primer? Biophys J 87:4056–4064

Mukrasch MD, Markwick P, Biernat J, Bergen M, Bernadó P, Griesinger C, Mandelkow E, Zweckstetter M, Blackledge M (2007) Highly populated turn conformations in natively unfolded tau protein identified from residual dipolar couplings and molecular simulation. J Am Chem Soc 129:5235–5243

Mylonas E, Hascher A, Bernadó P, Blackledge M, Mandelkow E, Svergun DI (2008) Domain conformation of tau protein studied by solution small-angle X-ray scattering. Biochemistry 47:10345–10353

Nairn KM, Lyons RE, Mulder RJ, Mudie ST, Cookson DJ, Lesieur E, Kim M, Lau D, Scholes FH, Elvin CM

(2008) A synthetic resilin is largely unstructured. Biophys J 95:3358–3365

Narayanan RL, Durr UHN, Bibow S, Biernat J, Mandelkow E, Zweckstetter M (2010) Automatic assignment of the intrinsically disordered protein Tau with 441-residues. J Am Chem Soc 132:11906–11907

Nørholm A-B, Hendus-Altenburger R, Bjerre G, Kjaergaard M, Pedersen SF, Kragelund BB (2011) The intracellular distal tail of the Na+/H+ exchanger NHE1 is intrinsically disordered: implications for NHE1 trafficking. Biochemistry 50:3469–3480

O'Brien DP, Hernandez B, Durand D, Hourdel V, Sotomayor-Pérez A-C, Vachette P, Ghomi M, Chamot-Rooke J, Ladant D, Brier S, Chenal A (2015) Structural models of intrinsically disordered and calcium-bound folded states of a protein adapted for secretion. Sci Rep 5:14223

Oh K-I, Jung Y-S, Hwang G-S, Cho M (2012) Conformational distributions of denatured and unstructured proteins are similar to those of 20 x 20 blocked dipeptides. J Biomol NMR 53:25–41

Otting G (2008) Prospects for lanthanides in structural biology by NMR. J Biomol NMR 42:1–9

Owens GE, New DM, West AP, Bjorkman PJ (2015) Anti-PolyQ antibodies recognize a short PolyQ stretch in both normal and mutant Huntingtin exon 1. J Mol Biol 427:2507–2519

Ozenne V, Bauer F, Salmon L, Huang J, Jensen MR, Segard S, Bernadó P, Charavay C, Blackledge M (2012) Flexible-meccano:a tool for the generation of explicit ensemble descriptions of intrinsically disordered proteins and their associated experimental observables. Bioinformatics 28:1463–1470

Palazzesi F, Prakash MK, Bonomi M, Barducci A (2015) Accuracy of current all-atom force-fields in modeling protein disordered states. J Chem Theory Comput 11:2–7

Pavlin MR, Brunzelle JS, Fernandez EJ (2014) Agonist ligands mediate the transcriptional response of nuclear receptor heterodimers through distinct stoichiometric assemblies with coactivators. J Biol Chem 289:24771–24778

Paz A, Zeev-Ben-Mordehai T, Lundqvist M, Sherman E, Mylonas E, Weiner L, Haran G, Svergun DI, Mulder FAA, Sussman JL, Silman I (2008) Biophysical characterization of the unstructured cytoplasmic domain of the human neuronal adhesion protein neuroligin 3. Biophys J 95:1928–1944

Pelikan M, Hura GL, Hammel M (2009) Structure and flexibility within proteins as identified through small angle X-ray scattering. Gen Physiol Biophys 28:174–189

Perevozchikova T, Stanley CB, McWilliams-Koeppen HP, Rowe EL, Berthelier V (2014) Investigating the structural impact of the glutamine repeat in huntingtin assembly. Biophys J 107:411–421

Pérez J, Nishino Y (2012) Advances in X-ray scattering: from solution SAXS to achievements with coherent beams. Curr Opin Struct Biol 22:670–678

Petoukhov MV, Svergun DI (2007) Analysis of X-ray and neutron scattering from biomacromolecular solutions. Curr Opin Struct Biol 17:562–571

Piana S, Klepeis JL, Shaw DE (2014) Assessing the accuracy of physical models used in protein-folding simulations: quantitative evidence from long molecular dynamics simulations. Curr Opin Struct Biol 24:98–105

Pizzi E, Frontali C (2001) Low-complexity regions in plasmodium falciparum proteins. Genome Res 11:218–229

Putnam CD, Hammel M, Hura GL, Tainer JA (2007) X-ray solution scattering (SAXS) combined with crystallography and computation: defining accurate macromolecular structures, conformations and assemblies in solution. Q Rev Biophys 40:191–285

Rambo RP, Tainer JA (2011) Characterizing flexible and intrinsically unstructured biological macromolecules by SAS using the Porod-Debye law. Biopolymers 95:559–571

Rambo RP, Tainer JA (2013) Super-resolution X-ray sacttering and its applications to structural systems biology. Annu Rev Biophys 42:415–441

Rauscher S, Gapsys V, Gajda MJ, Zweckstetter M, de Groot BL, Grubmüller H (2015) Structural ensembles of intrinsically disordered proteins depend strongly on force field: a comparison to experiment. J Chem Theory Comput 11:5513–5524

Receveur-Brechot V, Durand D (2012) How random are intrinsically disordered proteins? A small angle scattering perspective. Curr Protein Pept Sci 13:55–75

Rochel N, Ciesielski F, Godet J, Moman E, Roessle M, Peluso-Iltis C, Moulin M, Haertlein M, Callow P, Mély Y, Svergun DI, Moras D (2011) Common architecture of nuclear receptor heterodimers on DNA direct repeat elements with different spacings. Nat Struct Mol Biol 18:564–570

Rozycki B, Kim YC, Hummer G (2011) SAXS ensemble refinement of ESCRT-III CHMP3 conformational transitions. Structure 19:109–116

Ruskamo S, Chukhlieb M, Vahokoski J, Bhargav SP, Liang F, Kursula I, Kursula P (2012) Juxtanodin is an intrinsically disordered F-actin-binding protein. Sci Rep 2:899

Schwalbe M, Ozenne V, Bibow S, Jaremko M, Jaremko L, Gajda M, Jensen MR, Biernat J, Becker S, Mandelkow E, Zweckstetter M, Blackledge M (2014) Predictive atomic resolution descriptions of intrinsically disordered hTau40 and α-Synuclein in solution from NMR and small angle scattering. Structure 22:238–249

Schwarzinger S, Kroon GJ, Foss TR, Wright PE, Dyson HJ (2000) Random coil chemical shifts in acidic 8 M urea: implementation of random coil shift data in NMRView. J Biomol NMR 18:43–48

Shell SS, Putnam CD, Kolodner RD (2007) The N terminus of Saccharomyces Cerevisiae Msh6 is an unstructured tether to PCNA. Mol Cell 26:565–578

Shkumatov AV, Chinnathambi S, Mandelkow E, Svergun DI (2011) Structural memory of natively unfolded tau protein detected by small-angle X-ray scattering. Proteins 79:2122–2131

Sibille N, Bernadó P (2012) Structural characterization of intrinsically disordered proteins by the combined use of NMR and SAXS. Biochem Soc Trans 40:955–962

Stadler AM, Stingaciu L, Radulescu A, Holderer O, Monkenbusch M, Biehl R, Richter D (2014) Internal nanosecond dynamics in the intrinsically disordered myelin basic protein. J Am Chem Soc 136:6987–6994

Stanley CB, Perevozchikova T, Berthelier V (2011) Structural formation of huntingtin exon 1 aggregates probed by small-angle neutron scattering. Biophys J 100:2504–2512

Sterckx YGJ, Volkov AN, Vranken WF, Kragelj J, Jensen MR, Buts L, Garcia-Pino A, Jové T, Van ML, Blackledge M, van Nuland NAJ, Loris R (2014) Small-angle X-ray scattering- and nuclear magnetic resonance-derived conformational ensemble of the highly flexible antitoxin PaaA2. Structure 22:854–865

Stott K, Watson M, Howe FS, Grossmann JG, Thomas JO (2010) Tail-mediated collapse of HMGB1 is dynamic and occurs *via* differential binding of the acidic tail to the A and B domains. J Mol Biol 403:706–722

Stumpe MC, Grubmüller H (2007) Interaction of urea with amino acids: implications for urea-induced protein denaturation. J Am Chem Soc 129:16126–16131

Svergun DI, Koch MHJ (2003) Small-angle scattering studies of biological macromolecules in solution. Rep Prog Phys 66:1735–1782

Tamiola K, Acar B, Mulder FAA (2010) Sequence-specific random coil chemical shifts of intrinsically disordered proteins. J Am Chem Soc 132:18000–18003

Tidow H, Melero R, Mylonas E, Freund SMV, Grossmann JG, Carazo JM, Svergun DI, Valle M, Fersht AR (2007) Quaternary structures of tumor suppressor p53 and a specific p53 DNA complex. Proc Natl Acad Sci U S A 104:12324–12329

Tompa P, Schad E, Tantos A, Kalmar L (2015) Intrinsically disordered proteins: emerging interaction specialists. Curr Opin Struct Biol 35:49–59

Trakawa T, Takada S (2011) Multiscale ensemble modeling of intrinsically disordered proteins: p53 N-terminal domain. Biophys J 101:1450–1458

Tria G, Mertens HD, Kachala M, Svergun DI (2015) Advanced ensemble modelling of flexible macromolecules using X-ray solution scattering. IUCrJ 2:207–217

Uversky VN, Gillespie JR, Millett IS, Khodyakova AV, Vasilenko RN, Vasiliev AM, Rodionov IL, Kozlovskaya GD, Dolgikh DA, Fink AL et al (2000) Zn(2+)-mediated structure formation and compaction of the "natively unfolded" human prothymosin alpha. Biochem Biophys Res Commun 267:663–668

Uversky VN, Gillespie JR, Millett IS, Khodyakova AV, Vasiliev AM, Chernovskaya TV, Vasilenko RN, Kozlovskaya GD, Dolgikh DA, Fink AL et al (1999) Natively unfolded human prothymosin alpha adopts partially folded collapsed conformation at acidic pH. Biochemistry 38:15009–15016

Uversky VN, Li J, Souillac P, Millett IS, Doniach S, Jakes R, Goedert M, Fink AL (2002) Biophysical properties of the synucleins and their propensities to fibrillate: inhibition of alpha-synuclein assembly by beta- and gamma-synucleins. J Biol Chem 277:11970–11978

Varadi M, Kosol S, Lebrun P, Valentini E, Blackledge M, Dunker AK, Felli IC, Forman-Kay JD, Kriwacki RW, Pierattelli R, Sussman J, Svergun DI, Uversky VN, Vendruscolo M, Wishart D, Wright PE, Tompa P (2014) pE-DB: a database of structural ensembles of intrinsically disordered and of unfolded proteins. Nucleic Acids Res 42:D326–D335

Vestergaard B, Groenning M, Roessle M, Kastrup JS, van de Weert M, Flink JM, Frokjaer S, Gajhede M, Svergun DI (2007) A helical structural nucleus is the primary elongating unit of insulin amyloid fibrils. PLoS Biol 5:1089–1097

Vitalis A, Pappu RV (2009a) Methods for Monte Carlo simulations of biomacromolecules. Ann Rep Comput Chem 5:49–76

Vitalis A, Pappu RV (2009b) ABSINTH: a new continuum solvation model for simulations of polypeptides in aqueous solutions. J Comput Chem 30:673–699

Vitalis A, Wang X, Pappu RV (2008) Atomistic simulations of the effects of Polyglutamine chain length and solvent quality on conformational Equilibria and spontaneous Homodimerization. J Mol Biol 384:279–297

Wells M, Tidow H, Rutherford TJ, Markwick P, Jensen MR, E. Mylonas, Svergun DI, Blackledge M, Fersht AR (2008) Structure of tumor suppressor p53 and its intrinsically disordered N-terminal transactivation domain. Proc Natl Acad Sci U S A 105:5762–5767

Wishart DS, Bigam CG, Holm A, Hodges RS, Sykes BD (1995) (1)H, (13)C and (15)N random coil NMR chemical shifts of the common amino acids. I. Investigations of nearest-neighbor effects. J Biomol NMR 5:332

Wishart DS, Sykes BD (1994) The 13C chemical-shift index: a simple method for the identification of protein secondary structure using 13C chemical-shift data. J Biomol NMR 4:171–180

Wootton JC, Federhen S (1993) Statistics of local complexity in amino acid sequences and sequence databases. Comput Chem 17:149–163

Wright PE, Dyson HJ (1999) Intrinsically unstructured proteins: re-assessing the protein structure-function paradigm. J Mol Biol 293:321–331

Wright PE, Dyson HJ (2015) Intrinsically disordered proteins in cellular signalling and regulation. Nat Rev Mol Cell Biol 16:18–29

Wu K-P, Weinstock DS, Narayanan C, Levy RM, Baum J (2009) Structural reorganization of α-synuclein at low pH observed by NMR and REMD simulations. J Mol Biol 391:784

Yang S (2014) Methods for SAXS-based structure determination of biomolecular complexes. Adv Mater 26:7902–7910

Yang S, Blachowicz L, Makowski L, Roux B (2010) Multidomain assembled states of Hck tyrosine kinase in solution. Proc Natl Acad Sci U S A 107:15757–15762

Yang C, van der Woerd MJ, Muthurajan UM, Hansen JC, Luger K (2011) Biophysical analysis and small-angle X-ray scattering-derived structures of MeCP2-nucleosome complexes. Nucleic Acids Res 39:4122–4135

Zerze GH, Miller CM, Granata D, Mittal J (2015) Free energy surface of an intrinsically disordered protein: comparison between temperature replica exchange molecular dynamics and bias-exchange metadynamics. J Chem Theory Comput 11:2776–2782

Zhou H-X (2004) Polymer models of protein stability, folding, and interactions. Biochemistry 43:2141–2154

Zhou CZ, Confalonieri F, Jacquet M, Perasso R, Li ZG, Janin J (2001) Silk fibroin: structural implications of a remarkable amino acid sequence. Proteins 44:119–122

# What Can We Learn from Wide-Angle Solution Scattering?

# 8

Yujing Wang, Hao Zhou, Emre Onuk, John Badger, and Lee Makowski

**Abstract**

Extending collection of x-ray solution scattering data into the wide-angle regime (WAXS) can provide information not readily extracted from small angle (SAXS) data. It is possible to accurately predict WAXS scattering on the basis of atomic coordinate sets and thus use it as a means of testing molecular models constructed on the basis of crystallography, molecular dynamics (MD), cryo-electron microscopy or *ab initio* modeling. WAXS data may provide insights into the secondary, tertiary and quaternary structural organization of macromolecules. It can provide information on protein folding and unfolding beyond that attainable from SAXS data. It is particularly sensitive to structural fluctuations in macromolecules and can be used to generate information about the conformational make up of ensembles of structures co-existing in solution. Novel approaches to modeling of structural fluctuations can provide information on the spatial extent of large-scale structural fluctuations that are difficult to obtain by other means. Direct comparison with the results of MD simulations are becoming possible. Because it is particularly sensitive to small changes in structure and flexibility it provides unique capabilities for the screening of ligand libraries for detection of functional interactions. WAXS thereby provides an important extension of SAXS that can generate structural and dynamic information complementary to that obtainable by other biophysical techniques.

**Keywords**

WAXS • SAXS • Protein structure • Structural fluctuations • Molecular dynamics • Moderate throughput screening

Y. Wang • H. Zhou • E. Onuk • L. Makowski (✉)
Northeastern University, Boston, MA 02115, USA
e-mail: l.makowski@northeastern.edu

J. Badger
DeltaG Technologies, San Diego, CA, USA





B. Chaudhuri et al. (eds.), *Biological Small Angle Scattering: Techniques, Strategies and Tips*, Advances in Experimental Medicine and Biology 1009, DOI 10.1007/978-981-10-6038-0_8

## 8.1 Introduction

X-ray solution scattering studies of proteins produce data that can provide substantial insight into protein structure, flexibility and dynamics. Small angle scattering (SAXS) provides relatively low resolution information (~20 Å), whereas wide-angle scattering (WAXS) provides information about higher resolution features or motions. Scattering intensity in the WAXS regime is typically two or more orders of magnitude weaker than in the SAXS regime. However, over the past two decades, the development of SAXS/WAXS beam lines at high brilliance synchrotron sources has fostered rapid growth of solution scattering studies extending to wide angles (e.g. Allaire and Yang 2011; Fischetti et al. 2004a). The WAXS regime also extends to scattering angles where background scattering from buffer and sample chamber is considerable. Use of a synchrotron source can provide data of a quality that can overcome these challenges.

In both the SAXS and WAXS regimes, the scattered intensity distribution can be more valuable when combined with structural information generated from other techniques. Whereas x-ray crystallography and NMR produce high resolution 'snap shots' of protein structure and information about local motions, solution scattering has the capability of providing information about conformational changes, intermolecular interactions, large scale structural fluctuations, and slow, concerted, global motions. WAXS is particularly effective for study of large-scale motions that are difficult to characterize with other approaches. Like SAXS, it can be used to study virtually any macromolecule or molecular assembly that can be purified at concentrations of ~1 to 5 mg/ml. Its value is significantly enhanced when used in concert with crystallographic or NMR approaches, computational modeling and/or molecular dynamics simulations.

For purposes of this chapter, we will define the boundary between the SAXS regime and the WAXS regime as 20 Å spacing $\left(1/d = 0.05\,\text{Å}^{-1} = (1/20)\text{Å}^{-1}, \text{or}\, q = 0.3\text{Å}^{-1}\right)$ where $q = 4\pi \sin(\theta)/\lambda = 2\pi/d$ and $2\theta$ is the angle between incident and scattered x-rays. We choose this boundary because beyond 20 Å spacing internal fluctuations in electron density of a protein begin to contribute substantially to scattering. This distinction alters the nature of analyses possible in the two regimes. SAXS has been utilized for decades (Luzzati and Tardieu 1980) to estimate the radius of gyration ($R_g$), pair-distance distribution function, P(r), and oligomerization state of proteins (Putnam et al. 2007). The capability to generate three-dimensional molecular shapes directly from SAXS data (Svergun 1999; Walther et al. 2000) has dramatically increased the utility and utilization of these methods. Three-dimensional reconstruction from solution scattering data is limited to about 20 Å resolution. Beyond 20 Å resolution, solution scattering data from any protein will be consistent with multiple molecular shapes in part because of the contribution of internal structures to the observed scattering and in part due to the intrinsic limitation of information content in the measured intensities (more about this later). Thus, SAXS data can be used directly to calculate $R_g$ and a three-dimensional shape reconstruction. The pair-distribution function, P(r), can be calculated from data extending to any resolution. However, for virtually all other applications, WAXS intensities are used to test hypotheses or molecular models generated by other means such as crystallography, NMR, molecular dynamics (MD) or *ab initio* modeling. The use of WAXS for testing of molecular models of structure and dynamics is directly dependent on our ability to accurately predict WAXS data from atomic coordinate sets (Park et al. 2009).

For testing of models, WAXS may provide an advantage over SAXS data in detection of relatively small structural changes. WAXS intensities are highly sensitive to small structural changes (Fischetti et al. 2004b) and to changes in the magnitude of structural fluctuations (Makowski et al. 2011). On functional binding of a ligand, a protein may alter either its structure, its dynamics or both, and it may be a mute point to argue whether the structure or dynamics have changed. Strictly speaking it is virtually impossible to alter one without the other, so perhaps it is most appropriate to simply state that the structural ensemble has been altered. More to the point, any interaction that alters function is almost certain to trigger a change in structure and/or dynamics – and those changes will, in many cases be detectable using WAXS.

Solution scattering methods have evolved and matured over the past decade into a suite of highly informative probes of protein structure and activity that go well beyond a simple method for determining size and shape of the molecule. They now represent an approach to detailed characterization of biochemistry in the scattering volume. As such, increased focus must be given to the state of the sample. It is critical that the sample be well defined biochemically, absent precipitates. Wide-spread adaptation of SEC-SAXS/WAXS in which scattering patterns are collected from the output of a size-exclusion column reflects this trend. In conventional, static WAXS, background scattering from a precisely matched buffer is a critical aspect of any experiment as scattering contributions from even minor buffer constituents can be important. This is even more important in WAXS than in SAXS, because below q ~0.3 $Å^{-1}$ there is little scattering from buffer or sample chamber, whereas at higher angles these contributions exceed scattering from the protein.

Although early studies were limited to structurally homogeneous samples, solution scattering is now frequently used to study the ensemble of structural forms present in solution including, for instance, enzymes undergoing catalytic cycling (Onuk et al. 2015). WAXS can be used to generate information about changes in secondary, tertiary and quaternary structures (Doniach 2001; Hirai et al. 2002; Makowski et al. 2008a); conformational changes due to ligand binding (Fischetti et al. 2004b; Rodi et al. 2007; Zhou et al. 2015), or cofactor oxidation state (Tiede et al. 2002) and by amino acid substitutions (Makowski et al. 2011; Zhou et al. 2015), or protein folding (Hirai et al. 2004). The conformational ensemble of a protein in solution can also be studied with WAXS. WAXS has proven to be highly sensitive to changes in the ensemble due to protein concentration (Makowski et al. 2008b), mutations and ligand binding (Makowski et al. 2011; Zhou et al. 2015; Onuk et al. 2015). Time resolved (TR) studies can be carried out analogous to static studies and have been used to characterize light-triggered conformational changes occurring in nano- to milli-seconds (Cammarata et al. 2008; Cho et al. 2010). Although we will not explicitly consider TR studies in this chapter, all methods described can be applied to each diffraction pattern 'snapshot' of a TR data set.

## 8.2 WAXS Data

Collection of WAXS data simultaneously with SAXS data is challenging, even at state of the art beam lines (Zhang et al. 2000; Makowski 2010). It can be accomplished with a very small beam stop and large detector; or by using two detectors set at different sample-to-detector distances, the WAXS detector subtending only a portion of the wide angle region, but capturing enough intensity to provide good signal-to-noise ratio after merging with the SAXS data (e.g., Allaire and Yang 2011). Ideally, one would like to set a WAXS detector on axis and at relatively small sample-to-detector distance, but including a slot-shaped hole to allow passage of x-rays to a SAXS detector placed at a much higher sample-to-detector distance. Choosing a slot-shaped hole would generate a q-range in which data was collected at both detectors, providing adequate overlap for accurate scaling of data from the two detectors. This arrangement has not, as of yet, been implemented.

The scattered intensity, I(q), from a protein solution can be calculated, in principle, from the position of all atoms in the protein using the Debye formula,

$$I(q) = \sum_{i=1}^{n} \sum_{j=1}^{n} f_i f_j \frac{\sin 2\pi q r_{ij}}{2\pi q r_{ij}} \qquad (8.1)$$

where $f_i$ is the scattering factor from the $i$th atom, and $r_{ij}$ is the distance between atom $i$ and atom $j$. As will be discussed below, direct application of this formula fails when the protein is immersed in aqueous solution since this necessitates taking into account the impact of the shape of the region excluding solvent and the difference in water structure and density between hydration layer and bulk.

Whereas SAXS data to be used for calculation or $R_g$, P(r), or three-dimensional shape reconstructions requires measurements to small angles dictated by the maximum spatial extent of the scattering object (see other chapters for details), data used for testing of models does not necessarily need to extend to small angles. However, collection of SAXS data simultaneously with WAXS data provides important quality assurance tests for detection of the presence of aggregates or inter-particle interference effects which may be observed at higher concentrations. Interparticle interference effects, more likely as protein concentration increases, are usually limited to the small angle regime. When present, they can distort estimates of Rg, P(r), or three-dimensional shape reconstructions (see e.g. Inouye et al. 2016). They can usually be detected by comparing SAXS data collected at two or more protein concentrations. At high protein concentrations (say, >10 mg/ml) intermolecular crowding can suppress structural fluctuations in some proteins, resulting in a sharpening of wide-angle scattering features (Makowski et al. 2008b). More rigid proteins exhibit little reaction to changes in concentration. Amorphous aggregates (to be distinguished from multimers) can result in a sharp spike in scattering at very small angles but usually exhibit little wide-angle scattering except for potentially resulting in a small increase in diffuse background.

Figure 8.1 is an example of the impact of ligand binding on the WAXS scattering from a protein. Binding of substrate to hexokinase results in a relatively large conformational change – closing of the binding site cleft (McDonald et al. 1979). This alters the small angle scattering from the molecule, lowers $R_g$ and induces additional intensity changes in the wide angle regime.

## 8.3 Predicting WAXS Data from Atomic Coordinates

The ability to accurately predict WAXS data from atomic coordinate sets is key to the utility of WAXS, making it a sensitive method of assessing the accuracy of atomic-scale models. If proteins existed in a vacuum, calculation of

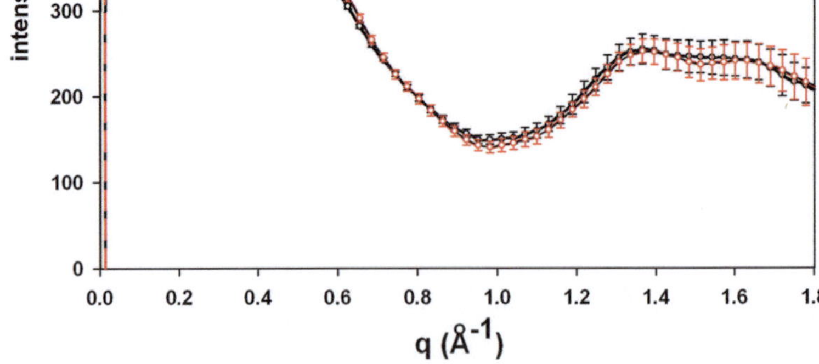


**Fig. 8.1** WAXS scattering from hexokinase in the presence and absence of substrate binding. Binding results in a closure of the ligand binding cleft altering $R_g$, as well as observed intensities in the SAXS and WAXS regimes. Error estimates increase at wide angles because of the increased intensity of buffer scatter in that regime

solution scattering would reduce to a simple application of the Debye formula (Eq. 8.1). However, proteins (or other macromolecules) are generally immersed in solvent making it essential to account for the exclusion of water in the volume occupied by the protein. One also has to model the hydration layer where the water takes on a density that may be as much as 10% greater than in bulk (Svergun et al. 1997). These effects were first taken into account in the iconic program CRYSOL (Svergun et al. 1995) that has transformed the use of SAXS for protein studies. In the WAXS regime, however, the approximations used in CRYSOL break down. In particular, CRYSOL underestimates the intensity of WAXS data by a factor of 2-3X relative to SAXS intensity when used with default parameters. This is due to the continuum representation of the hydration layer and method for representing excluded volume in CRYSOL (Bardhan et al. 2009). In the WAXS regime it is essential to utilize an explicit atom representation of water (Bardhan et al. 2009; Park et al. 2009; Grishaev et al. 2010). For precise modeling of intensity in the WAXS regime, CRYSOL may not be the most appropriate. Although CRYSOL refinement against experimental data often results in good agreement between calculated and observed, this may come as the result of non-physical values for adjustable parameters within CRYSOL (Barhan et al. 2009). Extensive experimental (Svergun et al. 1997) and computational tests indicate that the density of water in the hydration layer may be as much as 10% greater than bulk water, an amount detectable with solution scattering, and that these structural differences extend roughly 7 Å beyond the protein surface (Park et al. 2009). Once these issues are taken into account it is possible to calculate scattered intensities to within experimental error for most rigid proteins across both the SAXS and WAXS regimes (Park et al. 2009; Grishaev et al. 2010). This requires, however, MD simulation of the water in the hydration layer, a process that remains computationally laborious. Consequently, these calculations are not yet high throughput and the capability of carrying them out for large ensembles of representative structures has not yet been established. For this reason, CRYSOL remains the most widely used program for estimation of solution scattering from atomic coordinates.

Computational estimates of scattering intensity presented in this chapter utilize the software package XS as described (although not named) by Park et al. (2009). In XS, water molecules are positioned around a protein surface out to ~7 Å from the protein surface, and subject to 100 ps of MD simulation during which the protein atoms are held rigid. A 'snapshot' of the water positions is captured once each picosecond and WAXS intensity due to the protein plus water positions in each snap shot are calculated using the Debye formula and then averaged giving $I_{prot}$. Simulation of a 'droplet' of bulk water the same shape as the protein-containing droplet (including the 7 Å -thick hydration layer) is also carried out and snapshots of this droplet are used with the Debye formula to calculate WAXS patterns that are subsequently averaged to produce $I_{water}$. Subtraction of this bulk water intensity from the hydrated protein intensity results in approximation of 'excess intensity', $I_{XS}$. This excess intensity corresponds closely to the difference between scattering observed from protein-solution-filled and buffer-filled sample chambers,

$$I_{XS} = I_{prot} - I_{water} \approx I_{obs} - I_{buffer} \tag{8.2}$$

Out to ~5 Å spacing, excess intensity is virtually identical to $I_{prot}$. At wider angles, scattering from buffer is non-negligible. At these angles $I_{XS}$ differs from the more routinely calculated $I_{prot}$. Since protein usually occupies <1% of the total scattering volume, beyond 5 Å spacing (q ~1.2 $Å^{-1}$) the scattering from buffer is far more intense than that from the protein, and $I_{XS}$ will be negative. Thus, the moniker is 'excess intensity' rather than 'intensity' which is universally considered a positive number. Figure 8.2 is a comparison of the calculated and observed WAXS from ubiquitin. Intensity calculated using XS (Park et al. 2009) with no free parameters

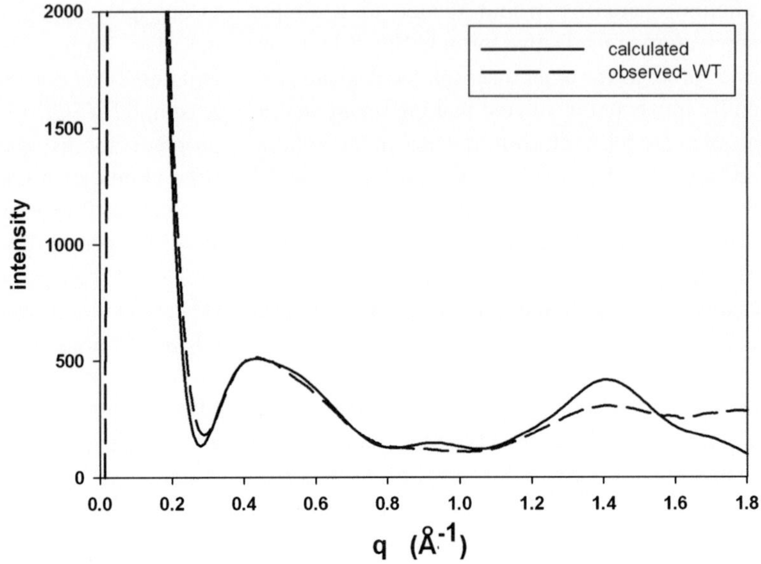


**Fig. 8.2** Comparison of WAXS intensity distribution, $I_{prot}$, from ubiquitin as calculated by the software package XS (*solid line*) and as observed in scattering from a 10 mg/ml solution (*broken line*)

results in an intensity distribution indistinguishable from observed out to a q ~1.2 $Å^{-1}$. In the region 1.2 < q < 1.6 $Å^{-1}$ the calculated intensity is greater than observed. This region corresponds to a spacing of ~4.7 Å and is generated largely from the inter-strand spacings of beta strands in the molecule. The comparison suggests that the strands are undergoing small structural fluctuations, leading to observed intensity somewhat lower than that calculated for a rigid molecule.

## 8.4 Size and Shape

The size and shape of a protein, other macromolecule or macromolecular complex can usually be determined from SAXS data. This is a topic well covered in other chapters of this book. Although WAXS data extends to much higher resolutions (scattering angles) it cannot be used to improve the accuracy of a radius of gyration or to enhance the level of detail in three-dimensional reconstructions of molecular shape. It is worth discussing the origins of these limitations.

The radius of gyration, $R_g$, literally, the average radius of scattering density from the center of mass can be estimated from data in the q-range where the Guinier approximation is valid ($qR_g$ <1.3 for most globular proteins). Intensities at higher scattering angles do not improve the estimate of $R_g$ because the Guinier plot is not, in general, linear at wider scattering angles. In fact, extending data to smaller angles is usually more important for accuracy of the estimate of $R_g$ than extending to higher angles. The arrangement of detector and beam stop required for WAXS data may place limits on minimum scattering angle at which data is collected. A hybrid SAXS/WAXS detector scheme tuned for collection of both simultaneously is used at a number of beam lines to overcome this problem.

SAXS data can also be used to reconstruct a three-dimensional shape of a macromolecule (Chacon et al. 1998; Svergun 1999; Walther et al. 2000; Svergun et al. 2001; Hura et al. 2009; other chapters in this book). The algorithms used to generate shape reconstructions from SAXS data implicitly assume the scattering density within the protein is roughly constant. For proteins, this is approximately true to ~20 Å resolution, but no higher. At spacings greater than (1/d) ~1/20 $Å^{-1}$ (q ~0.3 $Å^{-1}$), intensities are strongly influenced by internal structural features and extending data used to higher q may result in spurious features (although inclusion of higher angle data appears to stabilize some of the algorithms used for three-dimensional reconstructions without generating artefactual features). Combined SAXS-WAXS data should not be used for *ab initio* shape computation due to the breakdown of uniform scattering density model in the WAXS regime. It can, however, be used to test structural hypotheses. For instance, it can be used for evaluating the quality of rigid-body models derived from crystallographic structural information (Svergun et al. 2001; Zheng and Tekpinar 2011; Wen et al. 2014) or for refining the positions of (rigid body) domains. This has particular application to multi-domain proteins that may undergo large scale re-arrangements of domains in response to allosteric effectors or other interactions (Badger et al. 2016). Although most current studies utilize SAXS data, extending the approaches to WAXS has the potential to improve accuracy.

Validation of modeling efforts is not necessarily straightforward, there is potential for multiple solutions, and the calculation of uncertainty in optimized domain positions, while possible, has not usually been reported in published studies. At the very least, the use of WAXS for rigid body refinement of domain positions will produce testable hypotheses about the functional significance of domain movements.

Resolution of shape reconstructions is also limited by uniqueness (Volkova and Svergun 2003). The amount of information required for a three-dimensional reconstruction goes up roughly as $q^3$. The amount of information in a solution scattering pattern goes up proportional to q. At some limiting q value, the amount of information required for unique shape determination will exceed that contained in the scattering pattern. Another way of conceptualizing this is by considering a molecular shape as a sum of spherical harmonics (Lattman 1989). At very small angles, only a small number of spherical harmonics contribute to the observed intensities. At increasing scattering angle, increasing numbers of spherical harmonics contribute. The capability of three-dimensional reconstruction is only made possible by the oversampling of the continuous (spherically averaged) intensity distribution. At some limiting q, the amount of information required to estimate the intensity associated with all contributing spherical harmonics is greater than the amount of information within the pattern. At that point, estimation of the three-dimensional shape becomes an ill-posed problem with multiple solutions.

WAXS data can contribute to the accuracy and resolution of P(r), increasing the level of detail contained in it to resolutions beyond those of the SAXS regime (Hong and Hao 2009). Among other things, this may make possible a more accurate estimate of the longest interatomic vector lengths in the protein. Intensity in a WAXS pattern is a band-limited function with the band-limit equal to the length of the longest interatomic vector in the protein, $D_{max}$. Larger proteins exhibit scattering patterns with sharper features (e.g. peaks and troughs) because the patterns include more higher frequency terms – corresponding to the longest interatomic vectors (i.e., patterns from larger proteins have larger band pass). An estimate of $D_{max}$ can be made from the pair-distribution function, P(r). Nevertheless, since the longest interatomic vectors contribute very little to the measured intensity it is often challenging to make an accurate estimate of $D_{max}$. Iterative procedures may be required (e.g. Putman et al. 2007). WAXS can provide improved accuracy for P(r) and consequently, $D_{max}$. An accurate estimate of $D_{max}$ contributes to more accurate three-dimensional shape reconstructions since most algorithms require it as input. Validation of SAXS-derived structures is often difficult and, as in many biophysical approaches depends to some extent on self-consistency and consistency of models with all available data. Because of the very well defined relationship between atomic coordinates and WAXS data (Eq. 8.1), WAXS can provide a very well defined test of models constructed on the basis of multiple biophysical probes.

## 8.5 Secondary Structure

The Debye formula – Eq. 8.1 – demonstrates that solution scattering is due entirely to the distribution of interatomic vector lengths within a sample. Secondary structures, by definition, have strong patterns of interatomic vector lengths, so we would expect them to contribute to solution scattering in distinctive ways. α-helices, for instance, pack roughly 10 Å apart and, not surprisingly, α-helical proteins have a considerable number of interatomic vectors about 10 Å in length. This usually results in a strong scattering peak at a spacing of ~(1/10) $Å^{-1}$ (q ~0.6 $Å^{-1}$). Analogously, β-sheets may also lie about 10 Å from one another, face-to-face, and may also exhibit relatively intense scattering in the 10 Å region. Furthermore, they are made up from β-strands that typically lie ~4.7 Å apart. This results in solution scattering patterns with a peak at a spacing of ~ (1/4.7) $Å^{-1}$ (q ~1.3 $Å^{-1}$). Strong scattering in the 10 Å and 4.7 Å regions can be observed in WAXS patterns from ubiquitin as seen in Fig. 8.2. Similarly, Fig. 8.3 includes scattering from two $I_{gg}$ molecules, one showing well-defined, strong peaks at ~10 Å and 4.7 Å spacing, and a second $I_{gg}$ that, due to significant conformational flexibility, exhibits only modest peaks in these regions, an example of the impact of fluctuations on WAXS data.

## 8.6 Tertiary Structure

As of yet there has been no experimental demonstration that WAXS data could be used to generate information about protein tertiary structure. However, a quantitative analysis of the information embedded within a WAXS pattern was used to demonstrate that this may be possible, in principle, if accurate intensities can be measured to ~2.0 Å spacing (Makowski et al. 2008a). WAXS patterns computed from atomic coordinates of 498 protein domains corresponding to the known fold space at that time (Hou et al. 2003) were used to construct a multi-dimensional space of WAXS patterns ('WAXS space') corresponding to these folds. Within WAXS space, each scattering pattern is represented by a single vector. A principal components analysis (PCA) identified directions in WAXS space corresponding to the greatest discrimination among WAXS patterns. Estimates of the abundances of secondary structures were made based on training sets derived from these data. This analysis led to estimates of α-helical content with average error of 11%; and of β-sheet content with average error of ~9%. The distribution of proteins that are members of the four global structure classes, α, β, α/β and α+β, are well

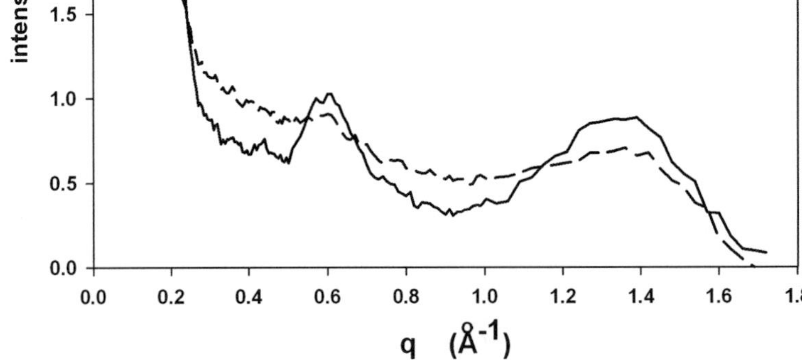


**Fig. 8.3** WAXS patterns from two $I_{gg}$ molecules, $I_{gg1}$ exhibiting strong WAXS scattering in the ~10 Å and 4.7 Å regions, consistent with a relatively rigid, well-formed immunoglobulin domain structure, and $I_{gg2}$ with muted intensities in the ~10 Å and 4.7 Å regions (q ~0.6 and 1.35 $Å^{-1}$, respectively), suggestive of a structural heterogeneity derived from relatively great conformational flexibility in solution

separated in WAXS space when data extending to a spacing of 2.2 Å are used, indicating that production of highly accurate WAXS data to high resolution has the potential for producing significant information on the structural class of any protein. By contrast, data limited to ~10 Å spacing exhibits little discriminatory power for classifying proteins according to secondary or tertiary structures.

## 8.7 Allosteric Proteins, Domain Organization and Quaternary Structure

One of the most promising areas for x-ray solution scattering is in the study of allosteric proteins, typically multi-domain and/or multi-subunit proteins that exhibit large-scale domain motions either as part of their function or in response to allosteric effectors or regulators. These re-arrangements of domains are difficult to study by crystallography because they typically involve movements that cannot be accommodated within a crystal lattice. Their characterization may require a new search for crystallization conditions for each allosteric effector studied or each structural configuration of functional importance. By contrast, domain motion results in large changes in solution scattering often within both the SAXS and WAXS regimes. For instance, Yang et al. (2010) studied the impact of peptide ligands and amino acid substitutions on the ensemble of structures exhibited by hck tyrosine kinase, characterizing large-scale re-arrangements of SH3, SH2 and kinase domain in response to different solution conditions. Badger et al. (2016) demonstrated the power of the approach by characterizing the re-arrangements of domains in abl kinase in response to amino acid substitutions that altered the activity of the protein. Their characterization of the T315I gatekeeper mutation (that exhibits resistance to all known drugs that target bcr-abl) revealed a novel configuration of the three domains of the abl core not previously characterized, suggesting the existence of multiple levels of regulation of abl kinase activity.

## 8.8 Ensembles

Protein solutions are not, in general, solutions of perfectly homogeneous macromolecular structures diffusing in an ideal buffer. Yes, it is possible to find small, relatively rigid proteins that will approach this ideal. But, these are not the most interesting cases. Much more frequently, proteins of interest may be large, flexible and capable of global internal motions of functional importance. Computational approaches to be used in concert with WAXS studies are developing rapidly and they may represent one of the most important applications of WAXS since they address issues difficult to resolve by other methods. It may seem counter intuitive that a single, one-dimensional intensity distribution could provide information about the relative abundances of multiple conformations within a solution. But if the structures of conformations that may be present can be hypothesized – on the basis of crystallographic, modeling or other information – then WAXS data represents a powerful test bed for determining which structures are, in fact, present and in what proportions (Konarev et al. 2003; Bernado et al. 2007; Tsutakawa et al. 2007; Yang et al. 2010; Petoukhov and Svergun 2007; Minh and Makowski 2013; Onuk et al. 2015). There are, of course, limits. In general, WAXS data seem capable of distinguishing relative abundances of three to ten distinct conformations. Contributions from conformations that are similar are more difficult to separate; dramatic structural differences far easier to distinguish.

OLIGOMER (Konarev et al. 2003) was originally conceived to separate out scattering from monomers; dimers and higher order oligomers when present together in a mixture. It has, however, found broader utility. It estimates the relative abundances of multiple constituents by solving a set of linear equations using nonnegative or unconstrained least-squares to minimize the difference between experimental and calculated scattering. It appears to adapt to WAXS data under conditions where the CRYSOL provides accurate estimates of scattered intensities (Onuk et al. 2015). Yang et al. (2010) introduced basis-set supported SAXS (BSS-SAXS) reconstruction, that combined solution scattering data with coarse-grained (CG) molecular dynamics to characterize the conformational states of Hck kinase in solution. In this approach, CG-MD simulations explore and sample conformational space; captured conformations are clustered into nine distinct conformational states and then used these as a basis set to analyze the scattering data. Onuk et al. (2015) took a somewhat different approach, using crystallographically determined structures of adenylate kinase as an initial basis set, clustering them into five distinct conformational classes and then used a maximum likelihood estimation (MLE) approach to generate estimates of relative abundances of these classes. It operates similar to OLIGOMER, but out-performs OLIGOMER when used with data having relatively low signal-to-noise ratio due to an accurate noise model (Onuk et al. 2015).

## 8.9 Priors

Addition of prior knowledge can greatly improve the accuracy or power of a calculation designed to characterize an ensemble of structures. That said, approaches to incorporation of priors are often non-trivial. Onuk et al. (2016) used a maximum a posteriori (MAP) approach to estimate the relative abundances of conformations in solutions of adenylate kinase. This enables estimates of the relative free energies of different conformations to be used to provide weights in the estimation of their relative abundances in solution (e.g., conformations with higher free energy are assigned lower weights than those with lower free energy). Computational tests indicated that prior knowledge improves estimation accuracy, and, not surprisingly, the stronger the prior constraints, the more accurate the resulting estimates of conformational abundances.

## 8.10 Modeling of Structural Fluctuations

It is not always convenient or informative to model structural fluctuations on the basis of an extensive ensemble of representative structures. This is particularly the case when proteins fluctuate about a single well defined conformation and characterization of the scattering can be made in terms of an average or consensus structure and the fluctuations about that structure. Increased flexibility leads to a broader structural ensemble that expresses itself in solution scattering patterns by filling in troughs in the scattered intensity and muting the intensity of peaks (Makowski et al. 2011). The range of motion of interatomic vectors can be estimated by comparison of the scattering pattern expected for a rigid protein with the observed scattering pattern (Makowski et al. 2011; Zhou et al. 2015). A formalism that makes it possible to predict the effect of these fluctuations on WAXS data has been developed and is called Vector Length Convolution (Makowski et al. 2011; Zhou et al. 2015). In this approach, the interatomic vector length of every atom pair in the protein is replaced by a distribution of vector lengths, and the breadth of that distribution is assumed to vary as a function of length. Not unreasonably, it has been found that small interatomic vectors exhibit smaller fluctuations than longer interatomic vectors (Zhou et al. 2016). Scattering from proteins undergoing this kind of fluctuation is predicted by (i) choosing a reference or consensus structure; (ii) calculating the scattering from the reference structure using XS; (iii) replacing each interatomic length in the pair correlation function, P(r), of the reference structure by a distribution of vector lengths – which amounts to a generalized convolution (see below); (iv) and re-calculating the intensity from the altered P(r). The resulting intensity function can then be compared with observed and parameters adjusted until a reasonable fit is achieved. Model ensembles with distinctly different properties can be generated by varying the way in which the fluctuations vary with interatomic vector length. The pair correlation function corresponding to the model structural ensemble, $P_m(r)$, is computed from the convolution of the pair correlation function of the reference structure, $P_r(r)$, and a Gaussian of half width σ(r) which may be a function of the interatomic vector length, r, according to

$$P_m(r) = P_r(r)^* \exp\left(-\sigma(r)^2/2r^2\right)$$

The '*' in the equation denotes convolution. Early applications of the method (Makowski et al. 2008b, 2011; Zhou et al. 2015) used a two parameter model for the radial variation of σ , $\sigma(r) = cr^e$, where c and e are free parameters, and varied the parameters to achieve an optimal fit to the observed data. More recently, σ(r) has been calculated directly from MD trajectories and used, with the model $P_m(r)$ to predict scattering.

The impact of structural flexibility is to generate a heterogeneous ensemble of protein conformations. This ensemble can be modeled through vector length convolution of P(r) to predict the impact of fluctuations on the WAXS pattern from a protein. Figure 8.4 includes the predicted scattering from ubiquitin assuming rigid conformation (solid line) and in the presence of fluctuations with a magnitude of $\sigma(r) = 0.7\ r^{0.5}$ Å. These predictions are compared to the scattering from a mutant ubiquitin (L50E) in which a hydrophobic core residue is replaced by a charged residue, disrupting structure and leading to flexibility and heterogeneity. Scattering from L50E is consistent with structural fluctuations of nearly 20% in interatomic vectors 10 Å in length – corresponding to very substantial structural heterogeneity.

Note that the 4.7 Å peak, due to the arrangement of β−strands is essentially gone in scattering from the mutant, indicating the complete or nearly complete disruption of β-sheets in the structure.

## 8.11 Unfolding

Folding (and unfolding) of proteins in response to environmental changes results in significant alterations in WAXS scattering. During the alcohol-induced unfolding of β–lactoglobulin the largely β-structure has been reported to transform into an open α-helical structure (Hirota et al. 1997; Kumar et al. 2003). However, WAXS patterns from β-lactoglobulin in increasing concentrations of ethanol suggest the preservation of β-structure during the transformation. Figure 8.5 includes WAXS data from

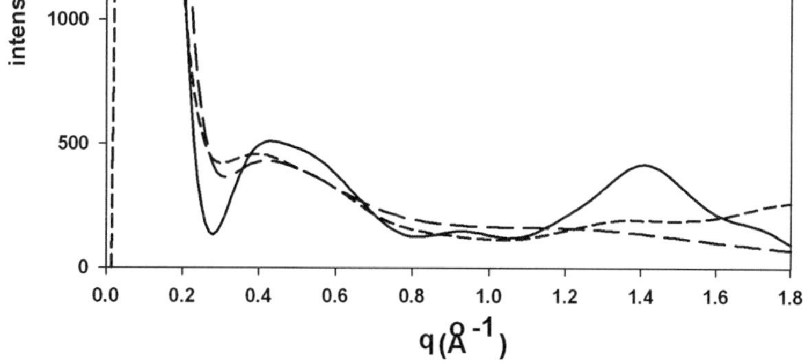


**Fig. 8.4** Predicted WAXS patterns for rigid ubiquitin (*solid line*) and a highly flexible ubiquitin molecule (*long dashes*) compared to observed scattering from a ubiquitin mutant (L50E) that is highly flexible (*short dashes*)

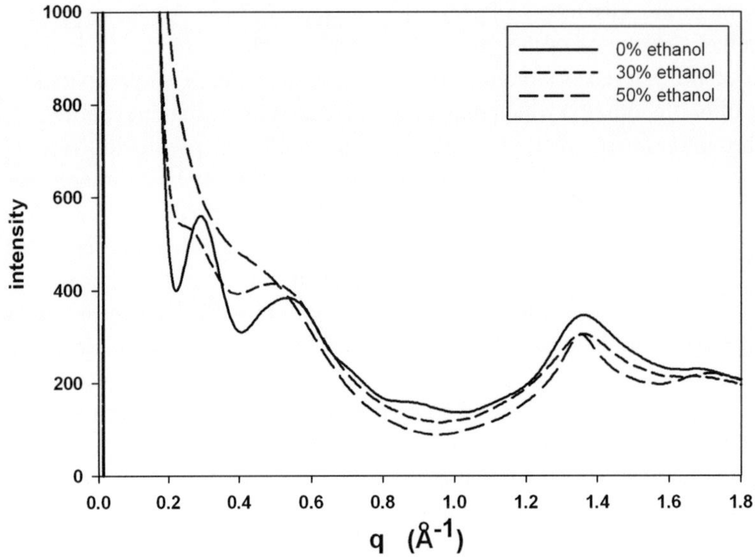


**Fig. 8.5** Unfolding of β-lactoglobulin in ethanol. WAXS patterns from β-lactoglobulin in buffer; and in 30% and 50% ethanol. Strong features that correspond to tertiary structure begin to disappear in ethanol solutions >12%, and are almost completely gone in 50% ethanol. However, the 4.7 Å peak (q ~1.35 $Å^{-1}$) that corresponds to β-strand separation remains strong in 50% ethanol, indicating the preservation of at least a part of the β-sheet structure, even in the near complete absence of tertiary structure

β-lactoglobulin in the absence of alcohol, in 30% and 50% alcohol. The scattering in 30% alcohol suggests the preservation of at least a semblance of tertiary structure since scattering features at in the 10–20 Å regime are not completely removed. In 50% alcohol, these features are gone, suggesting a complete obliteration of tertiary structure. Unexpectedly, the strong 4.7 Å peak remains present in 50% alcohol, strongly suggesting that the β-sheets retain some structural integrity even in the virtual absence of tertiary structure.

## 8.12 Screening Ligand Libraries for Detection of Functional Interactions

Because modulation of function by a small molecule ligand is almost always accompanied by a structural change detectable by WAXS (Fischetti et al. 2003), WAXS is becoming a promising technology for screening of ligand libraries for functional interactions. Target-based affinity screens may be used to screen libraries of up to $10^6$ compounds, typically yielding $10^1$–$10^2$ candidate ligands (Stockwell 2000). Subsequent functionality tests of these candidate molecules represent a serious bottleneck in the drug discovery pipeline. In vivo screens may detect phenotypic changes due to ligand action but are complicated by potential for the ligand tested to bind to other targets or yield false negatives due to parallel pathways that duplicate assayed function. *In vitro* screens often require a custom, function-specific assay and these assays are not available for all functions. An alternative approach is to use a generic biophysical method to detect structural changes that almost universally accompany functional ligand binding. Unfortunately, many approaches have limited sensitivity to structural change. For instance, circular dichroism (CD) is largely sensitive to changes in secondary structure (Wallace and Janes 2003), and SAXS may be insensitive to changes that do not alter the radius of gyration or result in large re-organization of domain structure. WAXS can be used to detect a broad range of ligand-induced alterations in secondary, tertiary, or quaternary structure. The speed of data acquisition, use of label-free targets, and adaptability to a broad range of solution conditions, make WAXS an attractive method for moderate-throughput detection and analysis of protein-ligand interactions.

How small a structural change can be detected using WAXS? Fischetti et al. (2004b) demonstrated detection of ligand binding in four proteins that had been crystallized both in the presence and absence of known ligands. Addition of ligands to transferrin, maltose binding protein (MBP), alcohol dehydrogenase (ADH) and calmodulin resulted in changes in WAXS patterns that corresponded to those predicted from atomic coordinate sets. The variation in structures triggered by these experiments ranged from ligand-induced re-folding in calmodulin, to ligand-induced domain rotation in transferrin, hinge-binding motion in MBP and change in the shape of the binding cleft in ADH. Figure 8.6 provides an example of the change in WAXS scattering for the binding of a small molecule ligand, 2,4,6 tribromophenol, to human transthyretin. This ligand is a common environmental contaminant that binds competitively with the natural ligand (Ghosh et al. 2000) and has been implicated in disruption of the thyroid hormone system.

It is also possible for ligand binding to alter the flexibility of a protein or the spatial extent of its structural fluctuations. These changes may also result in modulation of function. WAXS has proven unexpectedly sensitive to changes in structural fluctuations (Makowski et al. 2008b). For example, when an inhibitor is bound to HIV protease, the flaps may fold down over the inhibitor much as they do when binding substrate. However, detailed analysis of the ligand induced changes in intensity observed by WAXS indicated that the average structure does not change significantly. Rather there is a decrease in the magnitude of structural fluctuations that the protein is undergoing (Zhou et al. 2015).

## 8.13 Establishing the Significance of Small Intensity Changes

When ligands induce small changes in WAXS scattering that may or may not indicate a statistically significant change in structure, it is useful to have a statistical measure of the difference

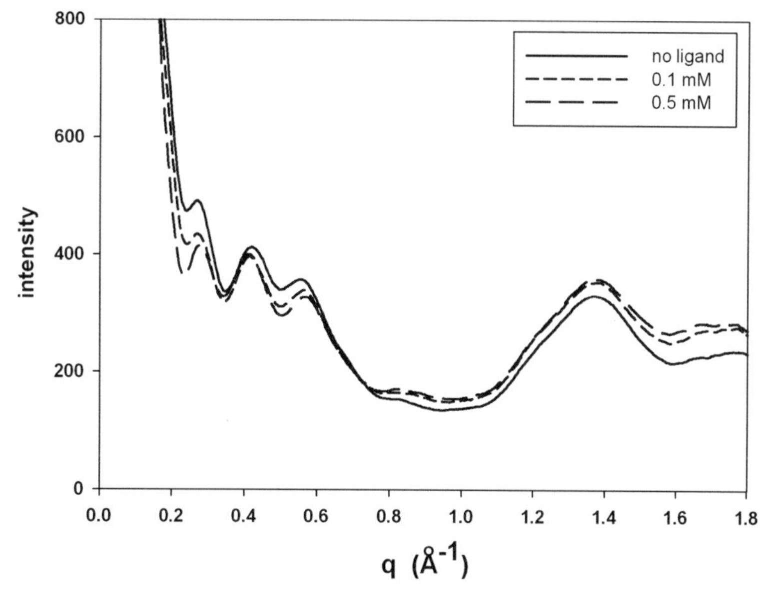


**Fig. 8.6** WAXS patterns from human transthyretin in solution, and in the presence of 0.1 and 0.5 mM of a 2,4,6 tribromophenol, a common environmental contaminant that competes with the natural ligand for binding to transthyretin

between two WAXS patterns. A chi-square measure has proven useful as a measure of statistical significance (Rodi et al. 2007). In our experience, a reduced chi-square (chi-square divided by number of degrees of freedom) $\chi_\nu > 1.0$ is indicative of a statistically significant difference between two WAXS patterns. The number of degrees of freedom is approximately equal to the number of independent measurements of intensity which is ~ $q_{max}d_{max}/\pi$, where $q_{max}$ is the greatest value of q for which data is used and $d_{max}$ is the longest interatomic vector in the structure (as estimated by a simple Shannon sampling theorem argument). However, some care needs to be taken in applying chi-square as a measure of significance of a structural change because estimates of standard deviation of intensities are difficult to make accurately and there remain questions about scaling of differences in the SAXS regime relative to differences in the WAXS regime due to the dramatic (two orders of magnitude) difference in their raw intensities.

Although it is quite easy to detect large domain motions with SAXS data, smaller motions may be detectable with WAXS. Investigation of the intensity changes generated by loop and side-chain re-arrangements (Fischetti et al. 2004b) suggested strongly that relatively minor movements can be detected with WAXS. Investigation of the impact of anomalous scattering on WAXS data (Makowski et al. 2012) suggested that differences corresponding to motion of even a few electrons can be detectable.

Establishing that two WAXS patterns are statistically distinguishable may be inadequate to address the biological question motivating the research. Comparison of scattering from several samples that represent impact of different ligands or amino acid replacements on the same molecule may require categorization or clustering of the patterns. Qualitative descriptions of differences may not elucidate the relationships among different structures or establish the structural origins of the differences observed. Use of a dimensionality reducing approach such as principal component analysis (PCA) to provide a quantitative classification of the patterns may make possible identification of features that most distinguish the patterns from one another and generate a foundation for establishing patterns underlying functional modulation.

## 8.14 Discussion

These examples indicate that WAXS provides enhanced sensitivity for detection of small structural changes relative to SAXS; that it can be used to test molecular models for protein structure, and provide insights into protein flexibility both quantitatively and qualitatively.

The need for generic approaches to screening ligands for functional binding has motivated consideration of WAXS as a moderate-throughput screen. Most intermolecular interactions that give rise to significant changes in structure or dynamics will modulate the function of a protein in some way. Since WAXS is sensitive to changes in secondary, tertiary or quaternary structure or domain motions, it provides a comprehensive option for these kinds of screens. Given that it is now possible to screen 10–20 samples per hour at a synchrotron source, a screen of several hundred candidate ligands is quite feasible. As a secondary screen, focused on ligands that were originally identified by a high-throughput affinity screen, WAXS can provide an attractive addition to the drug discovery pipeline.

Although WAXS cannot be used to calculate a molecular structure (it lacks adequate information content), it can be used to test molecular models, whether generated *ab initio*, or based on crystallography, cryoEM, NMR, MD or combinations of these methods. Development of more efficient computational approaches to modeling of the hydration shell and excluded volume would contribute substantially to these calculations. This does not represent a significant bottleneck for individual calculations. However, the increased focus on ensembles, and the consequent need for calculating patterns from a large number of protein conformations as a basis for characterizing the ensembles, places high priority on improved computational tools. Recent efforts to use solution scattering to refine structural models (Zheng and Tekpinar 2011; Roig-Solvas et al. 2017) would also benefit substantially from highly efficient computations taking into account the impact of hydration shell.

Characterization of flexibility is challenging for a number of reasons, not the least of which is the challenge of enumerating flexibility in a simple way. Utilizing the P(r) function as the basis for global characterization of structural fluctuations is advantageous because (i) it is relatively intuitive (ii) can be displayed as a simple one-dimensional plot and (iii) results in estimation of the dependence of the scale of structural fluctuation on interatomic vector length, $\sigma(r)$, a function that can be estimated from WAXS data or be calculated directly from an MD simulation trajectory (Zhou et al. 2016). Characterization in terms of the most abundant conformations of a structural ensemble (e.g. Yang et al. 2010) represents a highly informative, complimentary approach.

Time-resolved (TR) WAXS studies also have significant potential for investigating tertiary and quaternary conformational changes (Cammarata et al. 2008). When those changes can be induced by a short laser pulse time resolutions in the nanoseconds are possible. The methods outlined in this chapter are entirely applicable to each 'snap-shot' in a time series. Ligand-induced structural changes are much more difficult to track using TR WAXS because any structural change will be convoluted by variations in diffusion times – in other words, because each ligand will take a different amount of time to find and interact with a protein, it is impossible for all proteins to change structure synchronously.

New frontiers in the method have been suggested by the examples provided here. Efforts to collect highly accurate WAXS data to the highest possible resolutions (e.g. 2.0 Å) with the highest achievable signal-to-noise ratio have the potential to drive the method to the next level where structural changes induced by binding of a small molecule or ion, or even changes in the concentrations of buffer constituents could be observable and interpretable. Used in concert with computational approaches such as MD, these advances could increase the power of WAXS for characterization of the structure and structural fluctuations of macromolecules in solution and for comprehensive studies of biochemistry in the scattering volume.

## References


Allaire M, Yang L (2011) Biomolecular solution X-ray scattering at the National Synchrotron Light Source. J Synchrotron Radiat 18:41–44

Badger J, Grover P, Panjarian SB, Engen JR, Smithgall TE, Makowski L (2016) The c-Abl tyrosine kinase adopts multiple active conformational states in solution. Biochemistry 55:3251–3260

Bardhan JP, Park S, Makowski L (2009) SoftWAXS: a computational tool for modeling wide-angle x-ray solution scattering from biomolecules. J Appl Crystallogr 42:932–943

Bernado P, Mylonas E, Petoukhov MV, Blackledge M, Svergun DI (2007) Structural characterization of flexible proteins using small-angle x-ray scattering. J Am Chem Soc 129:5656–5664. (2007)

Cammarata M, Levantino M, Schotte F, Anfinrud PA, Ewald F, Choi J, Cupane A, Wulff M, Ihee H (2008) Tracking the structural dynamics of proteins in solution using time-resolved wide-angle X-ray scattering. Nat Methods 5:881–886

Chacon P, Moran F, Diaz JF, Pantos E, Andreu JM (1998) Low-resolution structures of proteins in solution retrieved from X-ray scattering with a genetic algorithm. Biop J 74:2760–2775

Cho HS, Dashdorj N, Schotte F, Graber T, Henning R, Anfinrud P (2010) Protein structural dynamics in solution unveiled *via* 100-ps time-resolved x-ray scattering. Proc Natl Acad Sci 107:7281–7286

Doniach S (2001) Changes in biomolecular conformation seen by small angle x-ray scattering. Chem Rev 101:1763–1778

Fischetti RF, Rodi DJ, Mirza A, Kondrashkina E, Irving T, Makowski L (2003) Wide angle x-ray scattering of proteins: effect of beam exposure on protein integrity. J Synch Res 10:398–404

Fischetti RF, Rodi DJ, Gore DB, Makowski L (2004a) Wide angle x-ray solution scattering as a probe of ligand-induced conformational changes in proteins. Chem Biol 11:1431–1443

Fischetti R, Stepanov S, Rosenbaum G, Barrea R, Black E, Gore D, Heurich R, Kondrashkina E, Kropf AJ, Wang S, Zhang K, Irving TC, Bunker GB (2004b) The BioCAT undulator beamline 18ID: a facility for biological non-crystalline diffraction and x-ray absorption spectroscopy at the advanced photon source. J Synchrotron Radiat 11:399–405

Ghosh M, Meerts IATM, Cook A, Bergman A, Brouwer A, Johnson LN (2000) Structure of human

transthyretin complexed with bromophenols: a new mode of binding. Acta Cryst D56:1085–1095
Grishaev A, Guo L, Irving T, Bax A (2010) Improved fitting of solution x-ray scattering data to macromolecular structures and structural ensembles by explicit water modeling. J Am Chem Soc 132:15484–15486
Hirai M, Iwase H, Hayakawa T, Inoue K (2002) Structural hierarchy of several proteins observed by wide-angle solution scattering. J Synchrotron Rad 9:202–205
Hirai M, Koizumi M, Hayakawa T, Takahashi H, Abe S, Hirai H, Miura K, Inoue K (2004) Hierarchical map of protein unfolding and refolding at thermal equilibrium revealed by wide-angle x-ray scattering. Biochemistry 43:9036–9049
Hirota N, Mizuno K, Goto Y (1997) Cooperative a-helix formation of P-lactoglobulin and melittin induced by hexafluoroisopropanol. Protein Sci 6:416–421
Hong X, Hao Q (2009) High resolution pair-distance distribution function P(r) of protein solutions. Appl Phys Letts 94:083903
Hou J, SIms GE, Zhang C, Kim S-H (2003) A global representation of the protein fold space. Proc Natl Acad Sci 100:2386–2390
Hura GL, Menon AL, Hammel M, Rambo RP, Poole FL, Tsutakawa SE, Jenney FE Jr, Classen S, Frankel KA, Hopkins RC, Yang S-J, Scott JW, Dillard BD, Adams MWW, Tainer JA (2009) Robust, high-throughput solution structural analyses by small angle X-ray scattering (SAXS). Nat Methods 6:606–612
Inouye H, Houde D, Temel DB, Makowski L (2016) Utility of solution x-ray scattering for the development of antibody biopharmaceuticals. J Pharm Sci 105:3278–3289
Konarev PV, Volkov VV, Sokolova AV, Koch MH, Svergun DI (2003) Primus: a Windows PC-based system for small-angle scattering data analysis. J Appl Crystallogr 36:1277–1282
Kumar S, Modig K, Halle B (2003) Trifluoroethanol-Induced β -> α transition in β-Lactoglobulin: hydration and cosolvent binding studied by $^{2}$H, $^{17}$O, and $^{19}$F magnetic relaxation dispersion. Biochemistry 2003 (42):13708–13716
Lattman EE (1989) Rapid calculation of the solution scattering profile from a macromolecule of known structure. Proteins Struct Funct Genet 5:149–155
Luzzati V, Tardieu A (1980) Recent developments in solution x-ray scattering. Ann Rev Biophys Bioeng 9:1–29
Makowski L (2010) Characterization of proteins with Wide-angle X-ray Solution Scattering (WAXS). J Struct Funct Genom 11:9–19
Makowski L, Rodi DJ, Mandava S, Devrapahli S, Fischetti RF (2008a) Characterization of protein fold using wide angle x-ray solution scattering. J Mol Biol 383:731–744
Makowski L, Rodi DJ, Mandava S, Minh D, Gore D, Fischetti RF (2008b) Molecular crowding inhibits intramolecular breathing motions in proteins. J Mol Biol 375:529–546
Makowski L, Gore D, Mandava S, Minh D, Park S, Rodi DJ, Fischetti RF (2011) X-ray solution scattering studies of the structural diversity intrinsic to protein ensembles. Biopolymers 95:531–542
Makowski L, Bardhan J, Gore D, Rodi DJ, Fischetti RF (2012) Observation of multi-wavelength anomalous diffraction using Medium Angle X-ray solution scattering (MADMAX) from proteins in solution. Biophys J 102:927–933
McDonald RC, Steitz TA, Engelman DM (1979) Yeast hexokinase in solution exhibits a large conformational change upon binding glucose or glucose 6-phosphate. Biochemistry 18:338–342
Minh D, Makowski L (2013) Wide-angle x-ray solution scattering for protein-ligand binding: multivariate curve resolution with Bayesian confidence intervals. Biop J 104:873–884
Onuk AE, Akcakaya M, Bardhan J, Erdogmus D, Brooks DH, Makowski L (2015) Constrained maximum likelihood estimation of relative abundances of protein conformation in a heterogeneous mixture from small angle x-ray scattering intensity measurements. IEEE Trans Signal Process 63:5383–5394
Onuk E, Akcakaya M, Bardhan J, Erdogmus D, Brooks D, Makowski L (2016) Dirichlet priors for MAP inference of protein conformation abundances from SAXS. J. Signal Processing Sys (in press)
Park S, Bardhan JP, Roux B, Makowski L (2009) Simulated x-ray scattering of protein solutions using explicit-solvent molecular dynamics. J Chem Phys 130:134114
Petoukhov MV, Svergun DI (2007) Analysis of x-ray and neutron scattering from biomacromolecular solutions. Curr Opin Struct Biol 17:562–571
Putnam CD, Hammel M, Hura GL, Tainer JA (2007) X-ray solution scattering (SAXS) combined with crystallography and computation: defining accurate macromolecular structures, conformations and assemblies in solution. Q Rev Biophys 40:191–285
Rodi DJ, Mandava S, Gore DB, Makowski L, Fischetti RF (2007) Detection of functional ligand binding events using synchrotron x-ray scattering. J Biomol Screen 12:994–998
Roig-Solvas B, Bardhan J, Erdogmus D, Brooks D, Makowski L (2017) ProSFIT: a modular optimization framework for flexible reconstruction of protein structures from SAXS measurements. (in preparation)
Stockwell BR (2000) Chemical genetics: ligand-based discovery of gene function. Nat Rev Genet 1:116–125
Svergun DI (1999) Restoring low resolution structure of biological macromolecules from solution scattering using simulated annealing. Biophys J 76:2879–2886
Svergun D, Barberato C, Koch MHJ (1995) CRYSOL-a program to evaluate x-ray solution scattering of biological macromolecules from atomic coordinates. J Appl Crystallogr 28:768–773
Svergun DI, Richard S, Koch MHJ, Sayers Z, Kuprin S, Zacci G (1997) Protein hydration in solution:

experimental observation by x-ray and neutron scattering. Proc Natl Acad Sci 95:2267–2272

Svergun DI, Petoukhov MV, Koch MMHJ (2001) Determination of domain structure of proteins from x-ray solution scattering. Biop J 80:2946–2953

Tiede DM, Zhang R, Seifert S (2002) Protein conformations explored by difference high-angle solution x-ray scattering: oxidation state and temperature dependent changes in cytochrome C. Biochemistry 41:6605–6614

Tsutakawa SE, Hura GL, Frankel KA, Cooper PK, Tainer JA (2007) Structural analysis of flexible proteins in solution by small angle X-ray scattering combined with crystallography. J Struct Biol 158:214–223

Volkova VV, Svergun DI (2003) Uniqueness of *ab initio* shape determination in small-angle scattering. J Appl Crystallogr 36:860–864

Wallace BA, Janes RW (2003) Circular dichroism and synchrotron radiation circular dichroism spectroscopy: tools for drug discovery. Biochem Soc Trans 31:631–633

Walther D, Cohen FE, Doniach S (2000) Reconstruction of low resolution three-dimensional density maps from one-dimensional small angle x-ray scattering data for biomolecules in solution. J Appl Crystallogr 33:350–363

Wen B, Peng J, Zuo X, Gong Q, Zhang Z (2014) Characterization of protein flexibility using small-angle X-ray scattering and amplified collective motion simulations. Biop J 107:956–964

Yang S, Blachowicz L, Makowski L, Roux B (2010) Multidomain assembled states of Hck tyrosine kinase in solution. Proc Natl Acad Sci 107:15757–15762

Zhang R, Thiyagarajan P, Tiede DM (2000) Probing protein fine structures by wide angle solution x-ray scattering. J Appl Crystallogr 33:565–568

Zheng W, Tekpinar M (2011) Accurate flexible fitting of high-resolution protein structures to small-angle x-ray scattering data using a coarse-grained model with implicit hydration shell. Biophys J 101:2981–2991

Zhou H, Li S, Badger J, Nalivaika E, Cai Y, Foulkes-Murzycki J, Schiffer C, Makowski L (2015) Modulation of HIV protease flexibility by the T80N mutation. Proteins Struct Funct Bioinf 83:1929–1939. doi:10.1002/prot.24737

Zhou H, Li S, Makowski L (2016) Visualizing global properties of a molecular dynamics trajectory proteins. Struct Funct Bioinforma 84:82–91

# SAS-Based Studies of Protein Fibrillation 9

Carlotta Marasini and Bente Vestergaard

**Abstract**

Protein fibrillation is associated with a number of fatal amyloid diseases (e.g. Alzheimer's and Parkinson's diseases). From a structural point of view, the aggregation process starts from an ensemble of native states that convert into transiently formed oligomers, higher order assemblies and protofibrils and, finally, fibrils. The different species exist in equilibrium in solution leading to a high degree of sample heterogeneity. It is impossible to physically isolate any single species for structural analysis: separation will alter the equilibrium and potentially cause structural changes.

Small angle scattering is an optimal method for structural studies of the fibrillation process in order to further the knowledge of the associated diseases. The recorded scattering data include the scattering contribution of all the species in solution and must be decomposed to enable structural modeling of the individual components involved during the fibrillation, notably without physical separation of the species. In this chapter we explain how to optimize a small angle scattering analysis of the fibrillation process and the basic principles behind analysis of the data. We include several practical tips and highlight existing reports, exemplifying the wealth of information that can be derived from the method.

**Keywords**

Amyloid • Data decomposition • Mixtures • Protein fibrillation

## 9.1 The Fibrillation Process

### 9.1.1 What Is an Amyloid Fibril?

It is difficult to coin a general definition of amyloids, since several definitions exist, either based on a relation to disease (Friedrich 1859),

C. Marasini • B. Vestergaard (✉)
Department of Drug Design and Pharmacology, University of Copenhagen, Universitetsparken 2, DK-2100 Copenhagen, Denmark
e-mail: bente.vestergaard@sund.ku.dk





B. Chaudhuri et al. (eds.), *Biological Small Angle Scattering: Techniques, Strategies and Tips*, Advances in Experimental Medicine and Biology 1009, DOI 10.1007/978-981-10-6038-0_9

ability to stain with particular compounds (Divry and Florkin 1927; Cohen and Calkins 1959), a biophysical definition (cross-β pattern in fiber diffraction (FD), (Cohen and Calkins 1964; Eanes and Glenner 1968), and probably more. Here, we define amyloid fibrils as: insoluble fibrous protein aggregates that arise from an erroneous folding of a given disease-specific protein and sharing specific structural characteristics (Chiti and Dobson 2006; Fandrich 2007) including cross-β FD pattern and a highly elongated morphology. More than 20 amyloidogenic diseases such as Alzheimer's disease (AD), Parkinson's disease (PD), and prion–associated encephalopathies have fibril formation as a common symptom (Wong et al. 1985; Hardy and Higgins 1992; Lucking and Brice 2000; Ross and Poirier 2004). Although the proteins associated with each disease do not share sequence homology, they exhibit similar insoluble filaments and fibrillation responses (Selkoe 1994; Khurana et al. 2003). This suggests that the underlying fibril formation mechanisms may be common (Kayed et al. 2003).

Even if the amyloid-related disorders have a high worldwide incidence (as an example, nearly 44 million people have AD or related dementia) unfortunately until now, only symptomatic treatment and no cure have been developed. This evidences a significant interest in furthering our structural understanding of fibrillation. In addition, functional amyloids are widespread in nature (Otzen 2010) and amyloid is an attractive nanoscopic building material because its robust design and simple repetitive structure make for very durable and metabolically cheap material (Dueholm et al. 2011). To take advantage of this material, a firm control of the production of amyloid is required, and hence the molecular basis of the fibrillation process must be understood.

### 9.1.2 The Complexity of the Fibrillation Process

Proteins can fibrillate *in vitro* under controlled conditions, normally following a nucleation-dependent polymerisation mechanism exhibiting three distinct phases: a lag phase, a growth phase and a stationary phase (Lee et al. 2007) (Fig. 9.1). The lag phase is the initial part of the fibrillation, in which fibrillation prone conformations and the initial nuclei are formed, but no or only little protein assembles into fibrils (Harper and Lansbury 1997; Serio et al. 2000). This is followed by a growth phase where fibrils are formed by integrating proteins to the growing end of the elongating fibrils. Here, secondary nucleation (auto-catalysis) can take place, i.e. stimulating the growth phase further (Fodera et al. 2008). The stationary phase is finally reached when the solution is depleted from monomers (or monomers are at a sufficiently low concentration) (Jarrett and Lansbury 1993). In this phase, however, fibrils may re-organize and form more elaborate morphologies and/or maturate. As pointed out by Ferrone (1999) the presence of a lag phase is not enough to prove a nucleated polymerization. In fact, for many amyloids, the process is expected to be more complex, and indeed several monomeric species may co-exist, or the distribution of monomeric species may change, depending on even subtle changes in the experimental conditions (e.g. Curtain et al. 2015). It is thus important to realize, that during each phase, several competing processes co-exist (Fig. 9.1), and that all amyloidogenic structures have a probability of being present in varying relative amounts at any given time of the fibrillation process. The process, however, will be dominated by the presence of certain species at different times.

For following the fibrillation process, different methods can be used, such as e.g. absorbance, light scattering, changes in secondary structure or the evolving fluorescence of fibril-specific dyes. A widely used technique for the identification and quantification of amyloid fibrils *in vitro* is the monitoring of extrinsic fluorescence of the benzothiazole dye, thioflavin T (ThT) (Bolder et al. 2007; Hawe et al. 2008). This specific dye has the peculiarity to fluoresce when it is interacting with the cross-β quaternary structure of amyloid fibrils (LeVine 1993; Nilsson 2004) while free ThT in an aqueous environment shows only weak fluorescence (Fig. 9.1). All these methods, however, provide no, or only indirect,

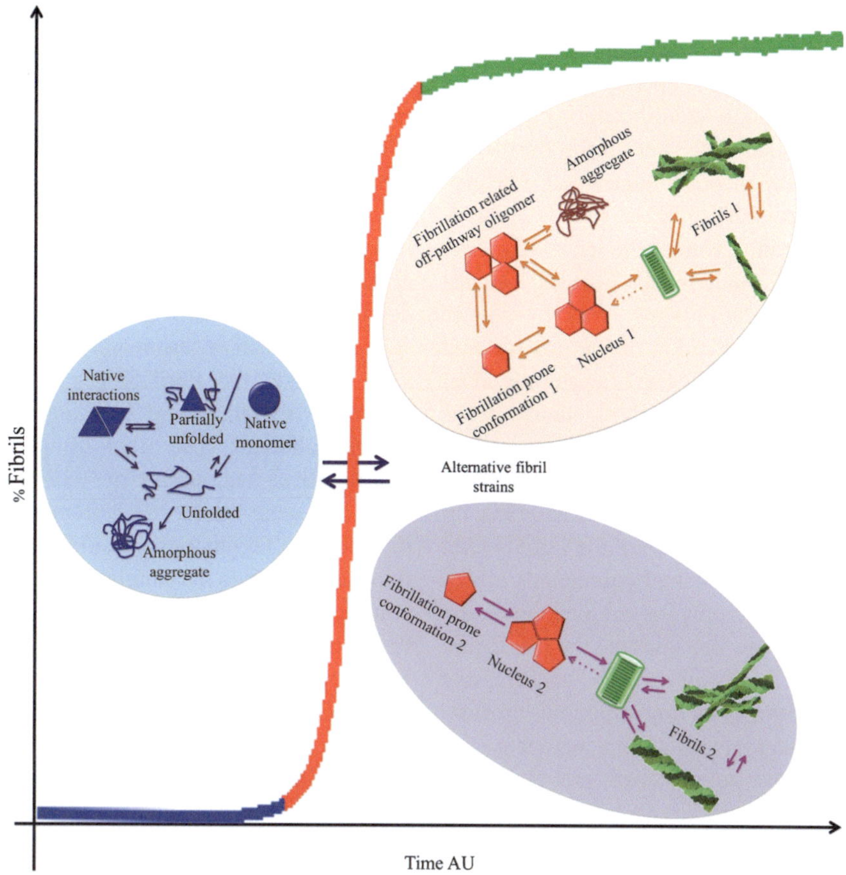


**Fig. 9.1** Schematic of the fibrillation process for a general partially unfolded or monomeric protein. An amyloid protein passes from its initial state (*blue circle*) to form fibrils (*green* species) passing through an intermediate state of transient/unstable oligomers (*red* components). Two important considerations have to be remarked: The physico-chemical properties of the protein are influenced by several parameters (temperature, pH, buffer, ionic strength, shaking…) and, as a consequence, both the structural state and protein:protein interactions during the fibrillation will be altered when experimental conditions change. This means that several potential fibrillation pathways co-exist (here indicated by the presence of the *purple* and *orange* areas). The percentage of fibrils can be followed e.g. by fluorescence spectroscopy using fibril-specific fluorescent dyes, and a schematic, typical for a nucleated polymerization reaction, is shown. The *curve* is colored in *blue* (lag-phase), *red* (elongation phase) and *green* (stationary phase), respectively. These correspond to the colors of the components that are particularly dominant during individual phases. It is stressed, however, that all three phases are highly heterogeneous, hence that all species may co-exist in varying amounts at all times of the fibrillation

structural measures for the formation of the aggregated species, but as we will describe here, SAS offers unique opportunities to investigate the complexity of the fibrillation process.

### 9.1.3 Why Is Fibrillation Analysis Difficult?

As we mentioned above, an adequate understanding of the structural and molecular basis behind fibrillation is still far from reached. Which initial conformational change is decisive for the onset of fibrillation? How is the internal structure of fibrils and oligomers? Do repeating monomeric or oligomeric units elongate fibrils? Which of the species involved is/are the toxic one(s)? What is the structural basis of the potentially toxic action? What is the molecular basis of primary and secondary fibril nucleation? These are just a few of several questions, which the research community is trying to answer.

Amyloid fibrillation is however a very challenging process to characterize. As outlined above, several different species co-exist in a constantly changing equilibrium (Fig. 9.1), meaning firstly that each state of the fibrillation (including the initial and the final state) is heterogeneous, and secondly, that one specific component cannot be extracted and purified. Purification of individual species equals disturbing the equilibrium, which induces response in the composition of the mixture and potentially alters the structure of each individual component. This inherent heterogeneity means that traditional structural analysis, using e.g. nuclear magnetic resonance or macromolecular crystallography, is difficult. In addition, all levels of the process (including the initial state) are sensitive to experimental parameters, which may be necessary to apply for sample preparation in different methods. For studying the fibrillation process, it is necessary to use non-invasive methods that allow obtaining structural information from the heterogeneous and developing mixture. As outlined below, SAS provides such opportunities.

## 9.2 Small Angle Scattering Captures the Structural Aspects of the Fibrillation Process

### 9.2.1 The Advantages of SAS from the Point of View of Fibrillation

As it is evident from this book, there are numerous reasons to use SAS for structural analysis. There are, however, three specific advantages of SAS that makes the method uniquely suitable for the analysis of protein fibrillation: Firstly, the samples are in solution. Secondly, SAS data are additive. And thirdly, SAS data cover a wide resolution range. As we will explain in this chapter, this makes the method particularly applicable to the study of heterogeneous macromolecular solutions with a spatiotemporal dynamic behavior, such as protein fibrillation.

The fact that samples are in solution with very few restrictions on the experimental conditions is indeed very advantageous. The protein fibrillation process is very sensitive to the experimental conditions hence changes in e.g. pH, salinity, temperature, co-solvents or protein concentration will influence fibrillation kinetics, and potentially alter the fibrillation pathway (Nielsen et al. 2001; Srinivasan et al. 2003; Arora et al. 2004; Morshedi et al. 2007; Wang et al. 2010). Almost any such experimental condition, necessary to study the fibrillation process, can be applied when collecting SAS data. This also allows collecting data from samples close to physiological conditions or, alternatively, to study the same fibrillating system as a function of different biophysical parameters. An additional advantage is the simplicity of sample preparation prior to the actual data collection, i.e. the fibrillating sample can be measured immediately when extracted from the reaction mixture, which is of evident importance, given the highly dynamic and developing nature of the samples.

The additive nature of SAS data (if structure factors can be ignored) means that the scattering curve measured from a mixture of species can be seen as a linear combination of scattering curves from the individual scattering components in solution, weighted by their relative volume fractions. It is thus principally possible to decompose the data into the pure scattering signals from each particular species and their relative volume fractions when data are available from several samples with varying compositions. The relevance for kinetic studies of amyloid processes is evident: the developing fibrillation process can be followed over time and the signal of each individual species can be isolated without altering the equilibrium. As simple as this principle may sound, the actual decomposition is challenging, due to the inherent ambiguity in such a decomposition process (Tauler et al. 1995; Tauler 2001; Jaumot et al. 2004).

The third particularly advantageous feature of SAS data is the wide q-range, which allows studying a process involving species of very different sizes. A typical fibrillation reaction includes monomers in the Ångstrom to nm length scales, intermediately sized oligomeric species, and mature fibril structures, with repeating units in the hundred-nm range. A typical synchrotron SAXS dataset spans all these distances, which is very different from other structural biology techniques. As we have recently demonstrated, this particular feature is central, if the method of decomposition is automated (Herranz-Trillo et al. 2017).

The three above-mentioned advantages are common for both SAXS and SANS. It should, however, be mentioned that neutrons also allow studies with the possibility to screen specific parts of the protein by recombinantly exchanging hydrogen with deuterium on one component (Gabel et al. 2002), just as neutron scattering enables very interesting studies of protein/fibril dynamics (Schiro et al. 2012; Fichou et al. 2015). Such studies will not be further discussed here.

### 9.2.2 SAS and Amyloids: Challenging Experiments

SAS data-collection is a non-invasive method unless the samples are damaged by the formation of free radicals in solution, caused by the often highly intense X-ray beams. If radiation damage is negligible it is possible to collect data continuously from one and the same fibrillating sample, such as it was done in the cases of Oliveira et al. (2009) and Chatani et al. (2015). In the first case a rotating anode SAXS setup was applied, hence, no radiation damage was expected. In the second case synchrotron data were collected from a 1 mm sample cuvette at 80 °C for two hours. In the latter case, it is difficult to test for radiation damage (distinguishing between radiation-induced aggregation and fibrillation is difficult). Furthermore, the fibrillation of proteins directly in the capillary/cuvette may generate other issues, e.g. precipitation (necessitating mixing) or evaporation of the samples during the measurement, in particular if applying incubation at high temperature and hence may not be generally applicable. A solution applied by us (Vestergaard et al. 2007; Giehm et al. 2011; Langkilde and Vestergaard 2012; Groenning et al. 2015) is to extract and measure samples from a fibrillation performed in parallel in the chemical laboratory. In this way, the risk of radiation damage is significantly reduced since every sample is exposed to minimal X-ray doses, and in addition it is possible to follow the fibrillation reaction by complementary measures (e.g. the ThT signal) in real time. The loading of the sample must be done with caution, ensuring that the handling of the sample, upon extraction from the reaction vial, does not alter the composition of the sample (e.g. if the sample changes temperature, see details below).

Neutron scattering does not cause radiation damage (Piccoli et al. 2007), but, in general, does not provide the same signal:noise as an X-ray experiment and demands longer exposure times (resulting in a corresponding poorer time resolution when resolving the kinetics of fibrillation). Additionally, if one wishes to take advantage of the powerful contrast matching techniques (Whitten and Trewhella 2009), new challenges are introduced related to the potentially altered protein solution behavior in deuterated media and the consequent risk of altered fibrillation kinetics/pathways.

### 9.2.3 The Central Step in SAS Analysis of Protein Fibrillation: Data Decomposition

Given all the advantages mentioned above, one may wonder why the method is not widely applied in the amyloid research community. One reason is definitely that both sample preparation and data collection are quite challenging, as outlined in more detail in Sect. 9.4. In addition, the data analysis is far from simple. As in all cases of SAS data evaluation, the initial steps of data reduction and basic analysis must be done with great care (Jacques et al. 2012) as elaborated elsewhere in this book. Here, we focus on the basic principles behind the decomposition process, which is the core peculiarity when analyzing SAS data from a fibrillation reaction. A scheme depicting these general principles is shown in Fig. 9.2. The first step is a singular

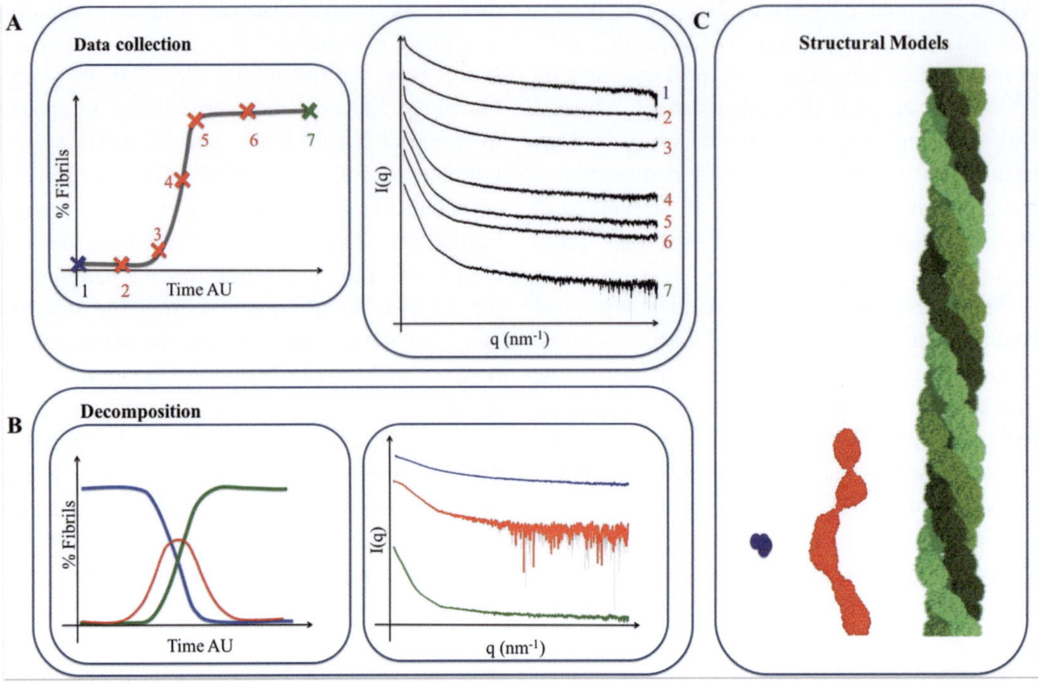


**Fig. 9.2** The basic strategy for a SAS analysis of the fibrillation process. (**a**) Data collection: (*left*) a ThT fluorescence spectrum exemplifies a method for following the fibrillation off-line. Samples are extracted at different time-points (here only seven time points are shown for clarity) and measured by SAXS, resulting in the seven corresponding SAXS curves (*right*). Each curve is the sum of scattering from the mixture of different scattering components. (**b**) Results from the decomposition: (*left*) the relative volume fractions of the three components present during the fibrillation and (*right*) the isolated SAXS curves of the three species. In blue the initial (often monomeric), in red the intermediate and in green the fibril state. (**c**) *Ab initio* models of the three dominating species existing during the fibrillation. Data are from (Giehm et al. 2011). Structural models are from (Vestergaard et al. 2007) and are illustrative only

value decomposition (SVD) or principal component analysis (PCA) (Hotelling 1933; Golub and Reinsch 1970) of the entire data set (i.e. the numerous data curves measured during the fibrillation reaction) (Fig. 9.2a). The SVD suggests how many independent scattering components contribute to the entire set of recorded data. The goal is to determine the isolated scattering curves from these individually contributing components, together with their relative volume fractions at the different time points of the fibrillation reaction (Fig. 9.2b). However, as mentioned above, the decomposition process is burdened by ambiguity, and this is a central challenge when analyzing fibrillation scattering data. This problem is significantly diminished if knowing either the volume fractions (from complementary biophysical experiments such as dynamic light scattering (Li et al. 2011) or quantitative analytical size exclusion chromatography (Hong et al. 2012) each protein species or the pure scattering curves of (some of the) individual species (either theoretically calculated from X-ray or NMR structures or experimentally measured scattering curves), beforehand. Most often, however, the volume fractions can not be estimated with acceptable accuracy, and, notably in the case of protein fibrillation analysis, the pure scattering curves of individual species cannot be measured, since the individual species are not stable in isolated form. In several published cases (Vestergaard et al. 2007; Groenning et al. 2015; Nors Perdersen et al. 2015) we have taken advantage of the assumption that the recorded scattering signal from the initial and the final point of the fibrillation process are coming from homogeneous solutions where just one (average) species contributes to the scattering curve: native (monomeric) state for the initial and the mature fibril state for the end point respectively. The second assumption is that these species represent two of the overall contributing species throughout the fibrillation process. Even in cases where the final fibril state is not homogeneous, then these assumptions are still valid if the relative composition of the final measured scattering curve corresponds to the relative composition of fibril species at earlier time points.

In the cases where these assumptions are valid, it is possible to isolate the scattering curve from an intermediately occurring species, even if this species is at no time-point the only contributing scattering species. This is crucially important for our approach, since it enables analysis of such transiently formed amyloidogenic species, without any prior physical separation from the reaction mixture. We argue that the highly dynamic nature of the protein:protein interactions, that cause the fibrillation, means that physical separation of the fibrillation species includes the risk to alter both the fibrillation pathway and the structural aspects of the individual components. That is, the intermediately formed oligomer may change structure if separating/purifying it from the reaction mixture. With the SAS approach advocated here, physical separation is avoided, hence, the 'in-reaction-intermediate' structure is elucidated (Fig. 9.2c).

Even with prior structural knowledge about two of the three components, the actual decomposition is not trivial. The bottleneck, in our experience, is the first approximation of a scattering curve representing the intermediate species. Different principles to do this have been applied both by others and us. Originally, in an analysis of insulin fibrillation (Vestergaard et al. 2007), we used the program SVDPLOT (Konarev et al. 2003) to estimate that the number of components was three. We assumed that two of these species were represented by the first and last measured experimental scattering curves, respectively. A large set of theoretical models, spanning from dimers to hexadecamers, was created by simply building various geometric oligomer conformations (e.g. elongated, compact, symmetric, asymmetric) in Pymol (The PyMOL Molecular Graphics System). These models hence represented a pool of possible conformations for the oligomeric species, and a theoretical SAXS curve was calculated for each model. Using the program OLIGOMER (Konarev et al. 2003) each of these theoretical curves were linearly combined with the initial monomeric and the final fibrils experimental SAXS curves, aiming to fit all the curves in the elongation phase. The resulting calculated volume fractions for all time points of the initial and final curves were then fixed, and the residual scattering isolated and averaged, resulting in a 'model-free' decomposition, which then is further refined, following the same procedure (Vestergaard et al. 2007).

Oliveira et al. (2009) used an experimental scattering curve recorded in the middle of the time development as a first approximation of the intermediate species, which subsequently was iteratively improved using weighted least-squares methods. A third approach, advocated by us (Giehm et al. 2011) makes an initial (deliberately erroneous) assumption that curves can be fitted by two components (monomer and fibril), knowing that there will then be large residuals at the time points where an intermediately formed species contributes strongly to the total scattering. The average residual scattering curves at these central time points are then a first approximation of the intermediate species.

Clearly, such a manual approach to the decomposition process can only be applied in simple cases, where, firstly, a low number of species are contributing to the total scattering curve, and secondly, prior information is available.

In some cases, the assumption that the initial and final experimental scattering curves are suitable representatives of the scattering contributors along the fibrillation pathway is not valid, and no or very little other prior information exists. In such cases, an objective and (semi-)automated rigorous method for the decomposition process is needed. We have recently developed such a method, and implemented it into the software named COSMiCS (Complex Objective Structural analysis of Multi-Component Systems) (Herranz-Trillo et al. 2017), applying principles from chemometrics analysis (Tauler 1995; Jaumot et al. 2005). Based on the popular MCR-ALS (multivariate curve resolution using alternating least squares) approach, we show that SAXS data curves can efficiently be decomposed and that the inherent ambiguity can be significantly reduced by multiplying the dataset with commonly used SAXS data representations, which zoom in on different q-ranges. Here, hence, the wide q-range feature of SAXS data, mentioned above, gives particular advantage in the decomposition process (Herranz-Trillo et al. 2017). The program enables decomposition of data sets, which (in our hands) could not be decomposed by the manual approach. Importantly, neither the initial nor the final states need to be homogeneous, and no prior assumptions are imposed, apart from simple closure and non-negativity restrains. In addition, the software can be expanded to handle simultaneous decomposition of data from orthogonal techniques (e.g. SAXS in combination with circular dichroism data, absorbance data or fluorescence spectra). The method is thus widely applicable, also to non-fibrillating systems. A prerequisite for the method is availability of adequate numbers of observations. If preparing for a fibrillation analysis, applying COSMiCS, a large number of data-points should hence be collected (Herranz-Trillo et al. 2017).

## 9.3 Current State of the Art

An early real-time study of the fibrillation process of beta-lactoglobulin was based on SANS data recorded for several hours from the monomeric state, incubated at 80 °C in a deuterated buffer, revealing the time dependence of the concentration of the monomeric and fibril protein respectively (Arnaudov et al. 2006). Our first SAXS-based study of insulin fibrillation represents the first ever structure determination of a transiently formed amyloid species (oligomer), investigated without perturbing the fibrillation mixture. The intermediately formed oligomeric species turned out to be an elongated, twisting structure, which existed in solution in surprisingly high relative concentrations (Vestergaard et al. 2007). A nucleus is per definition only present in minute amounts, but is expected to stabilize and accumulate at high protein concentrations (Powers and Powers 2006). We observed that the increase in fibril material and in oligomer concentration were correlated, suggesting that the intermediate oligomer is a building block of insulin fibrils (Vestergaard et al. 2007). Several other studies have come after, following the same basic principles, or variants hereof. Chatani et al. investigated insulin fibrillation under experimental conditions, different from those applied by us. As previously mentioned, using an in situ approach (Chatani et al. 2015), they observed the formation of early larger oligomers, preceding an actual nucleation of fibrillation. As previously noted, a potential role of radiation damage has not been investigated in that case. A very interesting work on glucagon fibrillation (Oliveira et al. 2009) shows how also in this case, intermediate prefibrillar species are formed at relatively high concentrations during the fibrillation (Oliveira et al. 2009). Two different initial protein concentrations were studied, proving that this does not affect the intermediate species involved in the fibrillation. In accordance with a previous study, where mature fibrils were described as helically twisted cylinders (Dong et al. 2006), the SAXS data from the fibrillation process were fitted using simple geometric bodies (cylinders with different length and radius).

An intermediately formed oligomeric structure was also revealed during the fibrillation of the PD related α-synuclein (aSN). Again, fibril mass accumulated most rapidly while large amounts of oligomers were present in solution, and observation of the relative dimensions of the oligomers and fibrils evoked the suggestion that fibril elongation can happen *via* oligomer stacking (Giehm et al. 2011). In a more recent study aSN fibrillation is stimulated by the addition of a 2-pyridone molecule (FN075) (Nors

Perdersen et al. 2015), a compound that had previously been shown to stabilize intermediately formed structures (Horvath et al. 2012). We showed that there is a direct correlation between the concentrations of oligomeric structures and fibrils, thereby strongly supporting the theory that oligomers incorporate into fibril structures (Nors Pedersen et al. 2015). Others have also investigated aSN oligomeric structures, which were formed by mechanic stress and isolated using size exclusion chromatography (Lorenzen et al. 2014). Clearly these oligomers are distinct from the oligomers that have been investigated without disturbing the fibrillation equilibrium, which emphasizes the importance of carefully choosing the experimental approach when investigating amyloid fibrillation. We do not intend to infer that the oligomers investigated by Lorenzen et al. are irrelevant, simply, we highlight that different sample treatment will also cause differences in the structures that are present during amyloid reactions. And finally, most recently, we have published the COSMiCS based decomposition SAXS data from the fibrillation pathway of a familial mutant of aSN (A53T), for which we show that the protein follows an alternative fibrillation pathway, passing through a structural state, distinct from the previously demonstrated oligomeric state. Rather, the intermediate state is an exceedingly large and apparently (partially) unstructured intermediate state, reminiscent of a membraneless organelle (Herranz-Trillo et al. 2017).

Transthyretin (TTR) is another amyloid protein responsible for several diseases as familial amyloid polyneuropathy and TTR amyloid cardiomyopathy During TTR fibrillation, and in contrast to the previously mentioned studies, our data are in accordance with an elongation model with a monomeric building block. This same monomeric state interchanges between the protofibril and soluble state, causing maturation of the final fibril structure (Groenning et al. 2015). This observation is potentially of great importance since such a monomeric dynamically interchanging structure could be responsible for the autocatalytic effect of fibrils.

A final SAXS-based study to be mentioned here is the study of the familial mutant $aSN_{A30P}$. The SAXS data clearly revealed that in this case, the assumption that the SAXS data are additive was not valid. A significant, gradual change in the scattering was observed, which did not derive from the occurrence of an additional structure, but rather, we suggested, reflected changes in the overall solvent:protein interface, as corroborated by several complementary biophysical observations (Nielsen et al. 2013). This is thus an example, where the complexity of the data makes it impossible to apply the decomposition approach, otherwise advocated here.

## 9.4 Experimental Procedures: Tips and Details

### 9.4.1 Suggestions for Sample Preparations

As in any SAS experiment, (see elaborate advice elsewhere in this book) sample preparation is a crucial step for a successful development of the experiment (Grishaev 2012; Skou et al. 2014a). Some specific considerations, however, apply when dealing with protein fibrillation, which we will dwell on here (Fig. 9.3). Overall, sample conditions must be chosen compromising between the fibrillation process and the SAS experiment and, of course, optimized for having reproducible samples. Buffers, protein concentration and temperature of fibrillation have to be selected such that the fibrillation process happens within a timeframe applicable within available beamtime (which evidently is less restrictive if collecting data on a home-source setup), but long enough to ensure a large number of measurements.

An optimal experimental strategy is shown schematically in Fig. 9.3 and basically it includes three different decisions:

(i) Which is the best buffer to use? In a SAS experiment, contents of the buffer can vary significantly, but not without restrictions. In general, the buffer scattering signal should

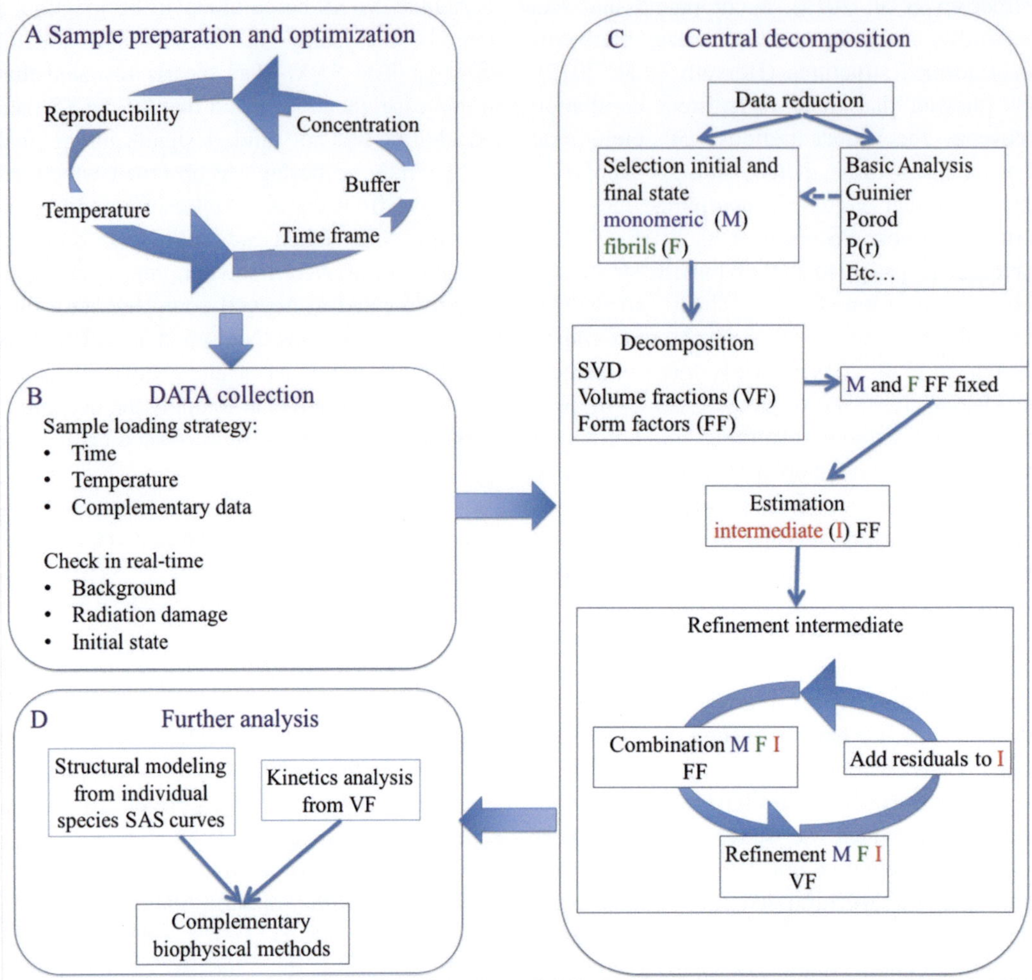


**Fig. 9.3** Scheme depicting the four steps that are necessary for studying a fibrillation process by SAS. (**a**) Sample preparation: buffers, concentrations, temperatures etc., must be chosen compromising between fibrillation time and the signal:noise of the scattering experiment while optimizing reproducibility. (**b**) Data collection: for avoiding any artifacts and misinterpretation of the data, it is necessary to check several parameters, incl. radiation damage, initial conditions, cleaning and more, during the data acquisition. (**c**) Decomposition: it is possible to separate the single curves of the species in solution performing a specific analysis (detailed in Sect. 9.2.3). If sufficient data points are known, the program COSMiCS can also be applied. (**d**) Further analysis: SAS information can be integrated with complementary data. The new program, COSMiCS, will widely improve this orthogonal analysis (Herranz-Trillo et al. 2017)

be minimized, relative to the protein signal, to ensure a good signal:noise. The concentration of strong scatterers (e.g. heavy atoms, phosphate) should be kept to a minimum. At the same time, buffers disfavoring structure factors (i.e. attraction and repulsion between the scatterers) are preferable (Zhang et al. 2007), since ignoring structure factors is a prerequisite for the decomposition process. The presence of lipids, detergents, and sugars (unless they are necessary for specific studies) should be closely controlled, since these compounds can assemble into micelles with a specific scattering signal (Kawaguchi et al. 1991; Thiyagarajan and Tiede 1994; Lipfert et al.

2007; Sanchez et al. 2011) (this phenomenon is well-known in studies of membrane proteins (Berthaud et al. 2012)). As all the above-mentioned conditions likely influence the actual fibrillation pathway and/or kinetics, these considerations must be made within the already mentioned demands on a relevant timeframe of fibrillation.

(ii) Which is the optimal protein concentration? A protein concentration series of fibrillation reactions has to be performed, compromising between the wish for an increased SAXS signal (higher concentration) and the risk for inducing structure factors or unwanted aggregation pathways (e.g. amorphous aggregation). Overall, the concentrations applied are relatively high, compared to the typical conditions applied in a complementary biophysical analysis of protein fibrillation. Taken together with the necessity of having several data points, this results in a considerable quantitative sample demand. It is desirable to avoid fibrillation at time zero, to ensure a proper measurement of the initial structural state, which normally implies that very high protein concentrations should be avoided. Again, the timeframe of the experiment must be optimized, which is highly concentration dependent. The quantitative demand is generally higher for a neutron experiment, in particular when performing contrast-match series, and due to the larger sample cells.

(iii) How will the experiment be performed? Several additional biophysical parameters influence fibrillation kinetics, and hence should be considered when preparing the experiment. The samples can be prepared either in a single volume, extracting samples from this vial at different time points, or, as advocated by us here, one may perform the fibrillation in a fluorescence plate reader (brought to the facility where the data are collected), and extract the samples from individual wells. Although both methods allow that one follows the developing ThT signal, this is certainly done with much greater ease in a plate reader, and an additional advantage is that the individual volumes in the wells are the same for all time points. When extracting from a single reaction vial, the volume:surface ratio gradually changes, and it is unforeseeable but likely, that this will also alter the fibrillation kinetics during the measurement time.

Tunable parameters influencing fibrillation kinetics, apart from the buffer composition, include mechanical stress (e.g. shaking) and temperature (Ikenoue et al. 2014), and these are easily varied in the fluorescence plate reader. The sensitivity to these parameters, however, poses an additional constrain on the experiment, since large temperature differences between the reaction vial and the SAS sample cell (and while transporting the sample between the two) must be avoided.

In addition, care should be taken, that the flow into the SAXS sample cell does not induce alignment of the fibrils in the sometimes highly viscous sample.

### 9.4.2 Suggestions for Data Collections

To optimize data quality, one should consider a few points during the data acquisition.

(i) Relative speed of conformational changes. The goal of the experiment is to follow the conformational changes of the protein during the fibrillation. Of course it is expected to have different signal at different time points but the structural changes investigated must be on timeframes that are significantly longer than the measurement times, such that one can assume that the recorded data represent a given state at a given time. The timeframe of the entire experiment must also be long enough to allow for collection of numerous data points, which is a prerequisite for a proper decomposition process. The lowest number of datapoints recorded in our own experience is 15, but a larger number (preferably >50 timepoints) is recommended. Particularly when applying the automated decomposition method COSMiCS, a large number of timepoints is recommended (Herranz-Trillo et al. 2017).

(ii) A homogeneous initial state. Since one of the crucial assumptions for the manual decomposition method is to consider the initial state as homogeneous, a control of the initial state is mandatory. If the starting state is not monodisperse, at least it should be polydisperse in a controlled and known manner. Furthermore, as discussed in Sect. 9.1.2, changes in the initial state can promote alternative fibrillation pathways (e.g. Curtain et al. 2015). Hence, it is recommended to include a thorough measurement of SAS data from a concentration series of the initial state prior to initiating the fibrillation reaction, and to investigate the initial state further using complementary data (e.g. dynamic light scattering, native gel electrophoresis, or size exclusion chromatography).

(iii) Fibrils are viscous and sticky. A general recommendation in SAXS experiments is to measure the buffer before and after each sample (Grishaev 2012), and this is particularly important for amyloid experiments. The fibril sample is often viscous and sticky, and frequent checks of the cleanness of the SAS capillary are recommended. Indeed it may be necessary to elaborate the cleaning process for avoiding protein residuals in the capillary.

(iv) Radiation damage may be confused with fibrillation. As mentioned previously, data can be recorded in situ (Oliveira et al. 2009; Chatani et al. 2015), but the risk for radiation induced sample damage must be taken into account, when applying synchrotron radiation (Kumta and Tappel 1961; Kuwamoto et al. 2004; Jeffries et al. 2015). At a first sight, it can be difficult to distinguish radiation induced amorphous aggregation from amyloid fibrils. Radical scavengers (e.g. DTT, TCEP or glycerol) can diminish the problem, but may alter the fibrillation pathway and hence must be included when initially optimizing the experimental conditions prior to the data collection. It is strongly recommended to use multi-frame data collection mode to monitor (and avoid) the potential radiation damage, i.e. since exposure times are short compared to the fibrillation time, the scattering signal should remain the same during these multiple exposures.

## 9.5 The Power of SAS in Hybrid Approaches to Study Protein Fibrillation

The synergy between SAS and other structural and biophysical methods is significant, as also addressed elsewhere in this book. Here, we will dwell only on hybrid approaches that in our opinion are particularly relevant for fibrillation studies.

### 9.5.1 Further the Decomposition Approach by Inclusion of Orthogonal Data

Circular dichroism (CD) and Fourier transform infrared spectroscopy (FTIR) are relevant methods for studying the secondary structural content of proteins in solution, and hence allow to follow the changes occurring during the fibrillation reaction (Bouchard et al. 2000; Arora et al. 2004; Librizzi and Rischel 2005; Miller et al. 2013). In addition, some synchrotrons now provide access to advanced laboratory facilities in the close vicinity of the beamlines (Boivin et al. 2016), enabling that these measurements are made on the same samples as those used for SAXS data collection. An alternative is to bring the equipment to the beamline, or e.g. use SRCD (synchrotron radiation CD) and SAXS from the same synchrotron facility, thereby avoiding batch variations. The advantage of obtaining data from the exact same samples is evident since this allows correlation of the occurrence of intermediate structures (assessed by SAS) with changes in e.g. secondary structural contents. COSMiCS further allows the incorporation of such complementary data sets, recorded along the same reaction coordinate (Herranz-Trillo et al. 2017), which will very efficiently reduce the ambiguity of the decomposition solution. Indeed, it should be possible also to decompose data from even more complex aggregating systems, e.g. in the presence of an additional protein co-factor, or in the presence of model lipid systems, potentially influencing the aggregation pathway. However, this has, to the best of our knowledge, not yet been demonstrated. Note, also, that if the investigated system is even more complex, additional data may be needed, either by collecting data from the same system but under different experimental conditions, or by adding data from complementary sources. This is also described in our recent publication about COSMiCS (Herranz-Trillo et al. 2017).

Another novel possibility lies in the coupling of high-resolution single particle cryo-electron microscopy (cryo-EM) and SAXS solution studies. The revolutionary hardware and software developments within cryo-EM now enable near-atomic resolution structure determination, and in addition enable high-resolution descriptions of mixed states (Scheres 2012; Nogales and Scheres 2015). It would be of evident interest to study fibrillation mixtures with cryo-EM. Solution studies under varying experimental conditions (potentially towards physiologically relevant conditions), could complement/validate the relevance and/or dominance of certain structural states, present in the mixtures under the experimental conditions applied during the cryo-EM analysis, with a significantly less laborious approach.

### 9.5.2 Use SAS as a Bridge Between Orthogonal Methods

SAS data contain information covering several orders of magnitude of length scales. This also means that the method brilliantly bridges from high- to low-resolution methods. One example is the study of the GNNQQNY peptide fragment from yeast prion protein (Langkilde et al. 2015). In this case, we combine SAXS, FD, negative stain transmission electron microscopy (TEM) and high-resolution X-ray crystallography. Information from the SAXS data facilitates detailed analysis of the diffraction data, and SAXS and TEM together elucidate the hierarchical nature of the ribbon formed fibrils. In addition, we show that the fibrillation of this peptide proceeds without accumulation of any intermediate oligomeric species, in contrast to e.g. the insulin, aSN and glucagon studies (Oliveira et al. 2009; Giehm et al. 2011; Groenning et al. 2015).

SAXS data also play a decisive role for the interpretation of functional data, on the interaction between aSN and lipid vesicles (van Maarschalkerweerd et al. 2014, 2015). Changes in vesicle morphology upon protein addition was followed in real-time by two-photon microscopy (2PM), SAXS and CD data, and the SAXS data reveal the occurrence of a new structural species, interpreted in the context of the 2PM data as a lipid:protein co-aggregate. It was demonstrated that early amyloidogenic states, and not the fibril state, potently disrupt vesicles. Evidently, SANS contrast variation data would enable further analysis of such a co-aggregated species.

In such a case, it could be relevant to attempt separation of this species from the reaction mixture, in order to improve the signal:noise of the SAS data. SAXS, coupled to size exclusion chromatography (SEC-SAXS) (David and Pérez 2009; Graewert et al. 2015; Malaby et al. 2015) can dramatically improve data quality from partially aggregated or polydisperse protein solutions and is particularly relevant when isolating species with limited stability. In principle, SEC-SAXS can also be used for trying to isolate the different species that are formed during the fibrillation, but here it is very relevant to consider whether the species of interest is an on-pathway/in-equilibrium species, which thus will either disappear or change structure upon isolation.

### 9.5.3 Moving Toward Higher Resolution in Time and Space

Within structural analysis of protein fibrillation, on-going development in e.g. detector technology, sample environments, and software (Round et al. 2008; Toft et al. 2008; Hura et al. 2009; Martel et al. 2012; Russel et al. 2012; Jain et al. 2013; Skou et al. 2014b; Blanchet et al. 2015; Graewert et al. 2015) promises for future increases in the resolution, both in terms of time and space. Indeed, these developments bridge towards other related technologies such as wide-angle X-ray scattering (WAXS), and the immense developments around the XFEL facilities. For fibrillation analysis, some of the relatively low-hanging fruits, which we envision could facilitate experiments resulting in significant scientific insight for the fibrillation community, would be further development of advanced sample environments, enabling simultaneous recording of data from several complementary sources (Bras et al. 2014; Haas et al. 2014; Blanchet et al. 2015; Graewert et al. 2015). Time will show what shall become possible, but expectations are high, that the community in the not too distant future will gain the insight necessary to structurally describe the molecular basis of protein fibrillation.

**Acknowledgements** The authors acknowledge financial support from the Copenhagen interfaculty collaborative program CONEXT, The Danish Council for Independent Research Sapere Aude project SAFIR, DANSCATT and the Lundbeck Foundation Initiative BRAINSTRUC. A warm thank you is extended to great support and interest from beamline staff at beamlines X33 and P12 (EMBL-Hamburg); BM29 (ESRF) and I911-4 (MAXlab) and to all collaborators on the referenced projects.

## References

Arnaudov LN, de Vries R, Stuart MAC (2006) Time-resolved small-angle neutron scattering during heat-induced fibril formation from bovine beta-lactoglobulin. J Chem Phys 124(8):084701

Arora A, Ha C, Park CB (2004) Insulin amyloid fibrillation at above 100 degrees C: new insights into protein folding under extreme temperatures. Protein Sci 13 (9):2429–2436

Berthaud A, Manzi J, Perez J, Mangenot S (2012) Modeling detergent organization around aquaporin-0 using small-angle X-ray scattering. J Am Chem Soc 134(24):10080–10088

Blanchet CE, Spilotros A, Schwemmer F, Graewert MA, Kikhney A, Jeffries CM, Franke D, Mark D, Zengerle R, Cipriani F, Fiedler S, Roessle M, Svergun DI (2015) Versatile sample environments and automation for biological solution X-ray scattering experiments at the P12 beamline (PETRA III, DESY). J Appl Crystallogr 48(Pt 2):431–443

Boivin S, Kozak S, Rasmussen G, Nemtanu IM, Vieira V, Meijers R (2016) An integrated pipeline for sample preparation and characterization at the EMBL@PETRA3 synchrotron facilities. Methods 95:70–7. doi:10.1016/j.ymeth.2015.08.003

Bolder SG, Sagis LM, Venema P, van der Linden E (2007) Thioflavin T and birefringence assays to determine the conversion of proteins into fibrils. Langmuir 23(8):4144–4147

Bouchard M, Zurdo J, Nettleton EJ, Dobson CM, Robinson CV (2000) Formation of insulin amyloid fibrils followed by FTIR simultaneously with CD and electron microscopy. Protein Sci 9 (10):1960–1967

Bras W, Koizumi S, Terrill NJ (2014) Beyond simple small-angle X-ray scattering: developments in online complementary techniques and sample environments. IUCrJ 1(Pt 6):478–491

Chatani E, Inoue R, Imamura H, Sugiyama M, Kato M, Yamamoto M, Nishida K, Kanaya T (2015) Early aggregation preceding the nucleation of insulin amyloid fibrils as monitored by small angle X-ray scattering. Sci Rep 5:15485

Chiti F, Dobson CM (2006) Protein misfolding, functional amyloid, and human disease. Annu Rev Biochem 75:333–366

Cohen AS, Calkins E (1959) Electron microscopic observations on a fibrous component in amyloid of diverse origins. Nature 183(4669):1202–1203

Cohen AS, Calkins E (1964) The isolation of amyloid fibrils and a study of the effect of collagenase and hyaluronidase. J Cell Biol 21:481–486

Curtain CC, Kirby NM, Mertens HD, Barnham KJ, Knott RB, Masters CL, Cappai R, Rekas A, Kenche VB, Ryan T (2015) Alpha-synuclein oligomers and fibrils originate in two distinct conformer pools: a small angle X-ray scattering and ensemble optimisation modelling study. Mol BioSyst 11(1):190–196

David G, Pérez J (2009) Combined sampler robot and high-performance liquid chromatography: a fully automated system for biological small-angle X-ray scattering experiemnts at the Synchrotron SOLEIL SWING beamline. J Appl Crystallogr 42:892–900

Divry P, Florkin M (1927) Sur les proprietes optiques de l'amyloid. Soc Biol 97:180–181

Dong MD, Hovgaard MB, Xu SL, Otzen DE, Besenbacher F (2006) AFM study of glucagon

fibrillation *via* oligomeric structures resulting in interwoven fibrils. Nanotechnology 17(16):4003–4009
Dueholm MS, Nielsen SB, Hein KL, Nissen P, Chapman M, Christiansen G, Nielsen PH, Otzen DE (2011) Fibrillation of the major curli subunit CsgA under a wide range of conditions implies a robust design of aggregation. Biochemistry 50 (39):8281–8290
Eanes ED, Glenner GG (1968) X-ray diffraction studies on amyloid filaments. J Histochem Cytochem 16 (11):673–677
Fandrich M (2007) On the structural definition of amyloid fibrils and other polypeptide aggregates. Cell Mol Life Sci 64(16):2066–2078
Ferrone F (1999) Analysis of protein aggregation kinetics. Methods Enzymol 309:256–274
Fichou Y, Schiro G, Gallat FX, Laguri C, Moulin M, Combet J, Zamponi M, Hartlein M, Picart C, Mossou E, Lortat-Jacob H, Colletier JP, Tobias DJ, Weik M (2015) Hydration water mobility is enhanced around tau amyloid fibers. Proc Natl Acad Sci U S A 112(20):6365–6370
Fodera V, Librizzi F, Groenning M, van de Weert M, Leone M (2008) Secondary nucleation and accessible surface in insulin amyloid fibril formation. J Phys Chem B 112(12):3853–3858
Friedrich N (1859) A. Zur amyloidfrage. Arch Pathol Anat Physiol Klin Med 16:50–65
Gabel F, Bicout D, Lehnert U, Tehei M, Weik M, Zaccai G (2002) Protein dynamics studied by neutron scattering. Q Rev Biophys 35(4):327–367
Giehm L, Svergun DI, Otzen DE, Vestergaard B (2011) Low-resolution structure of a vesicle disrupting alpha-synuclein oligomer that accumulates during fibrillation. Proc Natl Acad Sci U S A 108(8):3246–3251
Golub GH, Reinsch C (1970) Singular value decomposition and least squares solutions. Numer Math 14 (5):403.
Graewert MA, Franke D, Jeffries CM, Blanchet CE, Ruskule D, Kuhle K, Flieger A, Schafer B, Tartsch B, Meijers R, Svergun DI (2015) Automated pipeline for purification, biophysical and x-ray analysis of biomacromolecular solutions. Sci Rep 5:10734
Grishaev A (2012) Sample preparation, data collection, and preliminary data analysis in biomolecular solution X-ray scattering. Curr Protoc Protein Sci Chapter 17: Unit17 14
Groenning M, Campos RI, Hirschberg D, Hammarstrom P, Vestergaard B (2015) Considerably unfolded transthyretin monomers preceed and exchange with dynamically structured amyloid protofibrils. Sci Rep 5:11443
Haas S, Plivelic TS, Dicko C (2014) Combined SAXS/UV-vis/Raman as a diagnostic and structure resolving tool in materials and life sciences applications. J Phys Chem B 118(8):2264–2273
Hardy JA, Higgins GA (1992) Alzheimer's disease: the amyloid cascade hypothesis. Science 256 (5054):184–185
Harper JD, Lansbury PT Jr (1997) Models of amyloid seeding in Alzheimer's disease and scrapie: mechanistic truths and physiological consequences of the time-dependent solubility of amyloid proteins. Annu Rev Biochem 66:385–407
Hawe A, Sutter M, Jiskoot W (2008) Extrinsic fluorescent dyes as tools for protein characterization. Pharm Res 25(7):1487–1499
Herranz-Trillo F, Groenning M, van Maarschalkerweerd A, Tauler R, Vestergaard B, Bernado P (2017) Structural analysis of multi-component amyloid systems by chemometric SAXS data decomposition. Structure 25(1):5–15
Hong P, Koza S, Bouvier ES (2012) Size-exclusion chromatography for the analysis of protein biotherapeutics and their aggregates. J Liq Chromatogr Relat Technol 35(20):2923–2950
Horvath I, Weise CF, Andersson EK, Chorell E, Sellstedt M, Bengtsson C, Olofsson A, Hultgren SJ, Chapman M, Wolf-Watz M, Almqvist F, Wittung-Stafshede P (2012) Mechanisms of protein oligomerization: inhibitor of functional amyloids templates alpha-synuclein fibrillation. J Am Chem Soc 134 (7):3439–3444
Hotelling H (1933) Analysis of a complex of statistical variables into principal components. J Educ Psychol 24:417–441
Hura GL, Menon AL, Hammel M, Rambo RP, Poole FL 2nd, Tsutakawa SE, Jenney FE Jr, Classen S, Frankel KA, Hopkins RC, Yang SJ, Scott JW, Dillard BD, Adams MW, Tainer JA (2009) Robust, high-throughput solution structural analyses by small angle X-ray scattering (SAXS). Nat Methods 6 (8):606–612
Ikenoue T, Lee YH, Kardos J, Saiki M, Yagi H, Kawata Y, Goto Y (2014) Cold denaturation of alpha-synuclein amyloid fibrils. Angew Chem Int Ed Eng 53(30):7799–7804
Jacques DA, Guss JM, Svergun DI, Trewhella J (2012) Publication guidelines for structural modelling of small-angle scattering data from biomolecules in solution. Acta Crystallogr D Biol Crystallogr 68 (Pt 6):620–626
Jain R, Petri M, Kirschbaum S, Feindt H, Steltenkamp S, Sonnenkalb S, Becker S, Griesinger C, Menzel A, Burg TP, Techert S (2013) X-ray scattering experiments with high-flux x-ray source coupled rapid mixing microchannel device and their potential for high-flux neutron scattering investigations. Eur Phys J E Soft Matter 36(9):109
Jarrett JT, Lansbury PT Jr (1993) Seeding one-dimensional crystallization of amyloid: a pathogenic mechanism in Alzheimer's disease and scrapie? Cell 73(6):1055–1058
Jaumot J, Gargallo R, de Juan A, Tauler R (2005) A graphical user-friendly interface for MCR-ALS: a new tool for multivariate curve resolution in MATLAB. Chemom Intell Lab Syst 76 (1):101–110

Jaumot J, Marchan V, Gargallo R, Grandas A, Tauler R (2004) Multivariate curve resolution applied to the analysis and resolution of two-dimensional [H-1, N-15] NMR reaction spectra. Anal Chem 76 (23):7094–7101

Jeffries CM, Graewert MA, Svergun DI, Blanchet CE (2015) Limiting radiation damage for high-brilliance biological solution scattering: practical experience at the EMBL P12 beamline PETRAIII. J Synchrotron Radiat 22(Pt 2):273–279

Kawaguchi T, Hamanaka T, Kito Y, Machida H (1991) Structural studies of a homologous series of Alkyl sucrose ester micelle by x-ray-scattering. J Phys Chem 95(9):3837–3846

Kayed R, Head E, Thompson JL, McIntire TM, Milton SC, Cotman CW, Glabe CG (2003) Common structure of soluble amyloid oligomers implies common mechanism of pathogenesis. Science 300 (5618):486–489

Khurana R, Ionescu-Zanetti C, Pope M, Li J, Nielson L, Ramirez-Alvarado M, Regan L, Fink AL, Carter SA (2003) A general model for amyloid fibril assembly based on morphological studies using atomic force microscopy. Biophys J 85(2):1135–1144

Konarev PV, Volkov VV, Sokolova AV, Koch MHJ, Svergun DI (2003) PRIMUS: a windows PC-based system for small-angle scattering data analysis. J Appl Crystallogr 36:1277–1282

Kumta US, Tappel AL (1961) Radiation damage to proteins. Nature 191:1304–1305

Kuwamoto S, Akiyama S, Fujisawa T (2004) Radiation damage to a protein solution, detected by synchrotron x-ray small-angle scattering: dose-related considerations and suppression by cryoprotectants. J Synchrotron Radiat 11(Pt 6):462–468

Langkilde AE, Morris KL, Serpell LC, Svergun DI, Vestergaard B (2015) The architecture of amyloid-like peptide fibrils revealed by x-ray scattering, diffraction and electron microscopy. Acta Crystallogr Sect D-Biol Crystallogr 71:882–895

Langkilde AE, Vestergaard B (2012) Structural characterization of prefibrillar intermediates and amyloid fibrils by small-angle x-ray scattering. Methods Mol Biol 849:137–155

Lee CC, Nayak A, Sethuraman A, Belfort G, McRae GJ (2007) A three-stage kinetic model of amyloid fibrillation. Biophys J 92(10):3448–3458

LeVine H 3rd (1993) Thioflavine T interaction with synthetic Alzheimer's disease beta-amyloid peptides: detection of amyloid aggregation in solution. Protein Sci 2(3):404–410

Li Y, Lubchenko V, Vekilov PG (2011) The use of dynamic light scattering and brownian microscopy to characterize protein aggregation. Rev Sci Instrum 82 (5):053106

Librizzi F, Rischel C (2005) The kinetic behavior of insulin fibrillation is determined by heterogeneous nucleation pathways. Protein Sci 14(12):3129–3134

Lipfert J, Columbus L, Chu VB, Lesley SA, Doniach S (2007) Size and shape of detergent micelles determined by small-angle x-ray scattering. J Phys Chem B 111(43):12427–12438

Lorenzen N, Nielsen SB, Buell AK, Kaspersen JD, Arosio P, Vad BS, Paslawski W, Christiansen G, Valnickova-Hansen Z, Andreasen M, Enghild JJ, Pedersen JS, Dobson CM, Knowles TPJ, Otzen DE (2014) The role of stable alpha-synuclein oligomers in the molecular events underlying amyloid formation. J Am Chem Soc 136(10):3859–3868

Lucking CB, Brice A (2000) Alpha-synuclein and Parkinson's disease. Cell Mol Life Sci 57 (13–14):1894–1908

van Maarschalkerweerd A, Vetri V, Langkilde AE, Fodera V, Vestergaard B (2014) Protein/lipid coaggregates are formed during alpha-synuclein-induced disruption of lipid bilayers. Biomacromolecules 15(10):3643–3654

van Maarschalkerweerd A, Vetri V, Vestergaard B (2015) Cholesterol facilitates interactions between alpha-synuclein oligomers and charge-neutral membranes. FEBS Lett 589(19):2661–2667

Malaby AW, Chakravarthy S, Irving TC, Kathuria SV, Bilsel O, Lambright DG (2015) Methods for analysis of size-exclusion chromatography-small-angle x-ray scattering and reconstruction of protein scattering. J Appl Crystallogr 48(Pt 4):1102–1113

Martel A, Liu P, Weiss TM, Niebuhr M, Tsuruta H (2012) An integrated high-throughput data acquisition system for biological solution x-ray scattering studies. J Synchrotron Radiat 19(Pt 3):431–434

Miller LM, Bourassa MW, Smith RJ (2013) FTIR spectroscopic imaging of protein aggregation in living cells. Biochim Biophys Acta 1828(10):2339–2346

Morshedi D, Rezaei-Ghaleh N, Ebrahim-Habibi A, Ahmadian S, Nemat-Gorgani M (2007) Inhibition of amyloid fibrillation of lysozyme by indole derivatives – possible mechanism of action. FEBS J 274 (24):6415–6425

Nielsen L, Khurana R, Coats A, Frokjaer S, Brange J, Vyas S, Uversky VN, Fink AL (2001) Effect of environmental factors on the kinetics of insulin fibril formation: elucidation of the molecular mechanism. Biochemistry 40(20):6036–6046

Nielsen SB, Macchi F, Raccosta S, Langkilde AE, Giehm L, Kyrsting A, Svane ASP, Manno M, Christiansen G, Nielsen NC, Oddershede L, Vestergaard B, Otzen DE (2013) Wildtype and A30P mutant alpha-synuclein form different fibril structures. PLoS One 8(7):e67713

Nilsson MR (2004) Techniques to study amyloid fibril formation in vitro. Methods 34(1):151–160

Nogales E, Scheres SH (2015) Cryo-EM: a unique tool for the visualization of macromolecular complexity. Mol Cell 58(4):677–689

Nors Perdersen M, Fodera V, Horvath I, van Maarschalkerweerd A, Norgaard Toft K, Weise C,

Almqvist F, Wolf-Watz M, Wittung-Stafshede P, Vestergaard B (2015) Direct correlation between ligand-induced alpha-synuclein oligomers and amyloid-like fibril growth. Sci Rep 5:10422

Oliveira CLP, Behrens MA, Pedersen JS, Erlacher K, Otzen D, Pedersen JS (2009) A SAXS study of glucagon fibrillation. J Mol Biol 387(1):147–161

Otzen D (2010) Functional amyloid: turning swords into plowshares. Prion 4(4):256–264

Piccoli PMB, Koetzle TF, Schultz AJ (2007) Single crystal neutron diffraction for the inorganic chemist – a practical guide. Comments Inorg Chem 28(1–2):3–38

Powers ET, Powers DL (2006) The kinetics of nucleated polymerizations at high concentrations: amyloid fibril formation near and above the "supercritical concentration". Biophys J 91(1):122–132

Ross CA, Poirier MA (2004) Protein aggregation and neurodegenerative disease. Nat Med 10(Suppl):S10–S17

Round AR, Franke D, Moritz S, Huchler R, Fritsche M, Malthan D, Klaering R, Svergun DI, Roessle M (2008) Automated sample-changing robot for solution scattering experiments at the EMBL Hamburg SAXS station X33. J Appl Crystallogr 41(Pt 5):913–917

Russel D, Lasker K, Webb B, Velazquez-Muriel J, Tjioe E, Schneidman-Duhovny D, Peterson B, Sali A (2012) Putting the pieces together: integrative modeling platform software for structure determination of macromolecular assemblies. PLoS Biol 10(1): e1001244

Sanchez SA, Gratton E, Zanocco AL, Lemp E, Gunther G (2011) Sucrose monoester micelles size determined by Fluorescence Correlation Spectroscopy (FCS). PLoS One 6(12):e29278

Scheres SH (2012) RELION: implementation of a Bayesian approach to cryo-EM structure determination. J Struct Biol 180(3):519–530

Schiro G, Vetri V, Frick B, Militello V, Leone M, Cupane A (2012) Neutron scattering reveals enhanced protein dynamics in Concanavalin A amyloid fibrils. J Phys Chem Lett 3(8):992–996

Selkoe DJ (1994) Cell biology of the amyloid beta-protein precursor and the mechanism of Alzheimer's disease. Annu Rev Cell Biol 10:373–403

Serio TR, Cashikar AG, Kowal AS, Sawicki GJ, Moslehi JJ, Serpell L, Arnsdorf MF, Lindquist SL (2000) Nucleated conformational conversion and the replication of conformational information by a prion determinant. Science 289(5483):1317–1321

Skou S, Gillilan RE, Ando N (2014a) Synchrotron-based small-angle X-ray scattering of proteins in solution. Nat Protoc 9(7):1727–1739

Skou M, Skou S, Jensen TG, Vestergaard B, Gillilan RE (2014b) Microfluidic dialysis for biological small-angle x-ray scattering. J Appl Crystallogr 47 (Pt 4):1355–1366

Srinivasan R, Jones EM, Liu K, Ghiso J, Marchant RE, Zagorski MG (2003) pH-dependent amyloid and protofibril formation by the ABri peptide of familial British dementia. J Mol Biol 333(5):1003–1023

Tauler R (1995) Multivariate curve resolution applied to second order data. Chemom Intell Lab Syst 30 (1):133–146

Tauler R (2001) Calculation of maximum and minimum band boundaries of feasible solutions for species profiles obtained by multivariate curve resolution. J Chemom 15(8):627–646

Tauler R, Smilde A, Kowalski B (1995) Selectivity, local rank, 3-way data-analysis and ambiguity in multivariate curve resolution. J Chemom 9(1):31–58

The PyMOL Molecular Graphics System, Version 1.8 Schrödinger, LLC

Thiyagarajan P, Tiede DM (1994) Detergent Micelle structure and Micelle-Micelle interactions determined by small-angle neutron-scattering under solution conditions used for membrane-protein crystallization. J Phys Chem 98(40):10343–10351

Toft KN, Vestergaard B, Nielsen SS, Snakenborg D, Jeppesen MG, Jacobsen JK, Arleth L, Kutter JP (2008) High-throughput small angle x-ray scattering from proteins in solution using a microfluidic front-end. Anal Chem 80(10):3648–3654

Vestergaard B, Groenning M, Roessle M, Kastrup JS, van de Weert M, Flink JM, Frokjaer S, Gajhede M, Svergun DI (2007) A helical structural nucleus is the primary elongating unit of insulin amyloid fibrils. PLoS Biol 5(5):e134

Wang SS, Liu KN, Han TC (2010) Amyloid fibrillation and cytotoxicity of insulin are inhibited by the amphiphilic surfactants. Biochim Biophys Acta 1802 (6):519–530

Whitten AE, Trewhella J (2009) Small-angle scattering and neutron contrast variation for studying bio-molecular complexes. Methods Mol Biol 544:307–323

Wong CW, Quaranta V, Glenner GG (1985) Neuritic plaques and cerebrovascular amyloid in Alzheimer disease are antigenically related. Proc Natl Acad Sci U S A 82(24):8729–8732

Zhang F, Skoda MW, Jacobs RM, Martin RA, Martin CM, Schreiber F (2007) Protein interactions studied by SAXS: effect of ionic strength and protein concentration for BSA in aqueous solutions. J Phys Chem B 111(1):251–259

# High Resolution Distance Distributions Determined by X-Ray and Neutron Scattering

10

Henry Y.H. Tang, John A. Tainer, and Greg L. Hura


**Abstract**

Measuring distances within or between macromolecules is necessary to understand the chemistry that biological systems uniquely enable. In performing their chemistry, biological macromolecules undergo structural changes over distances ranging from atomic to micrometer scales. X-ray and neutron scattering provide three key assets for tackling this challenge. First, they may be conducted on solutions where the macromolecules are free to sample the conformations that enable their chemistry. Second, there are few limitations on chemical environment for experiments. Third, the techniques can inform upon a wide range of distances at once. Thus scattering, particularly recorded at small angles (SAS), has been applied to a large variety of phenomenon. A challenge in interpreting scattering data is that the desired three dimensional distance information is averaged onto one dimension. Furthermore, the scales and variety of phenomenon interrogated have led to an assortment of functions that describe distances and changes thereof. Here we review scattering studies that characterize biological phenomenon at distances ranging from atomic to 50 nm. We also distinguish the distance distribution functions that are commonly used to describe results from these systems. With available X-ray and neutron scattering facilities, bringing the action that occurs at



H.Y.H. Tang
Molecular Biophysics and Integrated Bioimaging, Lawrence Berkeley National Laboratory, 1 Cyclotron Rd, Berkeley, CA 94720, USA

J.A. Tainer
Molecular Biophysics and Integrated Bioimaging, Lawrence Berkeley National Laboratory, 1 Cyclotron Rd, Berkeley, CA 94720, USA

Department of Molecular and Cellular Oncology, The University of Texas M. D. Anderson Cancer Center, Houston, TX, USA

G.L. Hura (✉)
Molecular Biophysics and Integrated Bioimaging, Lawrence Berkeley National Laboratory, 1 Cyclotron Rd, Berkeley, CA 94720, USA

Department of Chemistry and Biochemistry, University of California, Santa Cruz, CA, USA
e-mail: glhura@lbl.gov






B. Chaudhuri et al. (eds.), *Biological Small Angle Scattering: Techniques, Strategies and Tips*, Advances in Experimental Medicine and Biology 1009, DOI 10.1007/978-981-10-6038-0_10

the atomic to the micrometer scale is now reasonably accessible. Notably, the combined distance and dynamic information recorded by SAS is frequently key to connecting structure to biological activity and to improve macromolecular design strategies and outcomes. We anticipate widespread utilization particularly in macromolecular engineering and time-resolved studies where many contrasting experiments are necessary for resolving chemical mechanisms through structural changes.

**Keywords**
SAXS • Pair distribution • Pair correlation • Protein structure • Resolution • Molecular ruler

## 10.1 Introduction

X-ray and neutron scattering from solutions provide information at atomic to intra-organelle distances for which microscope based visible light techniques cannot be applied. In addition scattering can be applied at any biologically relevant solution condition excellently complimenting techniques that provide greater detail but only under restrictive conditions. Significant advances in the collection and analysis of high quality data have been made over the last ten years leading to several important results. Recent examples include studies on polyketide synthase (Edwards et al. 2014), elastin (Baldock et al. 2011), photosynthesis (Stingaciu et al. 2016), chromatin (Andresen et al. 2013; Falk et al. 2015) and microbial chromatin-like systems (Hammel et al. 2016). Recent advances in genomic sequencing and purification have led to a large increase in the number of targets that would greatly benefit from increased structural understanding. Demand for structural information is outpacing the capabilities of current main stream experimental techniques. Scattering reliably provides results on most samples under nearly any solution condition and is capable of high-throughput and multi-condition analyses (Hura et al. 2009, 2013a), so the impact of these techniques can potentially be very profound.

A primary asset of X-ray and neutron scattering is the ability to elucidate relatively detailed or high resolution structural features. This fact is in contradiction to what is often stated about SAXS and SANS (collectively SAS), as a web search will quickly attest. The perception that the techniques are low resolution largely stems from the use of SAS to estimate 3D shape from macromolecules. 3D shape is a point of convergence between structural methods of macromolecular crystallography (MX), electron microscopy (EM) and nuclear magnetic resonance (NMR) techniques, so it is quite natural to compare results at this level. The resolution of SAS shapes is lower relative to other structural techniques but not because high resolution features do not contribute. SAS results do not preserve directional information and the resulting ambiguity reduces the precision of shape determination. The loss of directional information is the price paid for the ability to work effectively and efficiently in solution. Despite the lack of directional information, shapes from SAS experiments are calculated by supplementing data with assumptions about compactness and density connectivity that have proven to be valid and are valuable because of key advantages of working in solution (Franke and Svergun 2009). However, this may not be the most important asset of SAS.

A better illustration of the resolution capabilities of scattering data comes from examining differences between scattering profiles collected from a macromolecule undergoing a small structural change. To demonstrate the accuracy and multiple scales at which scattering techniques can measure, we will start by developing the formalism that is applied for most macromolecules through use of the pair distribution function or P(r). Comparisons of the P(r), attained through Fourier transformation of the primary data, provides insight into the resolution possible. Many examples of the value of this function are described in the literature on particles that are 1–10 nm in maximum dimension. Below we highlight the precision possible but rather than dwell on this length scale, for which there are several excellent tools (Semenyuk and Svergun 1991; Nielsen et al. 2009; Bergmann et al. 2000), we will explore either side of this scale. We will review scattering as a tool to measure the distances between water molecules in bulk water at atomic length scales (0.1 nm). We will also describe the use of metallic labels to characterize small changes over large distances on the order of 50 nm.

We aim to highlight how these broad length scales (0.1–50 nm) are bridged by scattering. We believe such a treatise is important since instruments with sufficient flux and large detectors are increasingly available, providing access to these regimes in a single experiment on time scales of 1 s or less. We endeavor to enable investigators to begin extending their X-ray or neutron scattering analysis to understand multiple important phenomena. In addition, we seek to clarify the differences between distance distributions extracted from each approach which can be confusing because they describe related quantities. The measurement of structural changes a macromolecule undergoes as part of its mechanism has wide spread application and scattering provides this capability with sub-nanometer resolution.

## 10.2 General Formalism of Extracting Distances from Scattering and Diffraction

Reconstructing a macromolecular structure from its coherent scattering starts with two fundamental equations. Equations 10.1 and 10.2 describe the relationship between measured scattering intensities I(**q**) and the structure of the scattering material ρ(***r***). Bold face variables indicate vectoral quantities. The quantity ρ(***r***) will either describe the electron density for X-rays or nuclear scattering length density for neutrons as a function of position from some origin r (Fig. 10.1).

$$f(\boldsymbol{q}) = \int_V \rho(\boldsymbol{r}) e^{i(\boldsymbol{r}\cdot\boldsymbol{q})} d\boldsymbol{r} \tag{10.1}$$

$$I(\boldsymbol{q}) = \sum_j \sum_k f_j(\boldsymbol{q}) f_k{}^*(\boldsymbol{q}) e^{i(\boldsymbol{r}_j - \boldsymbol{r}_k)\boldsymbol{q}} \tag{10.2}$$

In Eq. 10.1. $\boldsymbol{q} = 2\pi(\mathbf{s_o} - \mathbf{s})/\lambda$ is the vectoral change in momentum between a photon or neutron prior to its interaction with material and after its interaction with the material. The quantity ***q*** is therefore often referred to as momentum transfer. We are interested in the coherent scattering where momentum changes are solely directional,

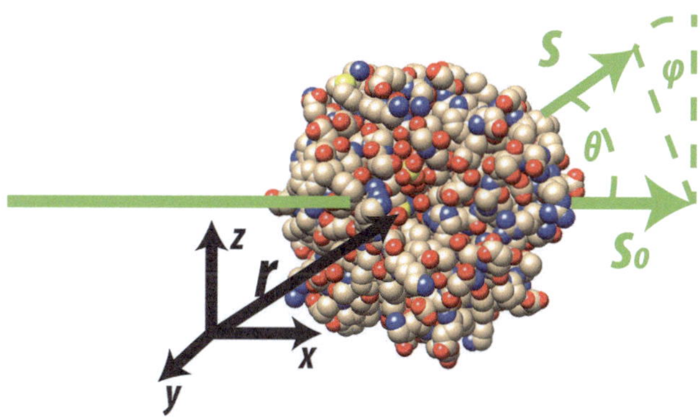


**Fig. 10.1** Coordinates for describing the scattering from an object. The coordinate system used to describe the scattering density is in *black*. The incident beam and outgoing scattering are described by the *green* vectors and coordinates

with the change in direction, $\mathbf{s_o} - \mathbf{s}$, between the incident and the outgoing angle. The vectors $\mathbf{s_o}$ and $\mathbf{s}$ are unit vectors describing the direction of the incident and outgoing scattering. The $f(\boldsymbol{q})$ function is often referred to as the "form" function and involves an integral over the volume V of the scattering particle (the portion of the sample illuminated by the incident X-ray beam).

In two common biological approaches that have been powerfully used together (Putnam et al. 2007), macromolecular crystallography and solution bioSAS from monodisperse samples, the scattering material is organized in two extreme ways. These are extreme in that with crystallography all particles are, ideally, perfectly ordered with respect to one another while in solution SAS, ideally, none of the particles are ordered in relation to one another. The organization of the material in these two different ways also distinguishes diffraction from scattering.

In most biological cases the total scattering from a macromolecule is not much stronger than the scattering of water regardless of whether neutrons or X-rays are used. Water of hydration is therefore an important contribution. However, for simplicity we neglect hydration now and will deal with some aspects later. For more complete detailed descriptions, we urge readers to consult several excellent resources (Schneidman-Duhovny et al. 2016; Koch et al. 2003; Poitevin et al. 2011).

In crystallographic diffraction, where all molecules are oriented identically, the vectoral distance of one atom relative to another is retained. In SAS, upon averaging all orientations of the macromolecules relative to the incident beam, the vectoral information is reduced to scalars. This split between the two techniques can be seen from the I($\mathbf{q}$) recorded from either type of sample. For crystallographic diffraction, $\boldsymbol{q}$ must be measured in two angular dimensions since the intensity of diffraction spots containing essential information vary in two dimensions.

In diffraction, a copy of an object located at $\mathbf{r_j}$ can be found at $\mathbf{r_k}$ oriented identically relative to the probing beam. The relation to the diffraction intensity can be made more useful when it takes into account the repeating regular translation of one scattering unit relative to another by describing $\mathbf{r_{jk}} = \mathbf{r_j} - \mathbf{r_k}$ in terms of the three crystal lattice dimensions. By replacing the general $\mathbf{r_{jk}}$ with lattice units, the equations describing crystallographic diffraction simplify. As we proceed and consider scattering from solutions, the same conditions do not hold and a different formalism is applied.

For bioSAS from dilute solutions of biomolecules, the scattering objects of interest are far enough away from one another that they may be treated as uncorrelated. There is no structure between individual particles. In this case the only term to consider in Eq. 10.2 is the self-scattering, shown in Eq. 10.3, where $(\mathbf{r_j} - \mathbf{r_k}) = 0$. With solution scattering, the components are in every orientation relative to the probing beam. Thus an extra mathematical operation is required to describe SAS data. The averaging over all orientations is indicated by the triangular brackets and is accomplished by integration over the traditional spherical coordinate system with axes r, θ and φ. This averaging results in loss of vectoral information but retention of scalar distance information. Therefore, only a one dimensional convolution of ρ($\boldsymbol{r}$) can be determined, the pair distribution function or P (r). The relationship between the intensity and P (r) is shown in Eq. 10.3.

$$I(q) = \langle f(\boldsymbol{q}) f^*(\boldsymbol{q}) \rangle = \int_0^{Dmax} 4\pi r^2 P(r) \frac{\sin qr}{qr} dr \qquad (10.3)$$

Since

$$\left\langle e^{i\boldsymbol{r}\cdot\boldsymbol{q}} \right\rangle = \frac{\int_0^{\pi} \sin\varphi d\varphi \int_0^{2\pi} e^{iqr\cos\theta} d\theta}{\int_0^{\pi} \sin\varphi d\varphi \int_0^{2\pi} d\theta} = \frac{\sin(q\,r)}{q\,r}$$

The momentum transfer $|\mathbf{q}| = q = (4\pi \sin(\theta/2))/\lambda$, now varies in one angular dimension, θ. Thus all information could be collected by a one dimensional strip detector. This is typically not done because by collecting on an area detector and averaging, the signal to noise is of greater quality.

Despite this mapping of three dimensional information on a one dimensional space, high resolution capability stems from the use of wavelengths ($\lambda$) on the order of 1 Å. When scattering is done with X-rays and neutrons, it is intrinsically capable of measuring scalar distances at the same resolution as crystallography because the wavelengths used are the same.

An often unappreciated advantage in solution scattering is that the scattering objects are not in contact as they are in crystallography. This lack of contact limits the integration of the distance r up to the maximum dimension of the particle $D_{max}$. In crystallography the boundary between scattering units can be difficult to decipher making the analogue to the P(r) function, called the Patterson function, less directly interpretable. Conversely the P(r) function is a histogram of pair distances of scattering density. Its properties are pictorially illustrated in Fig. 10.2 on a toy system. Note that changes in the position of one atom influences significant portions of the distribution.

For macromolecules in real systems, the P (r) can be very sensitive to conformational changes or modifications. Figure 10.3 illustrates the accuracy achievable on a protein where no atom changes more than 5 Å. This level of accuracy is achievable because (1) the protein is either in one state or another and (2) a substantial proportion of the molecule moves. Had the protein sampled a range of conformations in its apo state relative to a single conformation in the ligand-bound state, the P(r) functions for the apo and ligand-bound state would be more difficult to discern. Furthermore, if only a small portion of the molecule is changing, the P (r) function would be less sensitive to the change. For example, while the addition of a single amino acid to a terminus of lysozyme may be detectable in the P(r) function, the same cannot be said for one of the ribosomal proteins within a ribosome. While several factors influence the resolution of changes that can be determined, the P(r) function has successfully and uniquely detected subtle changes.

Until recently, existing methods for extracting the P(r) function from SAS data have been suitable. However, as data at higher angles are now routinely collected, particularly with X-rays, available tools are showing limitations. New methods are sure to arise for even greater definition and sensitivity of macromolecular structure. Of particular importance, as higher angles are

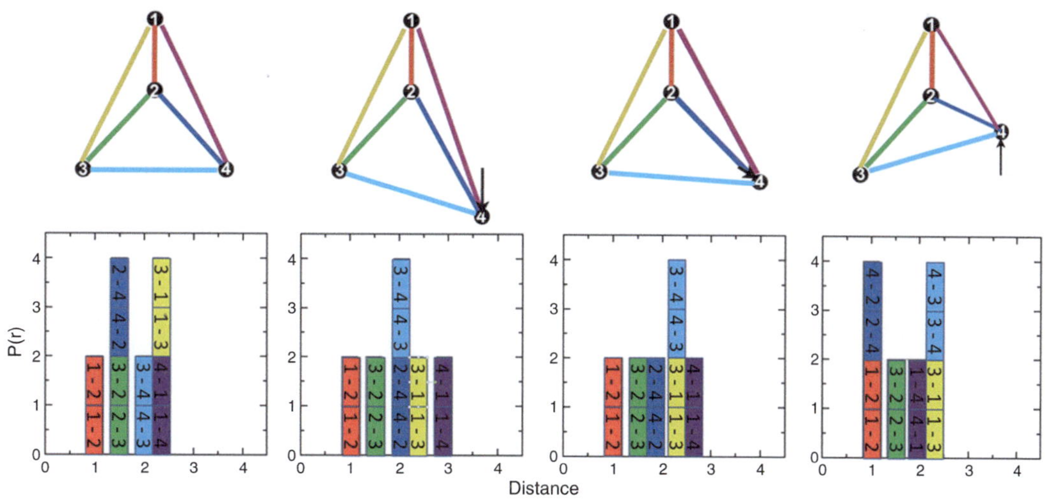


**Fig. 10.2** The pair distribution function, P(r), is a histogram of distances. A toy model of a scattering is composed of four labeled components. The P(r) function is shown below. As the spatial distribution changes within the system the P(r) function can be dramatically sensitive to these changes

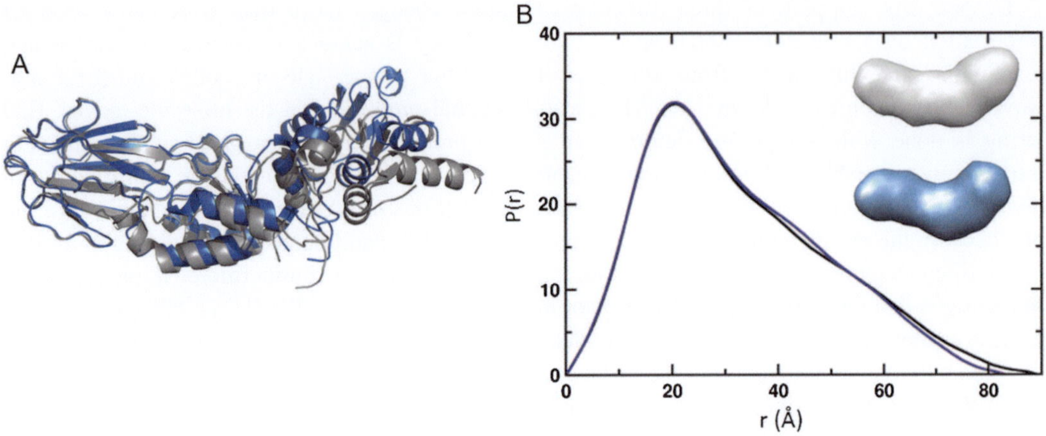


**Fig. 10.3** The P(r) function provides high resolution information. (**a**) The protein NBS1 undergoes is either extended (*black*) in an apo state or contracted (*blue*) when binding a small peptide (Williams et al. 2009). In this conformational change no atom moves more than 5 Å. (**b**) The difference can be detected in the P(r) function. The decrease in maximum dimension from 87 to 83 Å among other changes are visible. However these differences would be difficult to discern from SAS generated shapes on the two samples (inset)

measured, the influence of water which we consider below.

## 10.3 Measuring the Distance Between Water Molecules in the Liquid Phase

Liquids have structure as readily indicated by strong and non-monotonic features from their scattering. Nearly all scattering from biological samples will contain a significant signal from the structure of liquid water, as shown in Fig. 10.4a, b. Learning about the liquid water signal is of value if for no other reason than to de-convolute it from the signal of interest. Further investigation may be warranted though, as the structural interactions of water underlying this signal are also of important consequence for many processes in biology.

The molecules in liquid water are constantly translating and rotating due to thermal fluctuations. Thus the distance between any single water molecule and its instantaneous neighbor will vary with time. However, since all water molecules are identical, the influence of thermal energy is actually out-sized; in exchange for breaking bonds holding two waters together, another near equivalent bond with a different water is made. A water molecule will have a steady number of neighbors, shown in Fig. 10.4c. Water is famously polar giving a preferred directionality to interactions with neighbors. Consequently, certain spaces around a given water molecule are preferred relative to others. The strength of polarity competes with van der Waals interactions producing an optimal distance between water molecules. The influence of a water molecule goes beyond those in direct contact and extends to correlated second and third shells until thermal fluctuations dominate. These structural characteristics have a profound impact on the chemistry and biology of water. Furthermore, since water is a small molecule with a total of ten electrons it can be simulated in detail to provide insights into the quantum mechanics of inter-molecular bonds. Scattering measurements on water have played a fundamental role in these areas.

Interpretation of scattering data from molecular liquids focuses on the relationship between molecules rather than on the structure of the

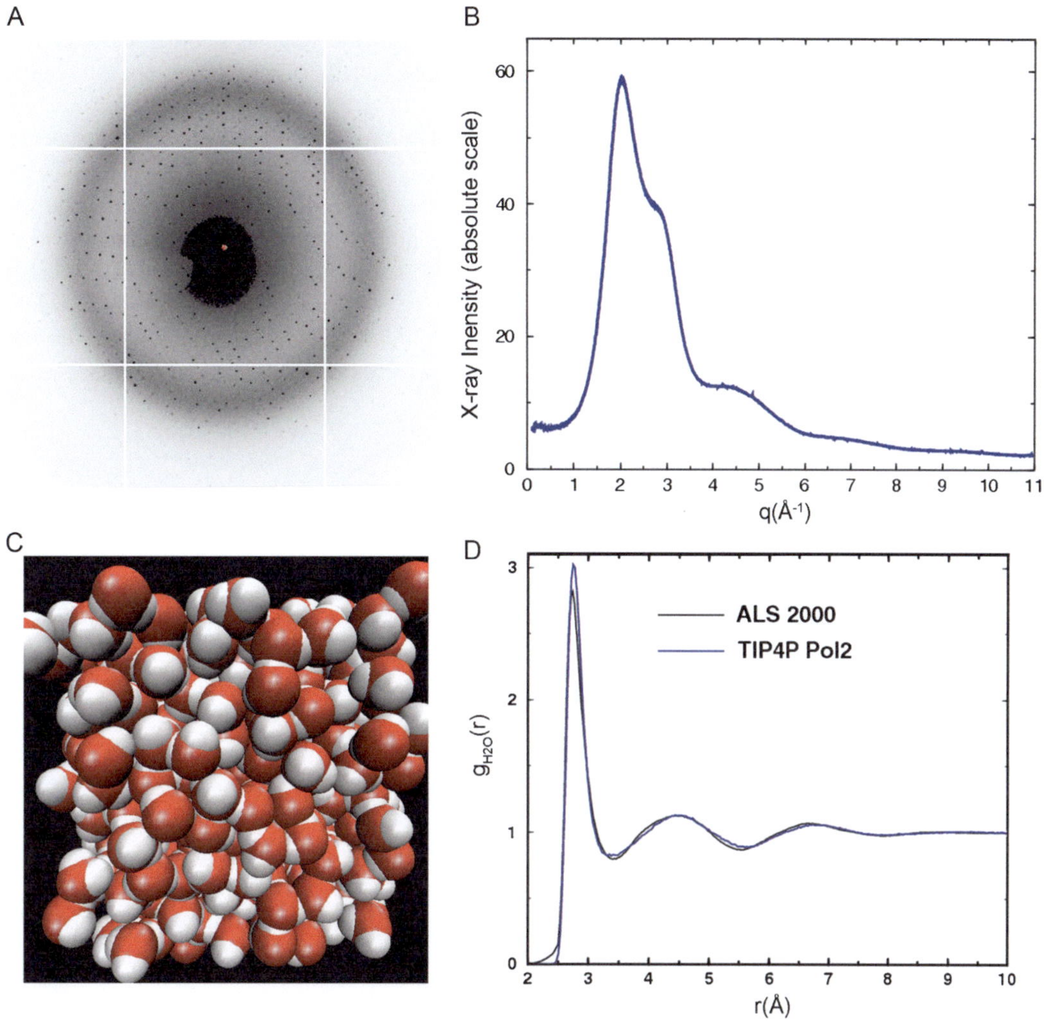


**Fig. 10.4** Scattering from water structure. (**a**) Underlying protein crystallographic data is a ring due to water. Because the protein is organized on a repeating lattice in the crystal, the diffraction spots must be characterized in two dimensions on the detector. In contrast the structures related to water are oriented in all directions relative to beam producing a symmetric ring that may be represented in one dimension (**b**) by integrating around central incident beam. (**c**) Water structure stems from several features including hydrogen bonding and van der Waals interaction. (**d**) The van der Waals interactions prevent water molecules from overlapping which can be seen in the radial distribution function $g_{H2O}(r)$ at small r up to 2.5 Å. The main correlation is with next nearest neighbors of which there are five. This can be determined by integrating the area under the peak of the first correlation shell which is centered at 2.8 Å

molecule itself. For this reason, the quantity of interest from Eq. 10.2 is the opposite of what it was for both crystallography and scattering from macromolecules. The $\rho_{H2O}(\boldsymbol{r})$, that describes a water molecule, in Eq. 10.1 is known and has been tabulated for both X-rays (Morin 1982; Hura et al. 2000) and neutrons. Using the known $\rho_{H2O}(\boldsymbol{r})$, a form factor for water has been calculated, $f_{H2O}(\mathbf{r})$. The scattering from water is thus composed of two parts, the scattering within a water molecule (intra-molecular scattering, a known quantity), and the scattering between different water molecules (inter-molecular scattering) where $r_j - r_k \neq 0$ as defined in the second part of Eq. 10.4. Intra-molecular scattering, $I_{intra}(q) = \langle f(q)f^*(q)\rangle$, can be calculated, however this calculated quantity can only be utilized when the measurement has been calibrated on an absolute scale. Absolute measurements can be challenging since few detectors are accurately calibrated. However, using calibrants or extremely high angles one can isolate the inter-molecular scattering $I_{inter}(q)$.

As in crystallography where a formalism is introduced that anticipates molecules will be on a lattice with extracted indices, for non-crystalline materials a construct has been created that anticipates the kind of inter-molecular structure we expect in liquids. This formalism is called the radial pair correlation function or $g(\mathbf{r_j}-\mathbf{r_k})$. The $g(\mathbf{r_j}-\mathbf{r_k})$ can be considered a probability weighting function that describes the probability that a neighboring water molecule has a specific orientation and distance relative to any other given water molecule. The quantity $\rho_o g(\mathbf{r_j} - \mathbf{r_k})d\mathbf{V}$ is the expected number of molecules that will be found at a distance $\mathbf{r_k}$ in a volume element $\mathbf{dV}$ from a molecule at $\mathbf{r_j}$. The constant $\rho_o$ is the average molecular density that is macroscopically measurable, calculable from ~1 g/cm$^2$ for water. By introducing this term, we can exchange one of the sums over all molecular pairs in the second part of Eq. 10.4 for an integral over the probability weighting function as in Eq. 10.5. This assumes that the population distribution of configurations has reached an equilibrium within the liquid.

The correlated structures in water will be in every orientation relative to the X-ray beam involving a spherical integration, with results similar to those described in the preceding section. All directional information is lost and the exponential term is further reduced to a "sinc" (sin(x)/x) function of scalars and g(r). The result of the spherical integration is shown in Eq. 10.5 (neglecting a term that is only of significance with strongly absorbing material). The function g(r) has several key properties. The value of g (r) describes the relative number of scatterers outside the bulk density having a center-to-center distance of r from a molecule, as shown in Fig. 10.4d. At small distances within the van der Waals diameter, g(r) will have a value of zero since this presents a no overlap zone. At large r, g(r) will be 1 as correlation has been lost and numbers have reached bulk average density, $\rho_o$. The vectoral $g(\mathbf{r_j} - \mathbf{r_k})$ can be calculated from a molecular dynamics calculation as can be its scalar form g(r).

$$I(q) = 2\langle f(q)f^*(q)\rangle + \left\langle \sum_j \sum_{k\neq j} f(\boldsymbol{q})f^*(\boldsymbol{q})e^{i(\boldsymbol{r_j}-\boldsymbol{r_k})\cdot\boldsymbol{q}} \right\rangle \tag{10.4}$$

$$I(q) = 2\langle f(q)f^*(q)\rangle + \left\langle \sum_k f(\boldsymbol{q})f^*(\boldsymbol{q}) \int_V \rho_o g(\boldsymbol{r_j} - \boldsymbol{r_k}) e^{i(\boldsymbol{r_j}-\boldsymbol{r_k})\cdot\boldsymbol{q}} d\boldsymbol{V_j} \right\rangle$$

$$I(q) = 2\langle f(q)f^*(q)\rangle \left(1 + 4\pi\rho_o \int_0^\infty r^2(g(r)-1)\frac{\sin(qr)}{qr}\,dr\right) \tag{10.5}$$

Water was almost certainly one of the first targets of X-rays and neutrons as it is both easy to attain and of critical importance. However, since water is of such great importance, there is a tremendous demand for precision. The g (r) function for water remains intensively studied and even debated (Brookes and Head-Gordon 2015; Clark et al. 2010; Gallo et al. 2016; Amann-Winkel et al. 2016). Each structural detail has wide spread implications. Though not presented here, further refinements have been made by starting at the atomic ($\rho_O(\boldsymbol{r})$ and $\rho_H(\boldsymbol{r})$) rather than the molecular level. Working from this basis allows an exploitation of a unique property of neutron scattering. Since deuterium scatters neutrons more strongly than hydrogen one can make use of a contrast change between deuterated water and hydrogenated water for extraction of the pair correlation function of hydrogens $g_{HH}(r)$. As X-rays are scattered by electrons, scattering experiments report mainly on the oxygens $g_{OO}(r)$. This is particularly true since oxygen electronegativity will draw electrons from hydrogen (Head-Gordon and Hura 2002).

A challenge for converting scattering data to real spatial information is that the two are related through a Fourier transform. For early investigators in particular but also of concern today are three features of scattering data that stymy Fourier transformation. (1) Noise in data adds unphysical Fourier terms, (2) sparsely sampled data from point detectors or other experimental factors may mean missing Fourier terms and (3) data are always truncated both at high and low angles with disastrous effects for Fourier transformation. When direct transformation is attempted on truncated data, an infinite set of non-physical Fourier terms are required. Modern detectors and bright sources have improved signal, increased the sampling and the angular range collected, greatly reducing challenges faced by early experimentalist. Thus an understanding of the structure of water as determined by scattering has been emerging.

General features that are agreed upon from analysis of g(r) is that a water molecule strongly influences its nearest neighbors. The g(r) as determined from X-ray scattering measurements from water is shown in Fig. 10.4d and has on average 5 nearest neighbors, that sit 2.8 Å from the center of any given water molecule. The presence of a water molecule perturbs structure as far as 10 Å away. This first coordinated shell is followed by a drop in density below bulk levels. Two more peaks are discernable in addition to the first, showing the minimum distance a water molecule maintains its influence. These length scales are larger than cavities within macromolecules affecting many important phenomena such as metabolite and drug binding. Thus accounting for the structural influence of water remains a major challenge for *in silico* based drug screening among other fields.

## 10.4 Measuring Changes in Distance Distributions Over Long Length Scales Using Labels

We now move to measuring changes in large scalar distances within a macromolecular assembly. Such measurements in solution can provide key biological insights. Outside of a crystalline state, there is almost always a population distribution of distances that cannot be adequately represented with a single structure. Biological systems may sample important states infrequently and thus these states occur in only a small subset of a population. The challenge for solution based techniques is to define this distribution and changes in this distribution as a function of some perturbation. A large variety of techniques have evolved to quantify these dynamic distributions, each with assets that are worth contrasting.

Like many other techniques, scattering can make use of labels to aid in following specific parts of macromolecules, increasing signal and accuracy. There are several types of labels and several ways of measuring the distance between them. This section is focused on the use of heavy metal labels in the context of X-ray scattering (SAXS) (schematically illustrated in Fig. 10.5a). Conceptually, the experiment is quite simple, however there is still a substantial investment into synthesis relative to other label based techniques for which synthesis has become more routine.

SANS based approaches have been used to measure distance distributions for many years. Selective deuteration of parts of a macromolecule adds extra scattering cross-section to that part. Collecting data from such samples in a mixture of $H_2O$ and $D_2O$ can make the non-deuterated portion invisible. This approach was applied to identify the relative distance between ribosomal components before atomic resolution structures were available. Several other interesting systems have been elucidated this way as discussed elsewhere in this book. Increased access to SANS instruments plus their increased brightness are certain to profoundly increase this type of application.

## 10.5 FRET and EPR for Measurement of Scalar Distances

Before we focus on SAXS we draw attention to important alternative non-scattering based techniques. When molecular rulers are required, FRET and EPR are often utilized with excellent effect. FRET and EPR labels are commercially supported, reducing sample preparation challenges. They have been powerfully applied in scenarios for which scattering techniques are difficult. For example, FRET can be applied *in vivo* and as a single molecule technique. EPR can be conducted at low concentrations and with membrane proteins. For these reasons FRET and EPR should be strongly considered for at least complementary information to scattering.

FRET and EPR also have specific challenges. Both techniques have an optimal range for distance measurements, beyond which they are no longer reliable. For FRET, this range is usually from 1.5 to 6 nm depending on the dynamics of the biomolecule and the size of the label (Lam et al. 2012). For EPR, the optimal range can extend from 1.5 to 2.5 nm for continuous-wave EPR, and up to 8 nm for pulsed wave EPR (Schiemann and Prisner 2007).

In addition to limitations in distance measurements, both FRET and EPR have specific experimental challenges. For FRET to accurately measure distance, both labels need to have the freedom to sample all rotational orientations. Limiting the rotation of one label relative to the other can increase measurement error significantly, from 10% at 5 nm to 50% at 1.5 nm. For EPR, the sensitivity of the spin label to the environment can be both a blessing and a curse. To increase the signal-to-noise for EPR, measurements are often taken at 50–80 K, sometimes for 10–12 h.

For systems and questions where these experimental limitations do not pose a challenge, both FRET and EPR will provide significant insights. Scattering remains inherently complementary. Scattering measurements can be conducted with and without FRET or EPR labels to determine the influence of labels.

## 10.6 Scattering from Metal Labels for SAXS Measurements

X-rays scatter from electrons, thus heavy elements scatter more strongly than lighter elements. Most biological SAXS measurements are difference experiments where the solvent is subtracted from the solution containing sample. When this is done, the relative signal of each contributor is scaled not by the square of the electron density but by the square of the electron density difference between the scattering object and solvent. The average electron density is 0.332, and 0.44 electrons/Å$^3$ for water and protein respectively. For gold, the electron density is approximately 4.6 electrons/Å$^3$. Using these values, in a difference experiment the scattering intensity at zero angle of a gold particle is 1650-fold larger than that of a protein of equivalent size (Fig. 10.5b). Differences of this order of magnitude are usually worthwhile to pursue despite challenges in preparation. For example, concentrations can be reduced to the nanomolar range or time scales can be reduced to milliseconds.

Gold labeling of biological systems has relied heavily on gold-sulfur bonding though other strategies are possible. Cysteines are a natural target in proteins while thiolated bases can be incorporated into DNA. As gold nanoclusters present a large surface, care must be taken to ensure the desired labeling ratio is eventually purified. Once the gold-sulfur bond is made, further reactions must be reduced. Thus the portion of the gold surface that is not involved in the desired bonding is protected by other chemical groups like thiolated polyethylene glycols. For control of bonding, the redox state of solutions during preparation must be carefully controlled as the gold-sulfur bond competes with potential disulfide bonding.

The larger the scattering from the label the simpler the subsequent analysis will be for the extraction of distances between two labels. However, any label can be disruptive to the biological system under investigation and a variety of heavy atoms and heavy atom clusters have been found useful including mercury (Vainshtein et al. 1980), lead (Grishaev et al. 2012), rubidium (Horkay et al. 2006), and terbium (Miakelye et al. 1983). Many of these studies also manipulated the scattering contrast by either using anomalous scattering properties of the metal atom or changing the solvent scattering to match that of the biological component. Depending on the strength of the signal and size of the label, data collection and analysis increase in complexity (Zettl et al. 2016; Mathew-Fenn et al. 2008a, b).

## 10.7 Extracting Length Information from the Scattering Curve

Contributions to the total scattering from labeled systems can be conveniently grouped. There are five types of contributions: the intra-label, intra-biomolecule, inter-label, inter-biomolecule, and finally scattering due to correlations between biomolecules and labels. Analysis of these terms vary in complexity. In the simplest case, the labels scatter so strongly that the scattering contribution of the biological macromolecule are negligible (Hura et al. 2013b). In this case only two terms are significant: the intra-label and inter-label terms.

With labels that scatter on the same scale as the biological macromolecule, all terms must be considered. In an effort to measure basic properties of DNA, small gold labels of 1.5 nm diameter have been used (Mathew-Fenn et al. 2008a, b). The intra- and inter-biomolecule scattering can be measured on the system without labels. The more difficult component is the scattering cross term due to correlations between the biomolecule and the label. Several strategies may be applied including the measurement of the system labeled at each point independently or modifying the scattering power of the label using resonant X-ray energies (Zettl et al. 2016) or varying label size or composition as reported in several studies.

In a system using equivalent labels at all labeling points and that scatter overwhelmingly, the analysis is similar to that used in the previous section for water. A distinction can be made between the experimental observable desired from labels relative to that from water. In the case of water, the absolute number of coordinating waters is an important experimental result. The number of coordinated labels is almost always known and if uncertain can be tested by varying the labeling strategy. For example one can test the agreement between the scattering of the label alone and the macromolecule labelled at one point. If these results are in poor agreement, the macromolecules may be multimerizing, adding additional label correlations that must be accounted for. Defining a distribution of labels relative to the bulk density is not necessary and so rather than work with g(r) which is a comparison to bulk density, a relative distribution is desired. Thus, starting from Eq. 10.4 we define a weighing distribution P(D) where D is the distance between label j and k. The labels all have the same f(q) so we can utilize Eq. 10.6.

$$I(q) = 2\langle f^2(q)\rangle + 2\langle f^2(q)\rangle \int_0^\infty P(D)\frac{\sin qD}{qD}dD \tag{10.6}$$

Rearranging terms to focus on the inter-label distance distribution and taking into account both concentration factors and instrumental parameters with two constants ($k_1$ and $k_2$) we arrive at the correlation scattering function (CSF), which is a Fourier transform of P(D).

$$CSF = \frac{I(q)}{k_1\langle f^2(q)\rangle} - k_2 = \int_0^\infty P(D)\frac{\sin qD}{qD}dD$$

Experimentally, $k_1$ and $k_2$ can be determined. However, they can also be treated as fitted parameters. The CSF should oscillate about 0 and at wide q the inter-particle contribution should be negligible as the inter-label distance must be larger than the label size. Drifts from an oscillation about 0 indicate either some level of aggregation in the labels.

We have applied this scenario to monitor protein mediated DNA repair (Hura et al. 2013b). Labeling both ends of damaged DNA with nominally 5 nm diameter gold labels, we followed the end-to-end distribution as proteins and metabolites in the repair pathway were added. We contrasted both short (31 base pair) and long (up to 71 base pair) DNA. The shorter DNA substrate accommodates a single protein footprint analogous to what can be done with FRET. The longer DNA substrate accommodates multiple proteins allowing the observation of cooperative effects common in DNA repair processes. Example results from this study are shown in Fig. 10.5c–e where we measure distances between labels that are 30 nm apart. These results can be extended to longer DNA strands or length scales as most modern SAXS instruments can sufficiently capture small angle data.

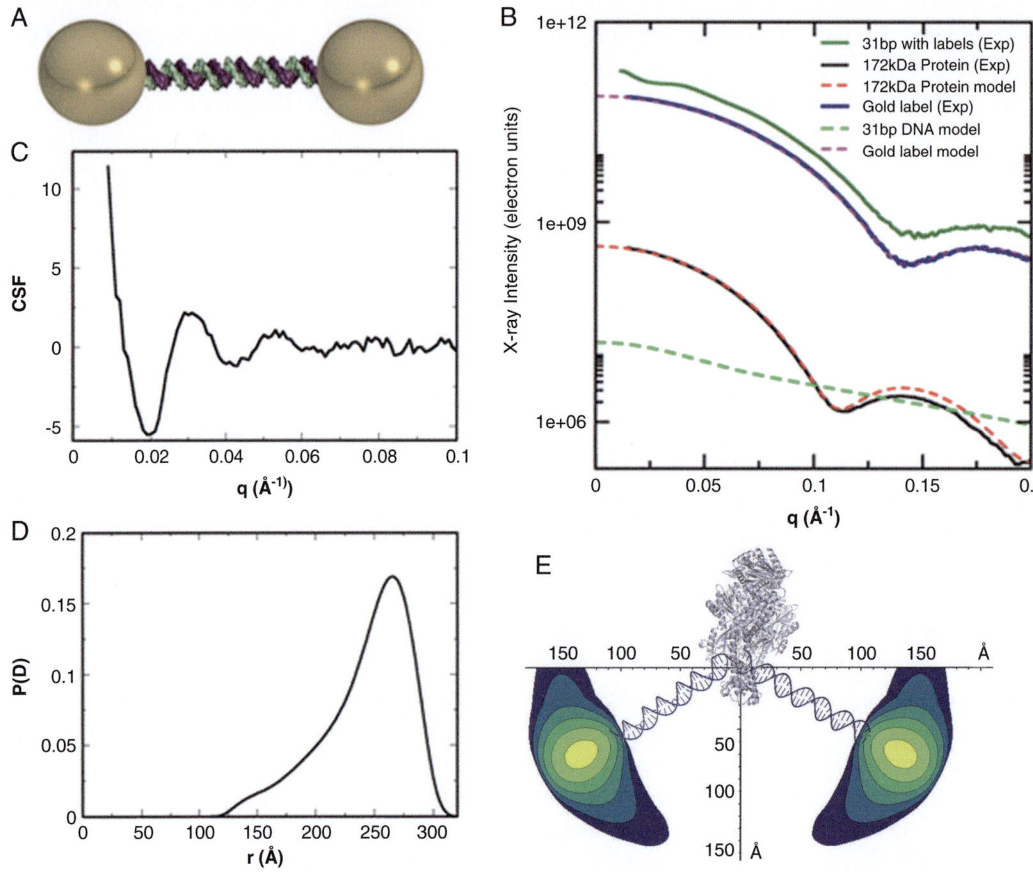


**Fig. 10.5** SAXS measures of long-distance distributions with gold labels on DNA. (**a**) Schematic representation of gold end-labeled DNA. (**b**) Experimentally measured scattering power of 5 nm gold is 5400 times higher than DNA and 500-fold greater than a globular protein of two times its diameter. (**c**) The CSF is derived from dividing two SAXS profiles, the labeled system through by the label alone. (**d**) The P(D) distribution characterizing the distances between labels as DNA is manipulated by a DNA processing enzyme. (**e**) Contour maps of the gold labels can be drawn based on the distribution shown in (**d**) combined with crystallographic information that the protein system bends DNA. The bending of the DNA is dynamic, sampling very dramatic bending angles that could not be deduced crystallographically

## 10.8 Conclusion

SAS as employed to study biological macromolecules in solution is a very flexible and powerful technology with widespread application. For most samples, a minimal amount of preparation is required to provide a comprehensive characterization of macromolecular structure. High flux sources and new detectors are capable of characterizing wide temporal and spatial scales – all with one sample and data collection. Here we worked through some details of analysis for probing distances between molecules in a liquid which can be pushed down to 0.1 nm resolution. Large area detectors provide access to the necessary angles that can be used to characterize the details of hydration layers around proteins. More work is required on tool development to utilize this information that has become routinely available.

For specialized cases where the organization or movement of subassemblies within a larger assembly is of central importance, samples may be modified so that these pieces have additional contrast. For X-rays, specific points may be labeled with metallic nanoclusters or for neutrons, regions may be deuterated. Since required sample quantities are already quite small and continue to decrease, the same sample preparation may be used to study the labeled system in a variety of contexts, providing unique insights into function. Here, we considered some of the detailed analysis required to extract information from labels separated by distances of 50 nm or greater.

While there have been big leaps in recent progress for EM and with the free electron laser for X-rays, advances in SAS data collection and analysis have been more wide spread and continuous. Access and utilization has grown, creating an ever larger community that contributes to analysis tools and interpretation. We anticipate that due to its widespread applicability and throughput, SAS will increasingly be looked to for complementary and unique structural information on a rapidly expanding set of targets from genomic and macromolecular engineering fields.

## 10.9 Few Assorted Experimental/ Computational Tips

SAS can monitor high resolution structural changes in solution when conducted as a relative measurement to other SAS data or an atomic resolution model.

Distance distributions are not always calculated from SAS in the same way and therefore do not always quantify the same scattering density distribution. Make sure you understand the assumptions that are part of a particular distance distribution.

In quantifying distances between scattering density, utilize any information that may be of value, whether it be bulk density or known semi-periodicity particular to the sample, to create the most intuitive distance distribution for your system.

Distance distributions are never direct Fourier transforms of SAS data as SAS data is finite and are often derived from fitting data. Always examine the quality of the fit of the Fourier transform.

## References

Amann-Winkel K, Bellissent-Funel MC, Bove LE, Loerting T, Nilsson A, Paciaroni A, Schlesinger D, Skinner L (2016) X-ray and neutron scattering of water. Chem Rev 116:7570–7589

Andresen K, Jimenez-Useche I, Howell SC, Yuan CL, Qiu XY (2013) Solution scattering and FRET studies on nucleosomes reveal DNA unwrapping effects of H3 and H4 tail removal. PLoS One 8:e78587

Baldock C, Oberhauser AF, Ma L, Lammie D, Siegler V, Mithieux SM, Tu Y, Chow JY, Suleman F, Malfois M, Rogers S, Guo L, Irving TC, Wess TJ, Weiss AS (2011) Shape of tropoelastin, the highly extensible protein that controls human tissue elasticity. P Natl Acad Sci USA 108:4322–4327

Bergmann A, Fritz G, Glatter O (2000) Solving the generalized indirect Fourier transformation (GIFT) by Boltzmann simplex simulated annealing (BSSA). J Appl Crystallogr 33:1212–1216

Brookes DH, Head-Gordon T (2015) Family of oxygen-oxygen radial distribution functions for water. J Phys Chem Lett 6:2938–2943

Clark GNI, Hura GL, Teixeira J, Soper AK, Head-Gordon T (2010) Small-angle scattering and the structure of ambient liquid water. Proc Natl Acad Sci U S A 107:14003–14007

Edwards AL, Matsui T, Weiss TM, Khosla C (2014) Architectures of whole-module and bimodular proteins from the 6-deoxyerythronolide B synthase. J Mol Biol 426:2229–2245

Falk SJ, Guo LY, Sekulic N, Smoak EM, Mani T, Logsdon GA, Gupta K, Jansen LE, Van Duyne GD, Vinogradov SA, Lampson MA, Black BE (2015) Chromosomes. CENP-C reshapes and stabilizes CENP-A nucleosomes at the centromere. Science 348:699–703

Franke D, Svergun DI (2009) DAMMIF, a program for rapid ab-initio shape determination in small-angle scattering. J Appl Crystallogr 42:342–346

Gallo P, Arnann-Winkel K, Angell CA, Anisimov MA, Caupin F, Chakravarty C, Lascaris E, Loerting T, Panagiotopoulos AZ, Russo J, Sellberg JA, Stanley HE, Tanaka H, Vega C, Xu LM, Pettersson LGM (2016) Water: a tale of two liquids. Chem Rev 116:7463–7500

Grishaev A, Anthis NJ, Clore GM (2012) Contrast-matched small-angle x-ray scattering from a heavy-atom-labeled protein in structure determination: application to a lead-substituted calmodulin-peptide complex. J Am Chem Soc 134:14686–14689

Hammel M, Amlanjyoti D, Reyes FE, Chen JH, Parpana R, Tang HY, Larabell CA, Tainer JA, Adhya S (2016) HU multimerization shift controls nucleoid compaction. Sci Adv 2:e1600650

Head-Gordon T, Hura G (2002) Water structure from scattering experiments and simulation. Chem Rev 102:2651–2669

Horkay F, Hecht AM, Rochas C, Basser PJ, Geissler E (2006) Anomalous small angle x-ray scattering determination of ion distribution around a polyelectrolyte biopolymer in salt solution. J Chem Phys 125

Hura G, Sorenson JM, Glaeser RM, Head-Gordon T (2000) A high-quality x-ray scattering experiment on liquid water at ambient conditions. J Chem Phys 113:9140–9148

Hura GL, Menon AL, Hammel M, Rambo RP, Poole FL II, Tsutakawa SE, Jenney FE Jr, Classen S, Frankel KA, Hopkins RC, Yang S-J, Scott JW, Dillard BD, Adams MWW, Tainer JA (2009) Robust, high-throughput solution structural analyses by small angle X-ray scattering (SAXS). Nat Methods 6:606–U683

Hura GL, Budworth H, Dyer KN, Rambo RP, Hammel M, McMurray CT, Tainer JA (2013a) Comprehensive macromolecular conformations mapped by quantitative SAXS analyses. Nat Methods 10:453–454

Hura GL, Tsai CL, Claridge SA, Mendillo ML, Smith JM, Williams GJ, Mastroianni AJ, Alivisatos AP, Putnam CD, Kolodner RD, Tainer JA (2013b) DNA conformations in mismatch repair probed in solution by X-ray scattering from gold nanocrystals. Proc Natl Acad Sci U S A 110:17308–17313

Koch MHJ, Vachette P, Svergun DI (2003) Small-angle scattering: a view on the properties, structures and structural changes of biological macromolecules in solution. Q Rev Biophys 36:147–227

Lam AJ, St-Pierre F, Gong YY, Marshall JD, Cranfill PJ, Baird MA, McKeown MR, Wiedenmann J, Davidson MW, Schnitzer MJ, Tsien RY, Lin MZ (2012) Improving FRET dynamic range with bright green and red fluorescent proteins. Nature Methods 9:1005. −+

Mathew-Fenn RS, Das R, Silverman JA, Walker PA, Harbury PAB (2008a) A molecular ruler for measuring quantitative distance distributions. PLoS One 3

Mathew-Fenn RS, Das R, Harbury PAB (2008b) Remeasuring the double helix. Science 322:446–449

Miakelye RC, Doniach S, Hodgson KO (1983) Anomalous x-ray-scattering from terbium-labeled parvalbumin in solution. Biophys J 41:287–292

Morin LRM (1982) Molecular-form factors and photon coherent scattering cross-sections of water. J Phys Chem Ref Data 11:1091–1098

Nielsen SS, Toft KN, Snakenborg D, Jeppesen MG, Jacobsen JK, Vestergaard B, Kutter JP, Arleth L (2009) BioXTAS RAW, a software program for high-throughput automated small-angle X-ray scattering data reduction and preliminary analysis. J Appl Crystallogr 42:959–964

Poitevin F, Orland H, Doniach S, Koehl P, Delarue M (2011) AquaSAXS: a web server for computation and fitting of SAXS profiles with non-uniformally hydrated atomic models. Nucleic Acids Res 39: W184–W189

Putnam CD, Hammel M, Hura GL, Tainer JA (2007) X-ray solution scattering (SAXS) combined with crystallography and computation: defining accurate macromolecular structures, conformations and assemblies in solution. Q Rev Biophys 40:191–285

Schiemann O, Prisner TF (2007) Long-range distance determinations in biomacromolecules by EPR spectroscopy. Q Rev Biophys 40:1–53

Schneidman-Duhovny D, Hammel M, Tainer JA, Sali A (2016) FoXS, FoXSDock and multiFoXS: single-state and multi-state structural modeling of proteins and their complexes based on SAXS profiles. Nucleic Acids Res 44:W424–W429

Semenyuk AV, Svergun DI (1991) Gnom – a program package for small-angle scattering data-processing. J Appl Crystallogr 24:537–540

Stingaciu LR, O'Neill H, Liberton M, Urban VS, Pakrasi HB, Ohl M (2016) Revealing the dynamics of thylakoid membranes in living cyanobacterial cells. Sci Rep 6:19627

Vainshtein BK, Feigin LA, Lvov YM, Gvozdev RI, Marakushev SA, Likhtenshtein GI (1980) Determination of the distance between heavy-atom markers in hemoglobin and histidine-decarboxylase in solution by small-angle x-ray-scattering. FEBS Lett 116:107–110

Williams RS, Dodson GE, Limbo O, Yamada Y, Williams JS, Guenther G, Classen S, Glover JNM, Iwasaki H, Russell P, Tainer JA (2009) Nbs1 flexibly tethers Ctp1

and Mre11-Rad50 to coordinate DNA double-strand break processing and repair. Cell 139:87–99
Zettl T, Mathew RS, Seifer S, Doniach S, Harbury PAB, Lipfert J (2016) Absolute intramolecular distance measurements with angstrom-resolution using anomalous small-angle x-ray scattering. Nano Lett 16:5353–5357

# A Successful Combination: Coupling SE-HPLC with SAXS

11

Javier Pérez and Patrice Vachette

**Abstract**

A monodispersed and ideal solution is a central (unique?) requirement of SAXS to allow one to extract structural information from the recorded pattern. On-line Size Exclusion Chromatography (SEC) marked a major breakthrough, separating particles present in solution according to their size. Identical frames under an elution peak can be averaged and further processed free from contamination. However, this is not always straightforward, separation is often incomplete and software have been developed to deconvolve the contributions from the different species (molecules or oligomeric forms) within the sample. In this chapter, we present the general workflow of a SEC-SAXS experiment. We present recent instrumental and data analysis improvements that have improved the quality of recorded data, extended its potential and turn it into a mainstream approach. We describe into some details two specific applications of SEC-SAXS that provide more than just separating associated forms from the particle of interest.

**Keywords**

Size exclusion chromatography • Small-angle X-ray scattering • Monodispersity • Membrane protein-detergent complex

## 11.1 Introduction

For meaningful subsequent 3D modelling of the studied particle, SAXS requires a monodispersed and ideal solution. In the case of a mixture, the composition may be fully defined by a given set of parameters that are determined by non-negative least-square approach in the case where form factors are known. When proteins are prone to aggregation, this can bar any valid

J. Pérez (✉)
Synchrotron Soleil, L'Orme des Merisiers, Saint-Aubin BP48, 91192 Gif-sur-Yvette Cedex, France
e-mail: javier.perez@synchrotron-soleil.fr

P. Vachette
Institut de Biologie Intégrative de la Cellule, UMR 9198, Université Paris-Sud, 91405 Orsay Cedex, France
e-mail: patrice.vachette@i2bc.paris-saclay.fr





B. Chaudhuri et al. (eds.), *Biological Small Angle Scattering: Techniques, Strategies and Tips*, Advances in Experimental Medicine and Biology 1009, DOI 10.1007/978-981-10-6038-0_11

SAXS study, even when trying to mitigate the effect of self-association by adding a small amount of detergent in the solution, using a slightly truncated construct or producing a fusion protein that does not aggregate any more. This has actually been the case in a recent controversy around two SAXS studies of PICK1, a BAR domain containing protein (Boczkowska et al. 2015; Erlendsson et al. 2015; Karlsen et al. 2015; Madasu et al. 2015). As stated in the conclusion of one of the letters exchanged "*These two models are radically different, and mutually exclusive, and should inspire scientists in the field, our laboratory included, to use alternative approaches to test their legitimacy. Lastly, we hope that as a result of this debate, aggregation issues in SAXS data analyses will receive the close scrutiny they deserve. The recent availability of in-line SEC-SAXS at an increasing number of beamlines worldwide should help mitigate aggregation problems, although this is not a universal solution.*" The present chapter is an attempt to present the implementation of the approach around an actual instrument and to describe data acquisition and processing.

To reach monodispersity in the biochemistry laboratory, Size Exclusion (aka Gel Filtration) Chromatography, is a routine and very efficient method. Its coupling on-line with the measuring cell on synchrotron radiation SAXS beamlines marked a clear breakthrough in biological SAXS. It was first introduced by "external users" at APS beamline BioCAT in 2004 (Mathew et al. 2004) and at the Photon Factory BL10C in 2008 (Watanabe and Inoko 2009) but it was only in 2008 that it became available to users as a routine mode of data acquisition on the SWING beamline at the SOLEIL synchrotron (David and Pérez 2009). Indeed, the instrument had been designed from its inception with an SE-HPLC setup as an integral part of the SAXS instrument. Since then this arrangement has spread to a number of other instruments worldwide (Table 11.1).

## 11.2 Features Required for a SEC-SAXS Setup

The SEC-SAXS combination basically resembles the popular SEC-MALLS combination now available in most biochemistry labs, but with the scattering of visible light "replaced" by the scattering of an X-ray synchrotron beam. A schematic simplified layout is represented in Fig. 11.1. For SAXS measurements, a thermalized flowing cell made of a cylindrical thin-walled (*ca.* 10 µm) quartz capillary (*ca.* 1.5 mm diameter) is the usual compromise between rigidity, low X-ray scattering background and cleaning ease. The capillary should best be inserted in a vacuum chamber to avoid shadow effects from air and window scattering. The chamber should itself be positioned on a horizontal motorized stage to select the zone of the capillary with the lowest scattering background. The tubing between the UV-Vis spectrophotometer of the SEC set-up and the downstream SAXS cell should be kept as short and as narrow as possible to minimize band broadening. There should be some communication between the chromatography instrument and the beamline control to synchronize protein elution and SAXS frames acquisition. As a minimum, the link is provided by the user who is in charge of launching both systems simultaneously. In more automated cases, electronic triggering signals are sent back and forth between the two set-ups, with no user intervention.

A multiple column system is useful to optimize duty cycle in case the SAXS acquisition finishes before elution completion. Switching columns allows injection and data collection on column B while column A completes the return to equilibrium with the elution buffer. An automated cleaning procedure of the SAXS measuring cell should be implemented and performed after each elution, often using a detergent emulsion followed by water rinsing (Fig. 11.1). A full compatibility with an

**Table 11.1** Beamlines worldwide where SEC-SAXS is available to users

| Beamline SR facility | Uniform resource locator | SEC-SAXS availability | Available SEC columns |
|---|---|---|---|
| P12-EMBL Petra-3, Germany | http://www.embl-hamburg.de/biosaxs/p12/ | On request | GE Healthcare Superdex 200 10/300 GL, 24 ml |
| B21 DLS, UK | http://www.diamond.ac.uk/Beamlines/Soft-Condensed-Matter/small-angle/B21.html | Routinely | GE Healthcare Superdex 200, 2.4 ml |
| | | | SHODEX KW-402, 4.6 ml |
| | | | SHODEX KW-403, 4.6 ml |
| | | | SHODEX KW-404, 4.6 ml |
| SWING Synchrotron SOLEIL, France | http://www.synchrotron-soleil.fr/Recherche/LignesLumiere/SWING | Routinely | Agilent Bio-Sec-3-300, 4.6 ml |
| | | | SHODEX KW 405, 4.6 ml |
| BM29 ESRF, France | http://www.esrf.eu/home/UsersAndScience/Experiments/MX/About_our_beamlines/bm29/beamline-setup/hplc.html | Routinely | User provided only |
| SAXS/WAXS Australian Synchrotron, Australia | http://www.synchrotron.org.au/aussyncbeamlines/saxswaxs | Routinely | User provided only |
| BL23A NSRRC, Taiwan | http://www.nsrrc.org.tw/www/eng/endstation/17b3/saxs/index.htm | Routinely | |
| BioSAXS MacCHESS, USA | http://www.macchess.cornell.edu/MacCHESS/inline_SEC.html | Routinely | GE Healthcare Superdex 200 3.2/300 Increase 2.4 ml |
| | | | GE Healthcare Superdex 200 5/150 GL, 3 ml |
| | | | GE Healthcare Superdex 200 10/300 GL, 24 ml |
| | | | GE Healthcare Superdex 75 3.2/300 (RNA work only), 2.4 ml |
| BL4-2 SSRL, USA | http://www-ssrl.slac.stanford.edu/~saxs/sample_environ/fplc_saxs.htm | On request | GE Healthcare Superdex 200 PC3.2 & Superose 6 PC Superose 6 PC 3.2 |
| BioCAT APS, USA | http://www.bio.aps.anl.gov/techniques/index.html | Routinely | Columns provided, but users are encouraged to bring their own column |

automated direct injection set-up is a strong asset for fast switching between both modes, the latter being mostly useful for better statistics at larger $q$-values. Finally, the chromatography set-up can be positioned on a lockable rolling table that allows easy removal and repositioning, which is especially convenient on those SAXS beamlines that are not uniquely dedicated to biological solutions (Fig. 11.2).

SEC-SAXS is often complemented by other parameter monitoring for a more complete sample characterization beyond the UV spectrometer at the column outlet. For instance, a refractometer has been installed downstream of the SAXS measuring cell on SWING in the case of membrane protein studies (see the dedicated section below, (Berthaud et al. 2012)). However, due to protein diffusion resulting in band broadening between the various measuring cells, the protein concentration and the quality of particle separation vary, which makes it difficult to combine the observations. In particular, the refractometer must usually be at the end of the circuit whereas the SAXS cell, with its large section, produces a marked band broadening. For these reasons, a second, parallel circuit can conveniently be used for other biophysical characterization. Hence, a significant step forward has recently

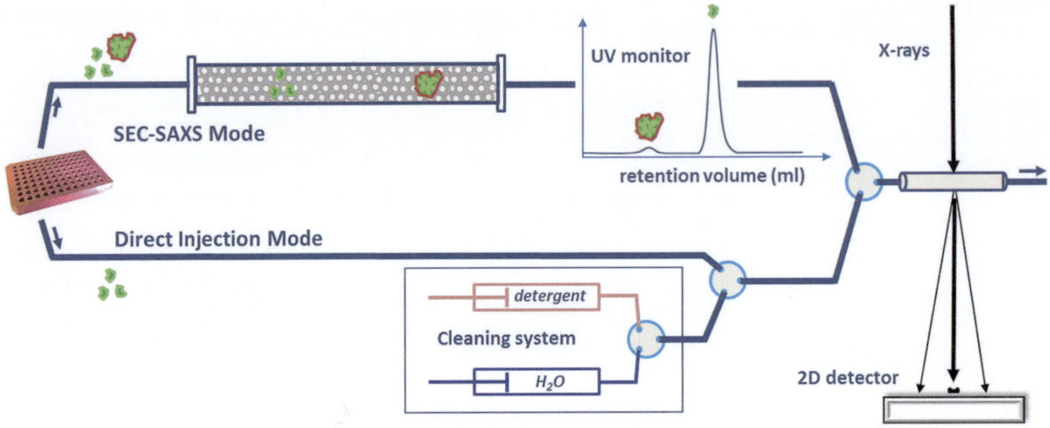


**Fig. 11.1** Schematic representation of a typical SEC-SAXS setup. Two alternative injection modes are represented, which can be selected by switching the various valves indicated in the diagram. The cleaning system is meant to remove any deposit on the walls of the SAXS measure cell. Cleaning, injection, and data acquisition are usually synchronized by a master control system to simplify the user task. Complementary features such as a second switchable column, or a splitted way for other biophysical characterizations, can be found at specific beamlines

been made at the EMBL Outstation in Hamburg by using a split valve at the outlet of the SE column creating two parallel and identical paths for the eluting sample, one flowing to the SAXS measuring cell while the other one flows successively through the U.V. cell, the refractometer and the viscometer (Graewert et al. 2015).

## 11.3 Columns

The first attempts at SEC-SAXS were the work of biochemist users who used lab columns with a 7.5 mm diameter, involving larger loaded protein amounts and longer elution times. Nowadays, 4.6 mm diameter columns are used practically everywhere. On SWING beamline, we typically favor matrices of polymer-coated silica beads, from Shodex™ [KW series] or from Agilent™ [BioSec series], due to their very high resolving power. However, for buffers with pH values higher than 7.5 or 8, organic polymer or carbohydrate based matrices have to be used. Column volumes should be minimized to avoid excessive dilution, e.g. around 5 ml. However, this should not be detrimental to resolution. Depending on the case (close peaks or not) resolution might be preferred and larger columns could be used. Column integrity should be regularly checked by eluting a mixture of reference samples. Filtering of buffers during elution is not as stringent a requirement as for MALLS, but it is still recommended to insert a 0.2 μm prefilter upstream of the pump. Even when columns are available on the beamline, users are encouraged to bring their own previously verified column. This limits problems of cross-contamination between various samples while avoiding the discovery on the beamline of unexpected interactions of the sample with the column matrix, a waste of resources and efforts.

## 11.4 Experimental Protocol

Choose the right column for the sample under study. The criteria for choice are fairly obvious: the range of molecular mass in which effective separation is achieved by the column compared with that of the sample; the sample hydrodynamic radius Rh, the pH to be used (silica based cannot stand pH higher than 7.5 or 8). If the column resolution is not enough to clearly separate two components, it is advisable to use a column with finer granulometry of the solid phase (the matrix). It can also be useful to combine two columns in series to improve the

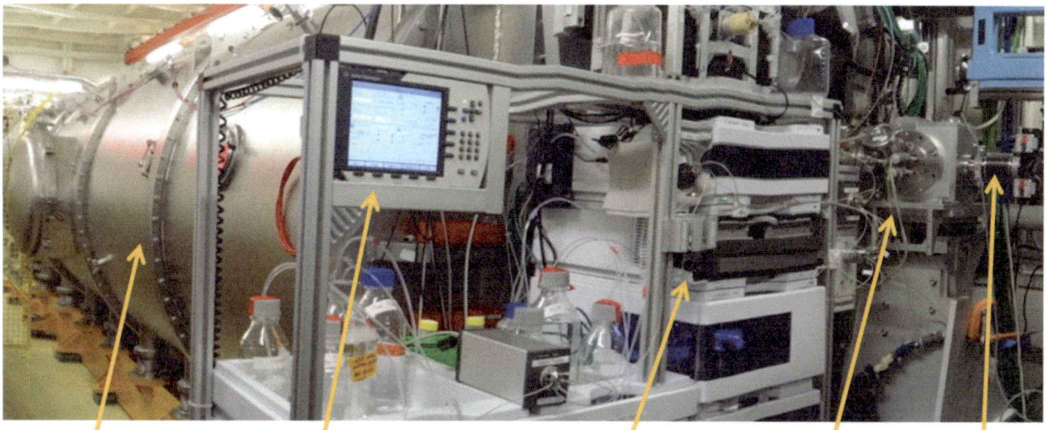


**Fig. 11.2** Photograph of the SEC-SAXS setup installed on the SWING beamline at the Soleil synchrotron

resolution, even if the resulting concentration at the peak top is slightly decreased.

Choose (if possible) the UV-Vis wavelength for elution monitoring. For instance 260 and 280 nm for a nucleoprotein complex are compulsory. A remote wavelength (400 nm) can be used to check the lamp stability. More specific wavelengths can be used in case chromophores are present to help distinguish particles containing them from the others and to better estimate their concentration.

Equilibration should be reached, usually checked by several stable UV-Vis profiles. The rule of thumb on SWING beamline for a 4.6 ml column is an absorbance variation at 280 nm lower than 5 mAU over 10 min at 200 μl/min.

At third generation synchrotrons, injection volumes and concentrations should be such that a concentration of the order of 1 mg/ml is reached at the elution peak head. Typically, the injection volume corresponds at most to 1 % of the column volume, beyond which resolution can be affected. For instance, 50 μl at 5 mg/ml can be loaded on top of a Biosec-3 300 Agilent 4.6 ml column, resulting in a fivefold dilution at the peak head in the case of a predominantly monodispersed solution with only a minor fraction of aggregates.

In general, the resulting concentration is low enough to make any potential coulombic repulsion effects negligible. However, when very high concentration solutions are loaded, it could be advisable to proceed to the "usual" dilution series (meaning the concentration at the top of the peak), and in that case, it is much simpler and effective to reduce the injected volume accordingly than to dilute the initial solution.

Buffer frames have to be collected before the void volume, and cover a significant fraction (typically 20–30%) of the time between injection and void volume to assess beam and elution stability and provide buffer scattering data with good statistics.

Data acquisition does not need to cover the entire elution process, typically lasting between 15 and 30 min. In particular, shining X-rays on aggregates only increases the risk of capillary fouling for no gain in useful information. It may be wiser to restrict the X-ray exposure to the time period of particle elution. In contrast it is good practice to collect data well after the last peak has eluted, in order to gain information on capillary fouling effects that could then be accounted for (see below).

If no preliminary run on the same column could be performed, the first elution should be monitored by the user and he/she may then have to trig manually the SAXS acquisition. Since the HPLC comprises its own UV monitor, usually positioned several tens of μl before the SAXS cell, there is some time left to check that the increase in optical density (OD) does indeed correspond to the expected peak and not to a spurious signal. Triggering at the rising side inflexion point of the expected peak is a good practice.

The acquisition frame rate should be compatible with the typical time widths of the elution peaks. About one frame per second allows more than adequate sampling of a 1-min broad peak.

It is also possible to collect data continuously with *ca* 10 ms frames when using photon counters as detectors. This "collect all" approach is simple and effective at the cost of (automatically) processing thousands of 2D images to yield 1D frames, most of which might contain no useful information. It can also produce a larger amount of undesired capillary fouling.

## 11.5 Data Processing

Data reduction, the first stage of data processing, is done using an automated pipeline converting 2D frames into 1D scattering profiles. Buffer frames are examined, their identity assessed by the absence of monotonic evolution or using some more general statistical tools such as CorMap (Franke et al. 2015) before averaging identical frames. The average buffer profile is subtracted from all frames of the elution dataset properly speaking. Then I(0) and $R_g$ are evaluated by applying the Guinier analysis to all buffer-subtracted curves. Two major cases can then be distinguished.

### 11.5.1 Simple Case: Well Resolved Peaks

In a favorable case, the species of interest elutes as a well-resolved peak far from the larger oligomers or aggregates that were present in the original solution. After performing the tasks mentioned above, one examines the evolution of $R_g$ with time (frame number) through the elution peak. The absence of monotonic variation across the peak suggests that the frames are identical. It is important to confirm this by a subsequent analysis of the ensemble of frames (e.g. using CorMap in PrimusQt (Petoukhov et al. 2012)) after scaling all frames to the same integral over a given $q$-range. Such an example is shown in Fig. 11.3. The series of frames around the peak maximum show no evolution of their $R_g$ value, and are practically identical after scaling. They can then safely be averaged for subsequent analysis and molecular modeling. The program DATASW (Shkumatov and Strelkov 2015) provides an option for automatic averaging using a sliding window to increase statistics if individual frames are too noisy and will yield for each peak the average of all frames under the peak and of frames $\pm 10\%$ around the maximum. This is fine as long as peaks are well resolved.

If one wishes to derive an estimate of the molecular mass from the value of I(0)/c at the peak top, one should take the band broadening effect into account due to the path between the UV detector of the HPLC instrument and the SAXS measuring cell. A simple and effective approximation is to divide the measured OD by the ratio SAXS-over-OD of the peak widths. More sophisticated approaches have been developed in data processing packages for the similar situation of SEC-MALLS data, in which the OD peak is convolved with a function mimicking the band-broadening process to match the MALLS peak shape. These approaches could easily be implemented for SEC-SAXS. The band broadening issue can also be addressed experimentally by directly measuring the OD within the SAXS measuring cell, but at the cost of a less precisely calibrated signal.

### 11.5.2 More Complex Cases

More complex situations are often encountered, in which peaks are not baseline resolved and most elution frames still contain contributions from two or more components. A first situation is shown in Fig. 11.4, where the main elution peak exhibits an unresolved shoulder on its rising

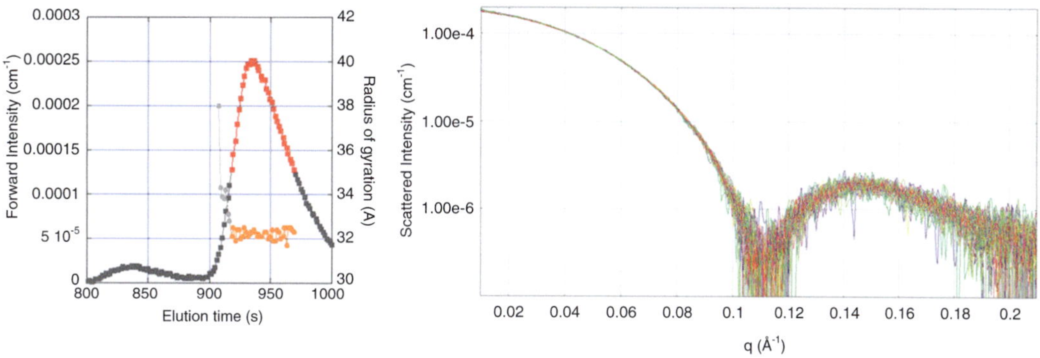


**Fig. 11.3** SEC-SAXS study of a protein well separated from aggregates. *Left graph*: time evolution of forward scattered intensity by a solution of urate oxidase (*squares, left side vertical axis*). The $R_g$-values for frames under the main elution peak are also shown (*filled circles, right side axis*). In orange are highlighted the values that are deemed stable and selected for subsequent frame averaging. The respective I(0) values are shown in *red*. All frames within full-width at half-maximum of the elution peak were kept, underlining the high monodispersity within the peak. *Right graph*: Superimposition of the 35 selected scattering frames, scaled to their integrated intensity in the range $0.07 < q < 0.10$ Å$^{-1}$. *Color code*: from *red* to *blue* with increasing time

edge, probably from a fraction of small oligomers. Accordingly, a systematic decrease of $R_g$ is observed throughout the peak, but still, frames with stable $R_g$ values are obtained on the descending side. These frames can then be used for averaging and further analysis, after thorough check that they are identical once scaled. However, since the data were extracted from the flank of a non perfectly monodispersed peak, it is crucial to confirm that the curve corresponds to the species of interest. This should be done by estimating its molecular mass, best using approaches not requiring the knowledge of the concentration (e.g. using SaxsMoW (Fischer et al. 2010)).

One may also want to retrieve more information from a set of unresolved patterns. To try and overcome the difficulties arising from poorly separated components, specific software was developed as a dedicated module within the US-SOMO package for macromolecular modeling using hydrodynamic and scattering data (Brookes et al. 2013, 2016). This module offers classical Singular Value Decomposition (SVD) analysis of the data set that informs the choice of the number of species m necessary to account for the ensemble of frames as well as filters noise by projection of all scattering frames on the subspace defined by the first m eigenvectors deemed meaningful. The specificity of this approach lies in its use of transposed vectors from the 2D scattering dataset I($q$,t) where chromatograms $I_q$(t) display the time evolution of intensity at a given $q$-value (Fig. 11.5). Elution peaks are modeled as gaussians and the entire dataset is globally fitted to yield the resulting m gaussians together with the m coefficients associated with each frame. Finally, the program extracts the m scattering patterns for each proposed species. This is illustrated in the article by the analysis of the data obtained from the elution of a BSA solution containing dimers and trimers beyond monomers (Fig. 11.6).

Very recently, an article of great interest was published that reports SEC-SAXS analysis of an allosteric protein in an inactive form and in an allosterically activated form (Meisburger et al. 2016). The authors show that, although only one elution peak is visible on the elution profile, the radius of gyration is not constant across the peak, showing that several species elute without being resolved. The authors utilize standard SVD analysis to determine the minimum number of species compatible with each dataset before using the so-called evolving factor analysis (EFA), a variant of SVD that determines the peak range associated with each species before retrieving the true SAXS pattern of each species

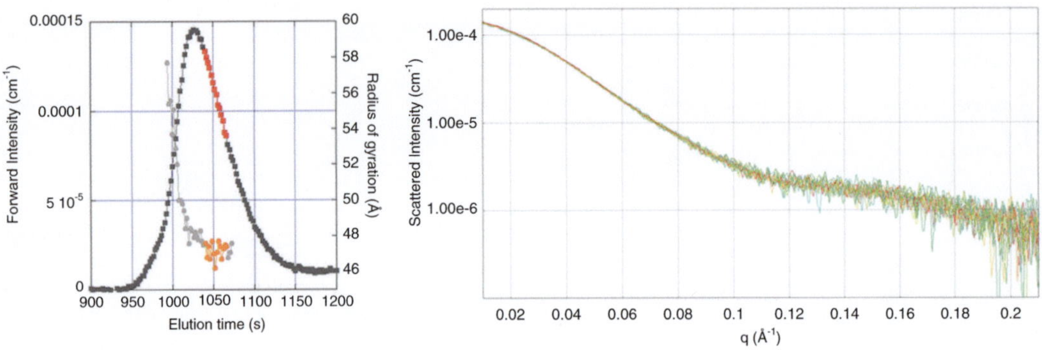


**Fig. 11.4** SEC-SAXS study of a protein poorly separated from oligomers. *Left graph*: time evolution of forward scattered intensity and radius of gyration by a solution of a complex of Smac/Diablo with a BIR2-BIR3 construct (Mastrangelo et al. 2015). The *symbols* and *color code* are the same as in Fig. 11.3. *Right graph*: superimposition of the 14 selected scattering frames, scaled to their integrated intensity in the range $0.07< q <0.10$ Å$^{-1}$. Only part of the frames from the *right-hand* side of the elution peak could be retained for further analysis, based both on the stability of the $R_g$ values and on the superposition of the scattering frames. In particular, the three last eluting frames with shown $R_g$ values were discarded due to small discrepancy (not shown) in their scaled scattering curves with respect to the ensemble, a difference not discernible from the $R_g$ values alone

from the singular vectors by imposing a zero concentration outside the peak range.

### 11.5.3 The Capillary Fouling Issue

Radiation damage can be a real issue and it is actually the object of a very interesting recent study (Hopkins and Thorne 2016) to which we refer the reader for an in-depth study of the phenomenon. Let us just mention a couple of simple advices: Tris-HCl acts as a radical scavenger and its wide use can only be recommended. We have also noted in some cases a tendency of phosphate buffer to increase radiation effect. Phosphate buffer should probably be avoided when planning a SAXS study (see Fig. 11.7). Finally, it has also become common practice to add from 2 to 5% (v/v) glycerol to sample solution as this limits the consequences of irradiation (Kuwamoto et al. 2004) but one should keep in mind that this could affect the conformation of the protein, primarily those that are partly unstructured as glycerol is known to stabilize compact, folded protein conformations.

The consequence of radiation damage on the SEC-SAXS data is the appearance of a small but increasing contribution to scattering at very low angles that can be directly attributed to progressive protein adsorption onto the irradiated part of the capillary wall. This parasitic scattering causes an increase of all apparent radii of gyration, and most spectacularly on frames after the elution peak, when the protein concentration of the circulating solution vanishes. This is directly visible in Fig. 11.7c showing early, top and late frames recorded in phosphate buffer.

This particular issue was recently addressed in the HPLC module of the US-SOMO package. In particular, the conversion of the experimental curves into chromatograms for each $q$-value clearly demonstrates that the parasitic intensity remains constant as soon as the protein elution is completed. This strongly supports that protein progressive deposition and not instantaneous aggregation is the origin of the unwanted scattering. The algorithm in the US-SOMO package estimates the time evolution of this effect at each $q$-value and corrects for it, finally delivering a set of corrected frames (Fig. 11.8). The experimental requisite for this algorithm to work is that data must be recorded well after the last peak elution. Recently, Kirby and coll. came up with a totally new design of a flow cell in which the

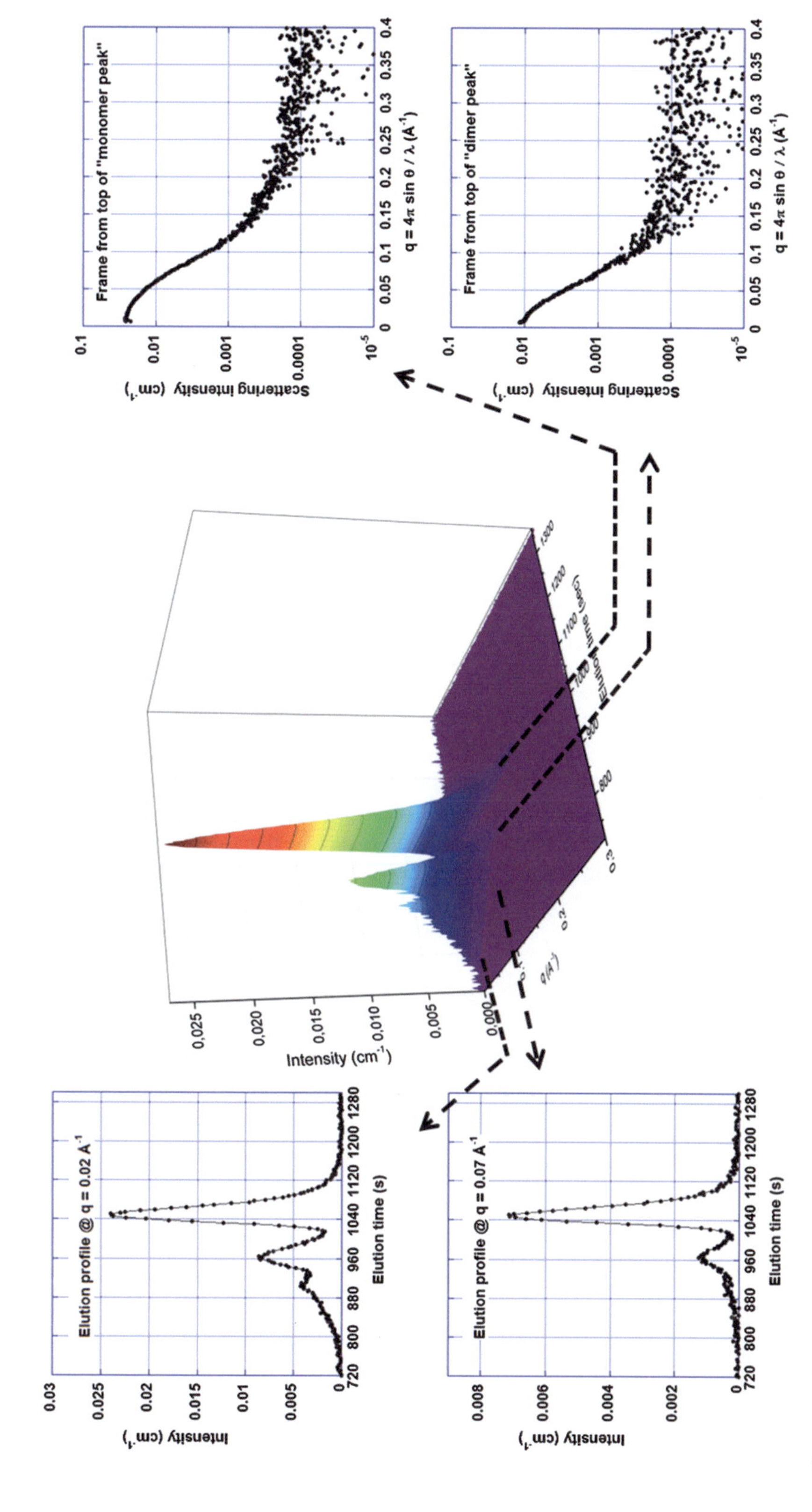


**Fig. 11.5** 2D representation of a SEC-SAXS dataset. A SEC-SAXS dataset consists on the collection of several SAXS frames collected at successive times. Such an ensemble recorded from a BSA solution is represented at the center of the figure as a 3D surface showing the scattered intensity I as a function of both the momentum transfer q and time t. On the *right-hand* side of the figure are shown two SAXS curves measured at different times during elution. The representation of the elution curves at fixed *q*-values shown on the *left-hand* part of the figure is typically put to good use in the HPLC module of the US-SOMO program that evaluates a possible parasitic baseline due to capillary fouling under irradiation, before attempting to deconvolve several species contributions from not fully resolved peaks

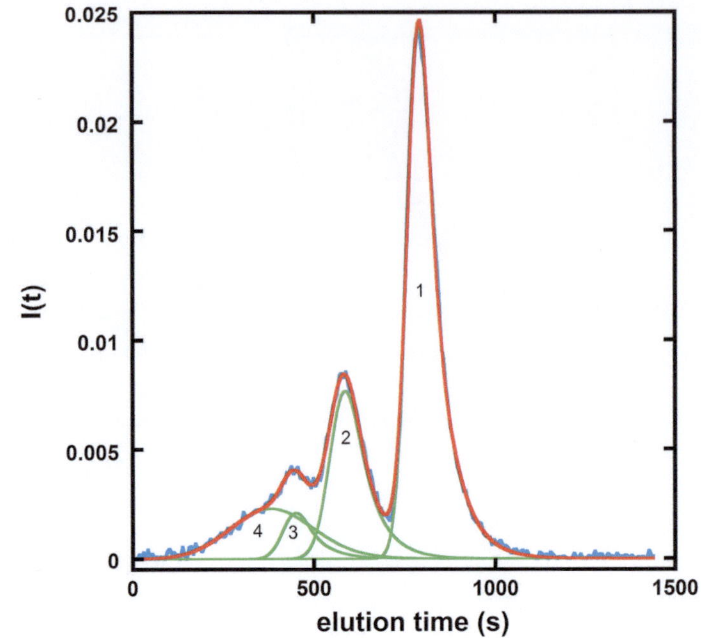


**Fig. 11.6** Gaussian decomposition of a chromatogram. Scattering intensities from a markedly polydispersed solution of BSA are represented as a function of time for a given $q$-value (here $q = 0.0205$ Å$^{-1}$). Experimental data in *blue* with the fit (*red line*) obtained by summing the contributions of the four Gaussian components shown in *green* and labelled *1–4*

main stream of protein solution is separated from the capillary walls by a thin laminar flow of buffer, which almost entirely eliminates any apparent radiation effect on the scattering pattern (Kirby et al. 2016). This new design has the potential to completely change the way SEC-SAXS data will be collected in the future.

## 11.6 Specific Applications

### 11.6.1 Case of Weak Complexes

Up to now, we have considered the use of SEC as a precious means of separating a protein from aggregates or various oligomeric forms through elution. We will consider now the opposite case where we wish to study a complex of limited or weak affinity without elution causing its dissociation. The origin of the problem is simply the law of mass action that makes dissociation unavoidable unless the complex is very tight, with a $K_D$ lower than a few nM. The solution to the problem is simple, at least in the case where one of the partners that is not the largest subunit in size can be produced in sufficient amount. In this favorable case, this partner is added to the elution buffer at a concentration that ensures stability of the complex (say, with a saturation level of 90 or 95% of its binding sites). The complex elutes in a milieu that contains the small partner and therefore will not significantly dissociate (Fig. 11.9). This approach has been used on several users' systems on the SWING beamline. An example in point is that of the study of PTPN4, a tyrosine phosphatase, with a PDZ domain upstream of the catalytic PTP domain separated by a short linker. Biochemical experiments had established that upon ligand binding on to PDZ, the PTP enzymatic activity that was inhibited by PDZ was practically restored to the level of isolated PTP domain. The system was investigated using a combination of AUC, NMR and SEC-SAXS (Maisonneuve et al. 2014). The two domain construct PDZ-PTP$^{WT}$ was studied free and in the presence of Cyto8-RETEV, the most affine peptide for the PDZ domain of PTPN4 ($K_D$ of 1 μM). However, even for the most affine peptide the complex will not survive SEC elution. Therefore 40 μM

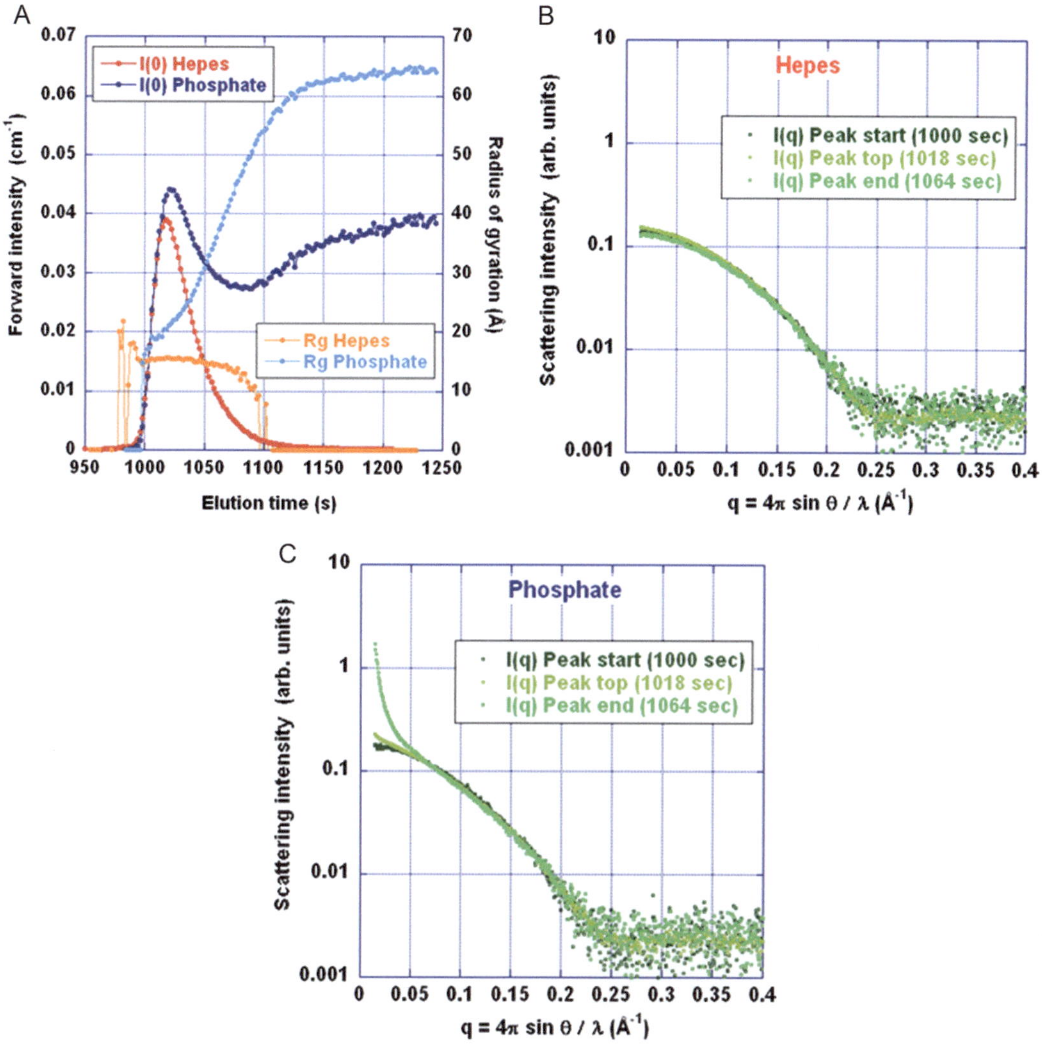


**Fig. 11.7** Comparison of capillary fouling with a lysozyme solution in two different buffers at same pH and ionic strength *Panel A* shows the evolution with time of both I(0) and $R_g$ obtained from a Guinier analysis on lysozyme datasets recorded in two different buffers at pH = 7, precisely Hepes: 20 mM, NaCl: 150 mM and Sodium Phosphate: 20 mM, NaCl: 150 mM. *Color code*: I(0) in *dark blue* (*phosphate*) or red (*Hepes*); $R_g$ in *cyan* (*phosphate*) or *orange* (*Hepes*). *Panel B* (*Hepes*) and *panel C* (*phosphate*) show three scattering profiles after scaling over the *q*-range (0.15 Å$^{-1}$–0.25 Å$^{-1}$) corresponding to the peak start (*dark green*), peak top (*kaki*) and peak end (*light green*). All three frames in *panel B* are nearly identical while the curve at peak *top* in *panel C* shows a clear upturn at small angles that is a dominant feature of the curve at peak end, a direct expression of a remarkably strong capillary fouling effect

of Cyto8-RETEV were added to the elution buffer before the experiment, a concentration that saturates 95% of PDZ binding sites. The two curves were then analyzed in detail and no meaningful difference was detected. It was thus essential to have ensured full saturation of PDZ binding sites. The authors conclude that "peptide binding modulates the overall stability/dynamics of the PDZ domain in the two-domain construct" without any detectable change in the average conformation of the protein as monitored by SAXS.

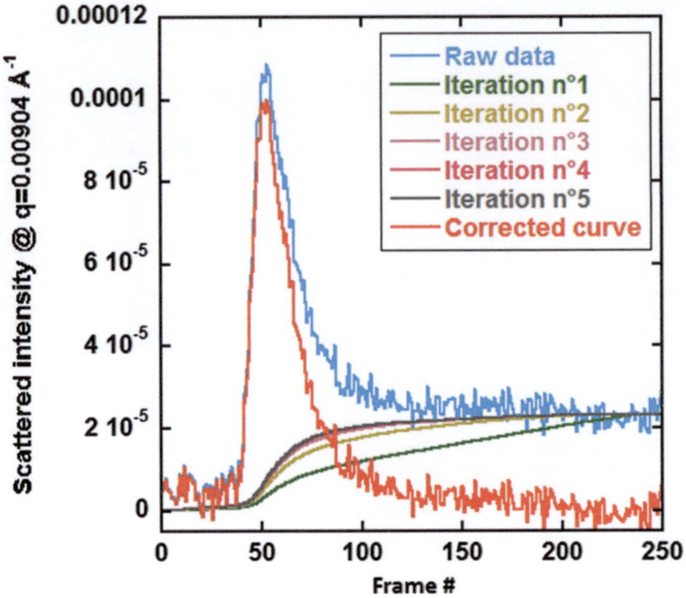


**Fig. 11.8** Baseline correction for capillary fouling effect. The *blue line* represents the intensity scattered by a protein solution as a function of elution time. The intensity does not return to zero well after the main protein peak. The US-SOMO program is used to determine and subtract a baseline accounting for capillary fouling. The five, grossly sigmoidal in shape profiles are the successive corrections obtained from an iterative process that has practically converged after four iterations. The *red line* shows the corrected profile after subtraction of the final correction

### 11.6.2 Case of Membrane Protein/Detergent Complexes

To be maintained folded in solution, membrane proteins have to be associated with amphiphilic molecules (in practice detergents), whose hydrophobic tails cover the hydrophobic transmembrane surface of the protein, thus hiding it from the (water) solvent. However, when their concentration is raised to values needed to dissolve membrane proteins, detergent molecules spontaneously associate into large micelles, with scattering power comparable to those of the protein. The resulting coexistence of different particles in the sample requires a specific strategy to analyze membrane protein structures in solution (Berthaud et al. 2012). SEC-SAXS must be used as the first step in this strategy, whereby the membrane protein is eluted through a column equilibrated with a buffer containing detergent molecules above the critical micellar concentration (cmc). In the previous process of concentrating the membrane protein solution, needed to obtain a reasonable data statistics, the concentration of free micelles is also often increased, to a level generally unknown with precision. The SEC process then ensures both that the level of free micelles around the protein is the same as in the buffer, and that the surplus of free micelles elutes with a different retention volume, usually higher, than the protein-detergent complex. It is a good idea to check that the chosen column can indeed separate the protein-detergent complex from the surplus of free micelles, by monitoring the elution process with a refractometer, since most detergents do not absorb light. The SAXS data collected under the protein peak can then be treated afterwards in the same way as for a soluble protein, i.e. by experimental subtraction of the detergent buffer data collected before the column void volume (Fig. 11.10, top panel).

The second step in the strategy requires modeling the detergent moiety around a known construct of the protein, e.g. using the program Memprot that models a detergent corona shaped as an elliptical torus (Fig. 11.10, bottom panel). The program was built on experimental data collected on the beamline SWING at Synchrotron Soleil from Aquaporin-0 solubilized in n-Dodecyl β-D-maltoside (DDM), a transmembrane protein of

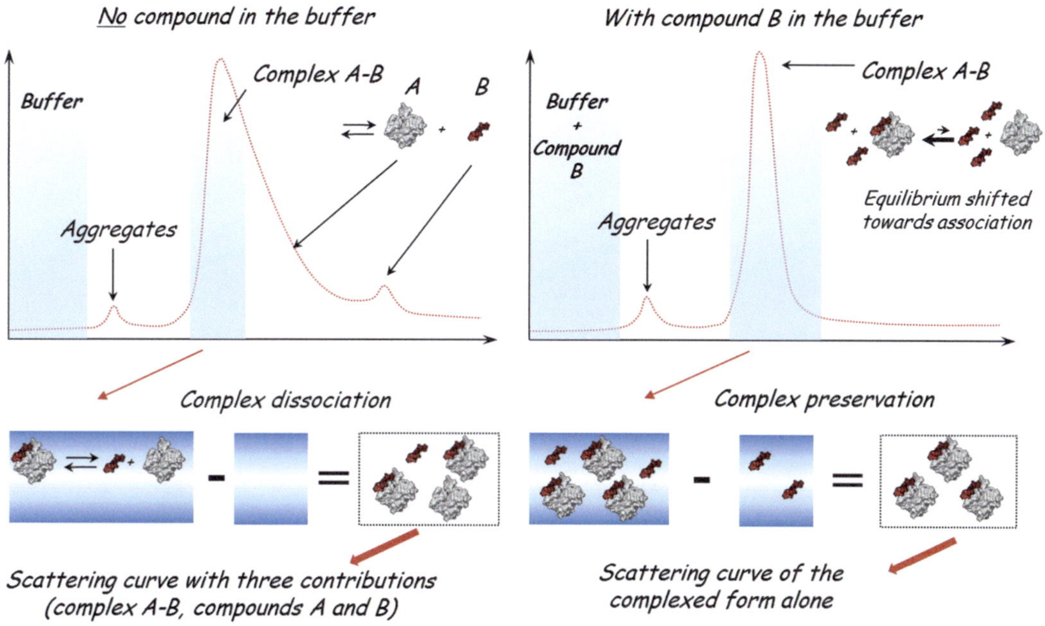


**Fig. 11.9** Stabilization of complexes during elution by addition of a ligand to the elution buffer

known crystallographic structure. It was shown that there is an optimal set of the parameters used to model the DDM torus, for which the experimental curve was perfectly fitted (Perez and Koutsioubas 2015). This comforted the approach, since it means that the hybrid structure can be considered as a reliable basis onto which further investigations can be made.

Hence, the third and final step consists in putting to good use the known hybrid structure to analyze its (unknown) interactions with soluble protein partners, whether these are additional domains within the same chain than the known (but partial) construct or separate chains that make a complex with the membrane protein (Fig. 11.11). An example of such a study was recently published by Wojtowicz et al. (2016). In this study, the interactions between the periplasmic signaling domain of the transmembrane heme transporter HasR and the main body of HasR could be directly investigated in the presence or not of the outer hemophore HasA. Given that HasR has no symmetry, the design of the detergent corona was improved to mimic the shape of the protein in the transmembrane plane, without increasing the number of parameters. The output of the study points to a position of the signaling domain closer to the main body when the latter is associated to HasA.

## 11.7 Conclusion

Whenever possible SEC-SAXS should be used in addition to or even instead of direct injection measurements. Despite the clear drawback of sample dilution during the elution process, the advantages in terms of monodispersity are overwhelming, and avoid drawing wrong conclusions from unreliable data. Several computing and analytical tools are currently being developed to help retrieve the structural information contained in these 2D datasets. These advantages are so evident that a group at the ILL (Grenoble, France), a high-flux neutron reactor facility is actively pursuing the installation of a similar set-up onto one of the SANS beamlines (Jordan et al. 2016). Moreover, the throughput of the method, though apparently much lower than that of "high-throughput", direct measurements, is in actual fact only a factor of the order of two to five times slower due to (i) the need to record several concentrations of a given molecule with the direct injection strategy, (ii) the often associated

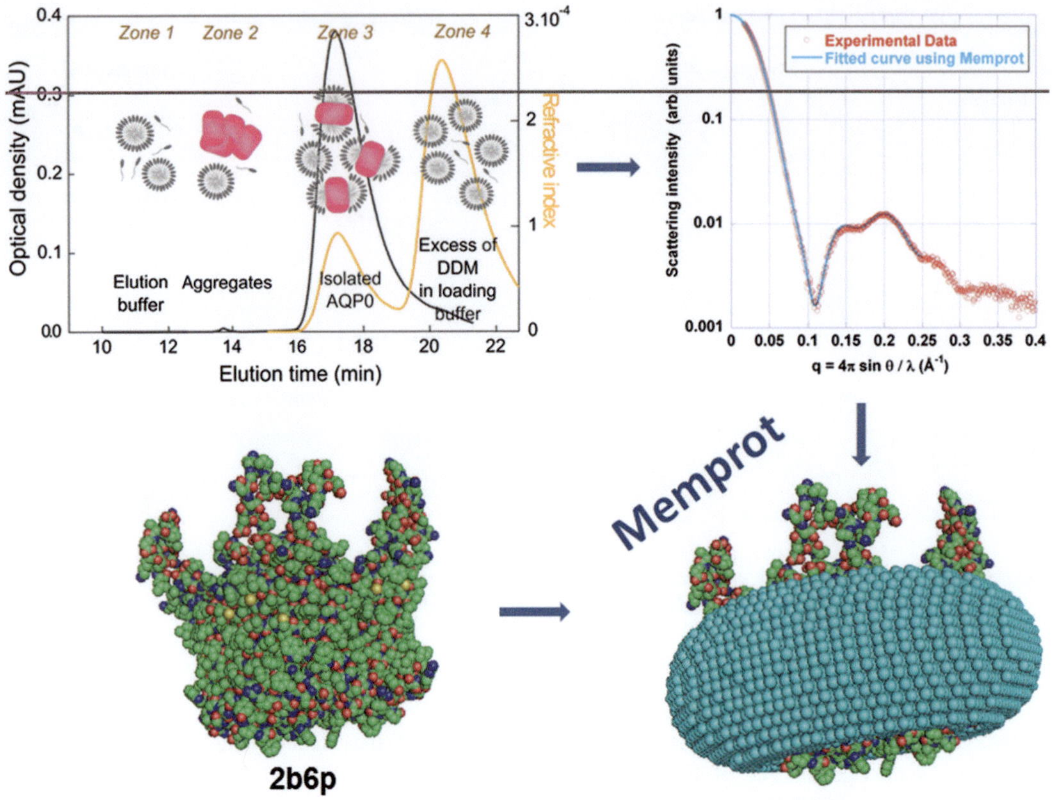


**Fig. 11.10** A protocol to model the detergent corona around a membrane protein. The first step shown on top consists in taking advantage of the SEC-SAXS set-up to extract the scattering curve from the protein-detergent complex, without being biased by the signal arising from free micelles. Partly reprinted with permission from Berthaud et al. (2012) (Copyright (Berthaud, Manzi et al.) American Chemical Society. The second step shown in *bottom row* consists in modeling the detergent corona around the known structure of the membrane protein by a coarse grained approach such as the one developed in the program Memprot (Perez and Koutsioubas 2015))

need of multiple buffer recordings interleaved between the various concentrations and (iii) the time needed to clean and dry the measuring cell after each acquisition, all factors that significantly increase the duty cycle for conventional measurements. Finally, Table 11.1 is the best illustration of the interest raised by SEC-SAXS measurements as the number of SAXS instruments offering this possibility is continuously increasing and as several of these instruments appear to be at the heart of methodological developments. We hope this short survey will have contributed to increase the awareness of SEC-SAXS potential among biochemists and molecular biologists.

A few general or specific tips

- SEC-SAXS should not be considered as the last purification step. The loaded solution should be as pure as possible for SEC-SAXS to be fully effective.
- Make sure to record buffer frames before the void volume and after protein elution is complete (return to a steady-state). This is necessary for baseline correction in the case of capillary fouling and for subsequent deconvolution (US-SOMO, Brookes et al. 2016).
- When studying membrane proteins, carefully check that the excess of free micelles elutes

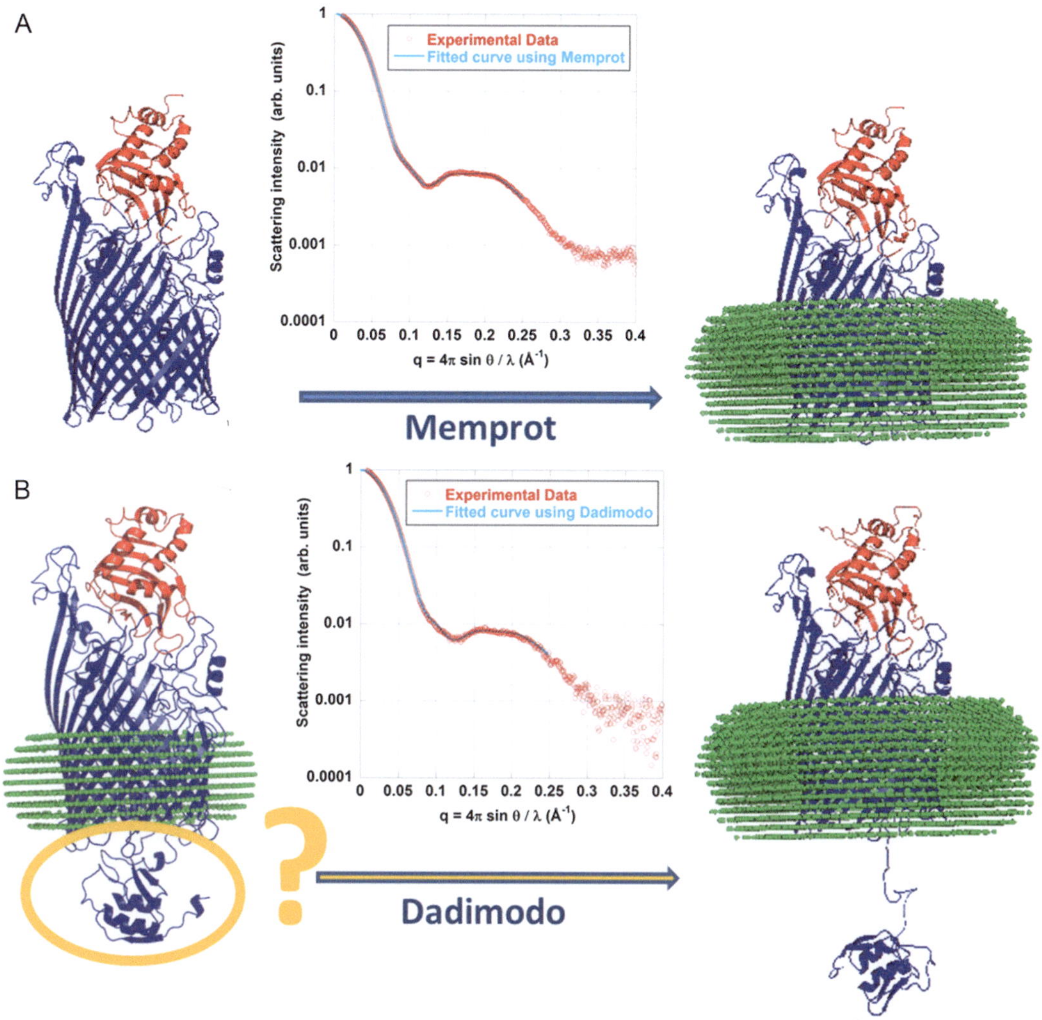


**Fig. 11.11** A strategy to derive structural information from membrane proteins. (**a**) First step: the protocol shown in Fig. 11.10 is followed to model a detergent corona around a transmembrane protein construct, whose atomic structure must be known. (**b**) Second step: the unknown position of a domain of known structure is optimized against the SAXS data from the whole construct, using the program Dadimodo (Evrard et al. 2011). Phi-psi angles of initially selected residues are stochastically modified until a good agreement with the experimental curve is reached. The program has been specifically modified to include the coarse grained representation of the detergent corona

separately from protein complex using refractometry monitoring.

- When studying a complex, if the $K_D$ is unknown or too large to be confident that the complex is preserved during elution, add one partner to the elution buffer at an adequate concentration.
- When attempting time profile (chromatogram) deconvolution using (modified)-Gaussian functions, start by fitting the right-most peak (last eluting species) to determine the distortion values that will subsequently be kept practically constant (US-SOMO, Brookes et al. 2016).

**Acknowledgments** We thank our colleagues in charge of SAXS beamlines at synchrotron facilities who kindly provided us with informations that constitute the basis of Table 11.1. We are very grateful to Dr. P. Roblin for help with Fig. 11.9. Finally, J. Pérez thanks the various groups of users who came to SWING with experimental challenges that prompted several of the reported developments.

## References

Berthaud A, Manzi J, Perez J, Mangenot S (2012) Modeling detergent organization around aquaporin-0 using small-angle X-ray scattering. J Am Chem Soc 134(24):10080–10088

Boczkowska M, Rebowski G, Dominguez R (2015) The challenges of polydisperse SAXS data analysis: two SAXS studies of PICK1 produce different structural models. Structure 23(11):1967–1968

Brookes E, Pérez J, Cardinali B, Profumo A, Vachette P, Rocco M (2013) Fibrinogen species as resolved by HPLC-SAXS data processing within the ultraScan solution modeler (US-SOMO) enhanced SAS module. J Appl Crystallogr 46:1823–1833

Brookes E, Vachette P, Rocco M, Pérez J (2016) US-SOMO HPLC-SAXS module: dealing with capillary fouling and extraction of pure component patterns from poorly resolved SEC-SAXS data. J Appl Crystallogr 49(Pt 5):1827–1841

David G, Pérez J (2009) Combined sampler robot and high-performance liquid chromatography: a fully automated system for biological small-angle X-ray scattering experiments at the Synchrotron SOLEIL SWING beamline. J Appl Crystallogr 42:892–900

Erlendsson S, Arleth L, Madsen KL (2015) Response to the challenges of polydisperse SAXS data analysis: two different SAXS studies of PICK1 produce different structural models. Structure 23(11):1969–1970

Evrard G, Mareuil F, Bontems F, Pérez J (2011) DADIMODO: a program for refining the structure of multidomain proteins and complexes against small-angle scattering data and NMR-derived restraints. J Appl Crystallogr 44:1264–1271

Fischer H, de Oliveira Neto M, Napolitano HB, Polikarpov I, Craievich AF (2010) Determination of the molecular weight of proteins in solution from a single small-angle X-ray scattering measurement on a relative scale. J Appl Crystallogr 43:101–109

Franke D, Jeffries CM, Svergun DI (2015) Correlation map, a goodness-of-fit test for one-dimensional X-ray scattering spectra. Nat Methods 12(5):419–422

Graewert MA, Franke D, Jeffries CM, Blanchet CE, Ruskule D, Kuhle K, Flieger A, Schafer B, Tartsch B, Meijers R, Svergun DI (2015) Automated pipeline for purification, biophysical and x-ray analysis of biomacromolecular solutions. Sci Rep 5:10734

Hopkins JB, Thorne RE (2016) Quantifying radiation damage in biomolecular small-angle X-ray scattering. J Appl Crystallogr 49(Pt 3):880–890

Jordan A, Jacques M, Merrick C, Devos J, Forsyth VT, Porcar L, Martel A (2016) SEC-SANS: size exclusion chromatography combined in situ with small-angle neutron scattering this article will form part of a virtual special issue of the journal, presenting some highlights of the 16th International conference on small-angle scattering (SAS2015). J Appl Crystallogr 49(6):2015–2020

Karlsen ML, Thorsen TS, Johner N, Ammendrup-Johnsen I, Erlendsson S, Tian X, Simonsen JB, Hoiberg-Nielsen R, Christensen NM, Khelashvili G, Streicher W, Teilum K, Vestergaard B, Weinstein H, Gether U, Arleth L, Madsen KL (2015) Structure of dimeric and tetrameric complexes of the BAR domain protein PICK1 determined by small-angle X-ray scattering. Structure 23(7):1258–1270

Kirby N, Cowieson N, Hawley AM, Mudie ST, McGillivray DJ, Kusel M, Samardzic-Boban V, Ryan TM (2016) Improved radiation dose efficiency in solution SAXS using a sheath flow sample environment. Acta Crystallogr Sect D 72(12):1254–1266

Kuwamoto S, Akiyama S, Fujisawa T (2004) Radiation damage to a protein solution, detected by synchrotron X-ray small-angle scattering: dose-related considerations and suppression by cryoprotectants. J Synchrotron Radiat 11(Pt 6):462–468

Madasu Y, Yang C, Boczkowska M, Bethoney KA, Zwolak A, Rebowski G, Svitkina T, Dominguez R (2015) PICK1 is implicated in organelle motility in an Arp2/3 complex-independent manner. Mol Biol Cell 26(7):1308–1322

Maisonneuve P, Caillet-Saguy C, Raynal B, Gilquin B, Chaffotte A, Perez J, Zinn-Justin S, Delepierre M, Buc H, Cordier F, Wolff N (2014) Regulation of the catalytic activity of the human phosphatase PTPN4 by its PDZ domain. FEBS J 281(21):4852–4865

Mastrangelo E, Vachette P, Cossu F, Malvezzi F, Bolognesi M, Milani M (2015) The activator of apoptosis Smac-DIABLO acts as a tetramer in solution. Biophys J 108(3):714–723

Mathew E, Mirza A, Menhart N (2004) Liquid-chromatography-coupled SAXS for accurate sizing of aggregating proteins. J Synchrotron Radiat 11 (Pt 4):314–318

Meisburger SP, Taylor AB, Khan CA, Zhang S, Fitzpatrick PF, Ando N (2016) Domain movements upon activation of phenylalanine hydroxylase characterized by crystallography and chromatography-coupled small-angle X-ray scattering. J Am Chem Soc 138(20):6506–6516

Perez J, Koutsioubas A (2015) Memprot: a program to model the detergent corona around a membrane protein based on SEC-SAXS data. Acta Crystallogr D Biol Crystallogr 71(Pt 1):86–93

Petoukhov MV, Franke D, Shkumatov AV, Tria G, Kikhney AG, Gajda M, Gorba C, Mertens HDT,

Konarev PV, Svergun DI (2012) New developments in the ATSAS program package for small-angle scattering data analysis. J Appl Crystallogr 45 (2):342–350

Shkumatov AV, Strelkov SV (2015) DATASW, a tool for HPLC-SAXS data analysis. Acta Crystallogr D Biol Crystallogr 71(Pt 6):1347–1350

Watanabe Y, Inoko Y (2009) Size-exclusion chromatography combined with small-angle X-ray scattering optics. J Chromatogr A 1216(44):7461–7465

Wojtowicz H, Prochnicka-Chalufour A, Cardoso de Amorim G, Roudenko O, Simenel C, Malki I, Pehau-Arnaudet G, Gubellini F, Koutsioubas A, Perez J, Delepelaire P, Delepierre M, Fronzes R, Izadi-Pruneyre N (2016) Structural basis of the signaling through a bacterial membrane receptor HasR deciphered by an integrative approach. Biochem J 473(14):2239–2248

# Applications of SANS to Study Membrane Protein Systems

# 12

Frank Gabel

**Abstract**

Small angle neutron scattering (SANS) is a powerful tool to obtain structural information on solubilized membrane proteins on the nanometer length-scale in complement to other structural biology techniques such as cryo-EM, NMR and SAXS. In combination with deuteration of components and/or contrast variation ($H_2O$:$D_2O$ exchange in the buffer) SANS allows to separate structural information from the protein and the detergent/lipid parts in solution. After a short historical overview on results obtained by SANS on membrane protein systems, this book chapter introduces the basic theoretical principles of the technique as well as requirements on samples. The two introductory sections are followed by an illustration of the specific consequences of sample heterogeneity of solubilized membrane proteins in the presence of detergent/lipid molecules on the interpretation of structural information by using simple, geometric models. The next sections deal with more sophisticated modelling approaches including *ab initio* shape reconstructions and full-atomic models in the presence of detergent/lipid and specific results obtained by these approaches. After a short comparison with the SAXS technique, this book chapter concludes with an overview of present and

F. Gabel (✉)
Université Grenoble Alpes, Institut de Biologie Structurale, 38044 Grenoble Cedex 9, France

Commisariat à l'Energie Atomique et aux Energies Alternatives, Direction de la Recherche Fondamentale, Institut de Biologie Structurale, 38044 Grenoble Cedex 9, France

Centre National de la Recherche Scientifique, Institut de Biologie Structurale, 38044 Grenoble Cedex 9, France

Large Scale Structures Group, Institut Laue-Langevin, 38042 Grenoble Cedex 9, France
e-mail: frank.gabel@ibs.fr





B. Chaudhuri et al. (eds.), *Biological Small Angle Scattering: Techniques, Strategies and Tips*, Advances in Experimental Medicine and Biology 1009, DOI 10.1007/978-981-10-6038-0_12

future developments and impact that can be expected by SANS on membrane structural biology in the coming years.

**Keywords**
Contrast variation • Heavy water • $D_2O$ • Deuteration • Form factor • MD simulation • Shape • Low resolution • Detergent • Lipid • Protein

## 12.1 Historical Overview

The application of small-angle neutron scattering (SANS) for the structural study of (weakly scattering) biological systems such as solubilized membrane proteins became technically feasible with the advent of high-flux neutron sources in the 1960s and 1970s. Earliest work, carried out at the Institut Laue-Langevin (Grenoble, France) and at the High Flux Beam Reactor (Brookhaven, USA) includes studies on several membrane protein systems purified directly from biological tissues: human serum low-density lipoprotein (Stuhrmann et al. 1975), bovine and frog rhodopsin (Osborne et al. 1978; Yeager 1976), the acetylcholine receptor from *Torpedo californica* (Wise et al. 1979) and porcine pancreatic colipase (Charles et al. 1980). Most of this early work was able to provide important model-free parameters of the protein-detergent complexes such as radii of gyration $R_G$ of both the complex and the individual partners, their stoichiometry, and molecular masses. The qualitative relative arrangements (in particular the distances) between protein and detergent moieties were obtained by a Stuhrmann analysis (Stuhrmann 1973), i.e. by interpreting the change of the measured $R_G$ at different contrast conditions ($H_2O$:$D_2O$ ratio in the solvent), occasionally combined with deuterated detergent to enhance contrast. In some cases, simple geometrical bodies were proposed for the protein and detergent moieties (Wise et al. 1979; Charles et al. 1980). Even though the functional interpretations of these pioneering results were limited in the absence of atomic-resolution structures, they allowed to probe and validate (or discard) basic working models of the membrane proteins studied. A more exhaustive overview on early SANS results from membrane protein systems can be found elsewhere (Timmins and Zaccai 1988).

In the second half of the 1980s the routine use of recombinant protein expression (Atkinson and Small 1986) made a wider range of membrane protein systems accessible to SANS, in particular in combination with an increasing commercial availability of deuterated detergent molecules. Concomitantly, more sophisticated modelling approaches using a number of beads to represent protein shapes were being developed (Perkins and Weiss 1983). The investigation of atomic-resolution membrane protein models became available in the 1990s with the first structures being deposited in the protein data bank (PDB) and the development of computer programs to back-calculate SANS curves efficiently from them (Svergun et al. 1998). At the same time, *ab initio* shape analysis was being developed (Svergun 1999; Chacon et al. 1998) and since the turn of the millennium, an increasingly versatile toolbox of computer programs for the interpretation of SANS data, including rigid body modeling and the comparison to cryo-EM data has become available (Zaccai et al. 2016; Chaudhuri 2015; Petoukhov et al. 2012). In parallel, ever more sophisticated labeling schemes (e.g. partial deuteration) became available for detergent/lipid molecules (Hiruma-Shimizu et al. 2016; Haertlein et al. 2016; Maric et al. 2015) to fine-tune contrast matching approaches.

## 12.2 Neutron Scattering Theory and Experimental Considerations

The general theory of small-angle scattering (SAS) as well as the specific theory and experimental setup for general SANS experiment have been introduced in previous sections of this book or in recent reviews and I will summarize here only briefly specific points related to membrane protein systems (Clifton et al. 2013).

The SANS signal measured from a solution of arbitrarily oriented particles $i$ (after solvent subtraction) corresponds to a one-dimensional intensity $I$ versus the modulus of the wave vector $\boldsymbol{q}$ and can be expressed as follows:

$$I(q) = \sum_i N_i \left\langle \left| \int \Delta\rho_i e^{i\vec{q}\cdot\vec{r}} dV_i \right|^2 \right\rangle, \quad q = \frac{4\pi}{\lambda}\sin\theta \tag{12.1}$$

Where $\lambda$ is the neutron wavelength and $2\theta$ the scattering angle. $N_i$ is the number of particles of a distinct species $i$. For solubilized membrane proteins these species correspond in practice to protein-detergent/lipid complexes (PDC), protein-free detergent/lipid aggregates (such as micelles, vesicles etc.) and single, free detergent and/or lipid molecules (Fig. 12.1). $\Delta\rho_i$ is the neutron scattering length density (SLD) contrast of specific particles with respect to the solvent. It varies in general for neutrons between protein and detergent/lipid head- and tail-groups (Timmins and Zaccai 1988; Jacrot 1976) (Fig. 12.2) and therefore has to be integrated over the whole particle volume $V_i$. The broken brackets correspond to a rotational average over all possible particle orientations.

Figure 12.2 represents the basics of contrast variation, that is the possibility by SANS to minimize the signal of either detergents/lipids or

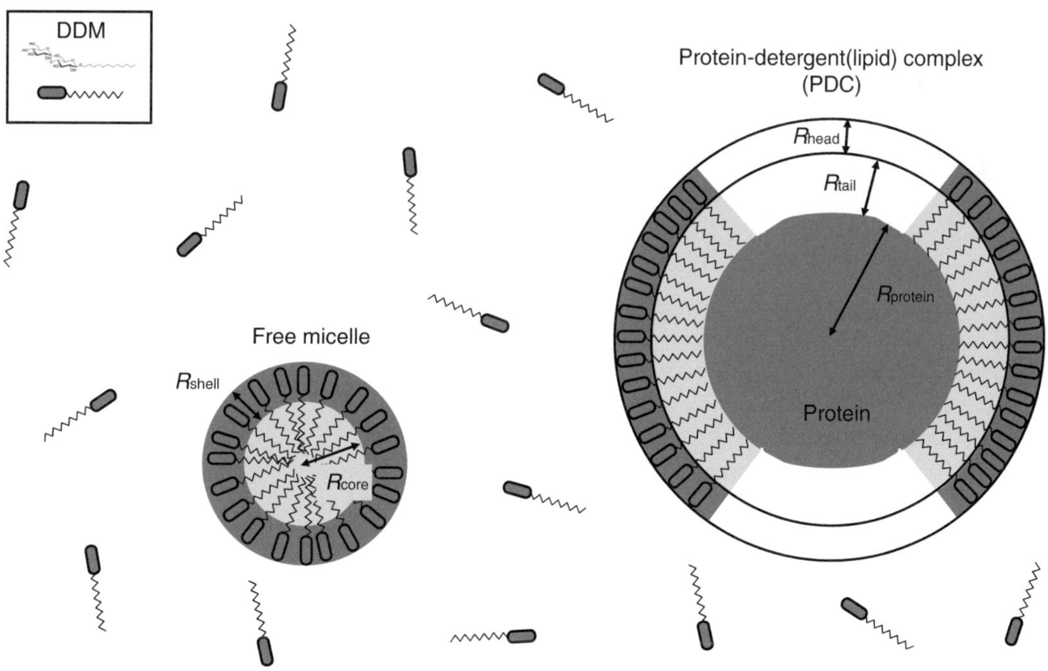


**Fig. 12.1** Schematic representation of compounds in a solubilized membrane protein sample. This schematic and strongly simplified overview shows a selection of individual particles in a sample of DDM (n-Dodecyl β-D-maltoside)-solubilized membrane proteins: free detergent molecules, free micelles, and the protein-detergent/lipid complex (PDC). The designations of geometric parameters (radii, thicknesses) in this figure correspond to the ones used in the following sections. *Different shades of grey* symbolize different contrasts $\Delta\rho$ (Table 12.1)

protein by measuring at an appropriate $H_2O$:$D_2O$ ratio and to focus on structural information from specific parts of a complex membrane protein system. Reasonable signal/noise on the most performant present-day SANS instruments can be obtained from membrane protein solutions in the (protein) concentration range 1–10 mg/mL for protein molecular masses in the range from ~20 kDa to several 100 kDa and for 100–200 μL sample volumes with exposure times varying between a few minutes and several hours, depending on contrast and instrumental setup.

As in the case of water-soluble proteins, mastery of the biochemistry and sample preparation prior to the experiment is of paramount importance for the interpretation of the experimental SANS curves in terms of structural parameters (Jacques and Trewhella 2010). For solubilized membrane proteins in particular, the monodispersity of the PDC and the detergent/lipid aggregates should be checked by preliminary complementary techniques including analytical ultracentrifugation (AUC) and size-exclusion chromatography (SEC) (Le Roy et al. 2015; le Maire et al. 2000). Ideally, the buffer subtracted from the sample containing the protein should match the latter in concentration of individual detergent/lipid molecules as well as free detergent/lipid aggregates (e.g. micelles) in order to eliminate their contribution to the SANS signal completely. If feasible, the buffer from the last purification (gel filtration) step should be used without further concentration if the detergent/lipid aggregate sizes and shapes depend on concentration.

Three major criteria should be respected when considering the choice of a detergent/lipid system for solubilizing membrane proteins for SANS experiments: (1) the detergent/lipid system should solubilize the membrane protein in a stable, monodisperse complex and in a functionally relevant state, (2) the contrast between protein and detergent/lipid (Fig. 12.2) should be chosen as high as possible and (3) protein-free detergent/lipid aggregates should be kept at a minimal concentration and as monodisperse as possible, i.e. in practice large aggregates such as vesicles, rod-like structures etc. should be avoided in order not to dominate the protein signal (see also following section). Point (3) requires a good knowledge of the detergent/lipid phase diagram and in particular the critical micellar concentration (CMC) (le Maire et al. 2000; Helenius and Simons 1975).

## 12.3 Influence of Membrane Protein Sample Heterogeneity on the SANS Signal: A Simple Geometrical Case Study

The extraction and interpretation of structural information by small angle neutron scattering according to Eq. 12.1 requires that the membrane proteins be solubilized in the presence of detergent and/or lipids. The required experimental conditions give rise to a composite system of (at least) three chemically different components with specific and distinct scattering length densities (SLDs): protein, detergent/lipids and (aqueous) solvent. In many cases, several distinct particulate species are formed (we exclude non-particulate systems such as lamellar phases, interconnected networks etc. (Qian and Heller 2015; Harroun et al. 2005; Seddon et al. 2004) and their respective scattering contributions need to be described separately: free (single) detergent/lipid molecules, protein-free detergent/lipid aggregates (micelles, vesicles etc.) and protein-detergent/lipid complexes (PDC) (Fig. 12.1). The general equation of the measured intensity (Eq. 12.1) can thus be rewritten as follows:

$$I(q) \propto N_{PDC}\left\langle\left|M_{prot}A_{prot}(q)+M_{lipid\_bound}\,A_{lipid\_bound}\,(q)\right|^2\right\rangle + N_{aggregates}\left\langle\left|M_{lipid\_aggregates}\,A_{lipid\_aggregates}\,(q)\right|^2\right\rangle + N_{free}\left\langle\left|M_{lipid\_free}\,A_{lipid\_free}\,(q)\right|^2\right\rangle \quad (12.2)$$

$N_{PDC}$, $N_{aggregates}$ and $N_{free}$ designate the number of particles of the respective species in solution. $A_{prot}$, $A_{lipid\_bound}$, $A_{lipid\_aggregates}$ and $A_{lipid\_free}$ correspond, respectively, to the scattering amplitudes (form factors) of the protein, the lipids/detergents bound to it, the lipids/detergents

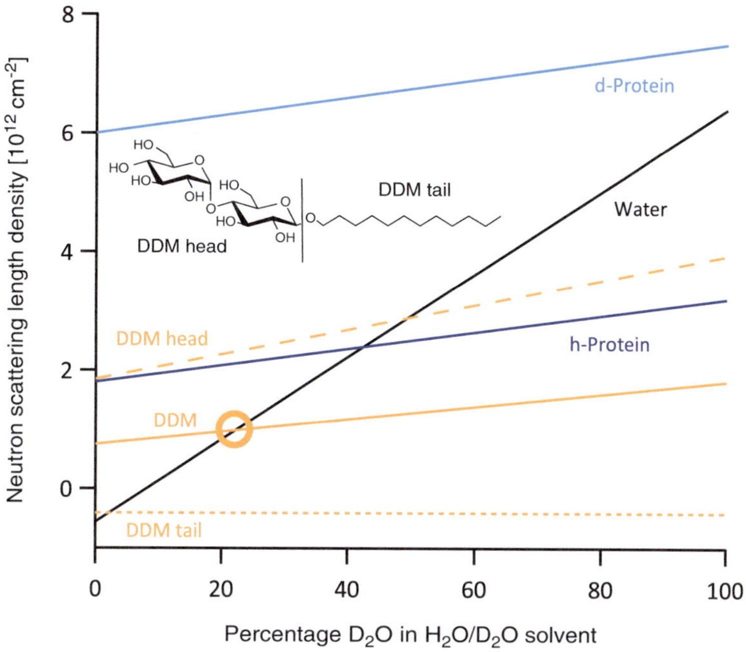


**Fig. 12.2** Neutron scattering length densities (SLD) $\rho$ of proteins and DDM detergent. The values are calculated from literature (Jacrot 1976; Breyton et al. 2013a). The contrast $\Delta\rho$ (Eq. 12.1) at a given $H_2O$:$D_2O$ ratio of each compound/moiety is defined as the difference between its own SLD and the one of the solvent (mixture of ordinary and heavy water). The *orange circle* indicates the contrast match point (CMP, i.e. $\Delta\rho = 0$) of an entire DDM molecule (head- plus tail-group) at ~22% $D_2O$. h-protein is a natural, hydrogenated protein, d-protein is a perdeuterated protein (all hydrogens replaced by deuterium). In general, all hydrogenated proteins have similar SLDs as the one depicted here while the SLDs for different detergents can deviate significantly from those of DDM (Breyton et al. 2013a)

associated in protein-free aggregates (micelles, vesicles etc.) and the free, single lipid/detergent molecules (for the sake of simplicity we use the index "lipid" in Eq. 12.2 for both lipid and detergent molecules). $M_i$ are the scattering masses of the respective parts of the systems and are given by $\int \Delta\rho_i \, dV_i$.

The significance and impact of Eq. 12.2 on the interpretation of SANS data is illustrated by a simplified model system consisting of a (hypothetical) spherical membrane protein covered entirely by a monolayer of detergent in addition to free, spherical micelles (Fig. 12.1). For the explicit calculations (see details in the Appendix), we assumed a hydrogenated protein and DDM (n-Dodecyl-β-D-maltopyranoside) detergent molecules at both 0 and 22% $D_2O$, the latter contrast corresponding to the overall contrast match point (CMP) of DDM detergent molecules (Fig. 12.2). The contrasts of the individual components are listed in Table 12.1 and in all cases random orientations of all individual components are assumed.

At 0% $D_2O$ (DDM not matched) the presence of bound detergent modifies both the intensity and the form factor of the protein (Fig. 12.3a, blue line vs continuous red line) and increases its apparent size (increase of $I(q = 0)$ intensity by a factor of 2.5 and radius of gyration $R_G$ by 24% from 31.4 to 38.8 Å). The relative contribution to the signal of free micelles (green line) with respect to the PDC complex (red line) is about 8% in intensity (at a stoichiometry 10:1) at small angles and reduces the $R_G$ of the PDC by 3% (to 37.7 Å). More importantly, their presence modifies the scattering curve from the complex significantly at intermediate $q$-ranges (0.07...0.25 Å$^{-1}$) (red continuous vs broken lines). Therefore, SANS curves of PDC complexes (in the absence or presence of additional free micelles) for non-matched detergent/lipids cannot, in general, be interpreted in terms of molecular mass or shape of the embedded proteins alone.

At 22% $D_2O$ (CMP of DDM), free micelles scatter very weakly at low angles ($q < 0.05$ Å$^{-1}$) with respect to the solubilized proteins and their contribution to $I(0)$ and the $R_G$ can be neglected, even in tenfold molar excess with respect to the PDC complexes (Fig. 12.3b, green and red broken lines). At intermediate $q$-values (0.07...0.25 Å$^{-1}$) the situation is similar to the one at 0% $D_2O$ with a modification of the signal from the isolated PDC complex (red lines). The situation is more complex for protein-bound detergent: even though matched on average at 22% $D_2O$ and therefore not modifying the $I(0)$ intensity and the apparent molecular mass of the protein, the form factor of the PDC complex is not the same as the one of the isolated protein alone (blue vs continuous red line) but increases the $R_G$ by 14% from 31.4 to 35.7 Å. This effect is due to the specific spatial arrangement and contrast of detergent moieties with respect to the protein: tail-groups are closer to the protein and have negative contrast while head-groups are further away from the protein and have positive contrast (Table 12.1). While their respective contributions to the scattered intensity (Eq. 12.1) cancels in the forward scattering direction $I(q = 0)$, they do not annihilate for $q > 0$.

While instructive regarding the sensitivity of SANS signal from complex, solubilized membrane protein systems regarding internal SLD heterogeneity and sizes of different particles, the analytical analysis presented here is heavily oversimplified with respect to several aspects: assumption of a spherical protein, covered homogeneously and completely by a double layer representing a head- and tail-group of detergent/lipid molecules, and a monodisperse and spherical population of micelles, composed of neatly separated homogeneous layers. More accurate descriptions of realistic systems would include non-spherical proteins, covered only partly by detergent/lipids and ellipsoidal micellar structures (Lipfert et al. 2007) as well as polydispersity in micellar size (Manet et al. 2011). These modifications would qualitatively lead to more smeared minima of the idealized calculated curves here but would not change the overall conclusions. Unfortunately, form factors of geometrical bodies that deviate even slightly from the spherical shape (e.g. ellipsoids, cylinders etc.) can no longer be written explicitly but are represented by mathematical integrals that need to be solved numerically (Pedersen 2002). Finally, it should be noted that the conclusions drawn here on micellar detergent/lipid aggregates as "contaminants" of the SANS signal of solubilized membrane proteins are even more pronounced when larger aggregates such as vesicles, liposomes or rods are present (Rubinson

**Table 12.1** SLD differences (i.e. contrast) $\Delta\rho$ of individual components shown in Figs. 12.1, 12.2, and 12.3

| **$\Delta\rho$component ($10^{10}$ cm$^{-2}$)** | **0% $D_2O$** | **22% $D_2O$** | | |
|---|---|---|---|---|
| Protein | 2.36 | 1.12 | | |
| DDM head | 2.41 | 1.34 | | |
| DDM tail | 0.15 | −2.51 | | |
| **Radii and thicknesses (Å)** | **Protein** | **DDM head** | **DDM tail** | **Whole particle** |
| Free DDM molecule | * | * | * | * |
| Free DDM micelle | * | 5.3 | 20 | 25.3 |
| Free protein | 40 | * | * | 40 |
| Protein-detergent complex (PDC) | 40 | 4.8 | 6 | 50.8 |

The geometrical components were chosen to approach published values on DDM molecules (Oliver et al. 2013) but also to yield a matched intensity ($I(0) = 0$) at 22% $D_2O$

* *ND*

et al. 2013; Qian and Heller 2011; Breyton et al. 2009; Hunt et al. 1997).

## 12.4 Strategies to Minimize/Homogenize the Detergent/Lipid SANS Signal and Shape Analysis

The previous section illustrates that the internal (SLD) heterogeneity of detergent/lipid molecules and their presence both in the protein-detergent/lipid complex and in the form of free micelles has a strong impact on the interpretation of membrane protein structures by SANS experiments. While it is in general reliable to extract molecular masses of membrane proteins from the $I(0)$ intensity at the detergent/lipid contrast match point (Fig. 12.3b, $I(0)$ intensities of blue and red curves) and therefore determine their oligomeric state (Compton et al. 2011), other elementary structural parameters such as the radius of gyration ($R_G$) cannot be interpreted as due to the protein alone. Moreover, *ab initio* shape analysis using single phases such as DAMMIN (Svergun 1999) will yield erroneous envelopes, encompassing both bound detergent/lipid molecules as well as a weighted contributions of micellar features from intermediate $q$-ranges (Fig. 12.3, deviations between red broken lines from red continuous lines). Several strategies have therefore been applied in literature to focus on the signal of the embedded membrane protein and to minimize the signal due to protein-bound detergent/lipid and/or free detergent/lipid aggregates:

1. **Reinforcing contrast by deuteration**: the relative contribution of internal SLD fluctuations of free detergent/lipid aggregates can be minimized by either increasing the signal of the protein by its deuteration in the presence of hydrogenated detergents/lipids or by working with deuterated lipids/detergents and hydrogenated proteins at elevated $D_2O$ percentages in the solvent (Compton et al. 2011; Gabel et al. 2014). In the former approach, the contrast of the deuterated protein (d-protein, Fig. 12.2) is much larger than the SLD fluctuations between head/tail of the detergent/lipid molecules which can be neglected in favorable cases, in particular for small differences between head/tail SLDs (Breyton et al. 2013a), by choosing detergent/lipids with low aggregation numbers close to their CMC (le Maire et al. 2000) or when only a few detergent/lipid molecules are attached to a protein complex of very large size (Efremov et al. 2015). Deuteration of lipids has the advantage of homogenizing the internal SLD variation for certain detergents/lipids (Timmins and Zaccai 1988; Breyton et al. 2013a) and, in addition, allows to work at high $D_2O$ concentrations which minimizes incoherent neutron scattering background (Gabel et al. 2002) and therefore improves signal/noise.
2. **Homogenizing internal contrast of detergents/lipids**: while internal SLD fluctuations between head- and tail-groups of detergent/lipid molecules can often be reduced by deuteration (see above), full deuteration unfortunately leads to SLD that can no longer be matched, even when working in 100% $D_2O$ (Breyton et al. 2013a). Alternatively, some detergent molecules have small SLD fluctuations in their natural (hydrogenated) state, e.g. fluorinated surfactants (Breyton et al. 2013b). Some of these compounds, however, have a match point close to 40% $D_2O$ and therefore require the use of deuterated proteins for structural studies by SANS. Another strategy consists in mixing hydrogenated and deuterated detergents/lipids at appropriate ratios in order to obtain a desired match point (Osborne et al. 1978; Clifton et al. 2012). Finally, more complex systems such as nanodiscs (mixed polypeptide/detergent/lipid particles) have been used recently to solubilize membrane proteins (Kynde et al. 2014; Skar-Gislinge et al. 2010; Bayburt and Sligar 2010; Nakano et al. 2009). In some cases (so-called "stealth nanodiscs"), the systems were designed to be matched out relatively homogeneously at

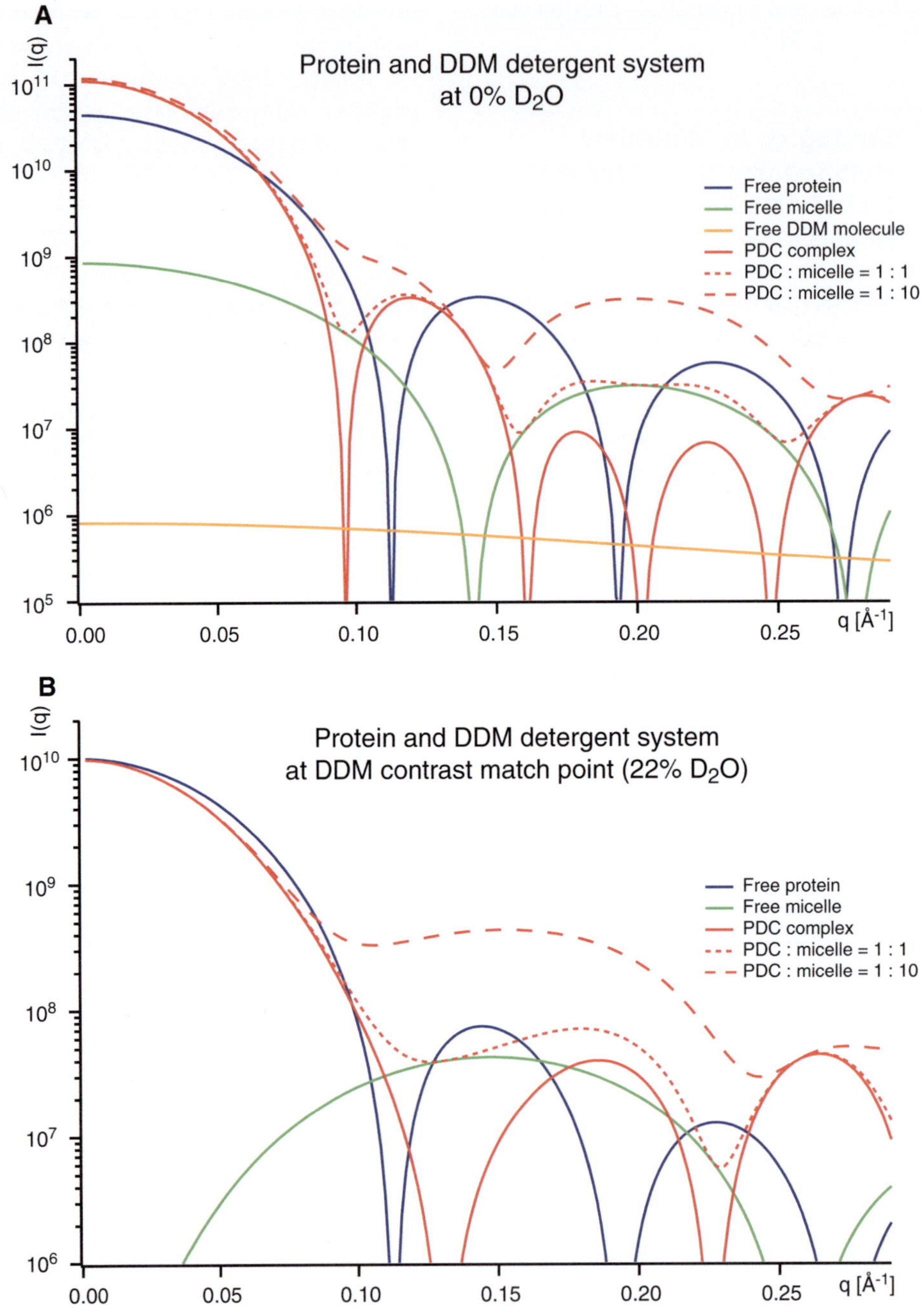


**Fig. 12.3** Theoretical SANS curves of simple, composite membrane protein systems at 0% (**a**) and 22% (**b**) $D_2O$. SANS curves calculated for free protein, free DDM micelle, isolated PDC complex and mixtures of PDC and micelles at two different stoichiometric ratios. The curves were calculated with Eq. 12.2 and the equations presented in the Appendix, by using neutron scattering length densities from literature (Jacrot 1976; Breyton et al. 2013a), listed in Table 12.1, and represent faithfully the relative intensities. The geometrical shapes of the micelle and PDC were assumed to be spherical as in Fig. 12.1, with the additional assumption that a detergent monolayer covers the protein completely (*black lines*, Fig. 12.1). The numerical values of the different parameters (protein radius, micelle core and shell radii and detergent head and tail lengths) are detailed in Table 12.1. The SANS curve of an individual DDM molecule was calculated with CRYSON (Svergun et al. 1998) using its atomic structure from the PDB (ligand ID entry "LMT")

about 100% $D_2O$ allowing thus the study of hydrogenated membrane proteins with very good signal/noise (Maric et al. 2014).

The strategies (1) and (2) allow, in a more or less accurate way, to assume that the detergents/ lipids can be considered to be homogeneous (in terms of SLD fluctuations between head- and tail-groups) and that the measured SANS signal at their contrast match point is due exclusively to the solubilized membrane proteins themselves. Under these conditions, the same approaches developed for water-soluble proteins have been applied on membrane proteins, in particular single- (Compton et al. 2011, 2014; Breyton et al. 2013b; Le et al. 2014; Sharma et al. 2012; Tang et al. 2010; Nogales et al. 2010; Cardoso et al. 2009; Zimmer et al. 2006; Johs et al. 2006; Bu et al. 2003) and multi-phase (Clifton et al. 2012) *ab initio* shape reconstructions. In the latter case, both detergents/lipids and protein are represented by two distinct envelopes and it is possible to determine their relative shapes and positions simultaneously by fitting SANS curves at multiple contrasts, e.g. with the program MONSA (Svergun 1999; Petoukhov and Svergun 2006).

## 12.5 Full-Atomic and Other Sophisticated Modelling Approaches

The previous section assumes that deuteration strategies allow to interpret SANS signals from complex systems (PDC plus free detergent/lipid aggregates) exclusively in terms of the protein structure at the detergent/lipid contrast match point. This is unfortunately not possible in all experimental cases and in general, a non-negligible contribution of bound detergent/ lipid as well as free detergent/lipid aggregates to the overall SANS signal needs to be taken into account. Size exclusion chromatography, coupled to SANS (SEC-SANS), which in analogy to SEC-SAXS experiments would allow the separation of the PDC from detergent/lipid aggregates such as micelles (see next section), has been developed recently (Jordan et al. 2016) but has not been applied, to our best knowledge, to membrane protein systems so far.

An accurate approach if SANS signals of detergent/lipid molecules cannot be neglected due to internal heterogeneity or due to perdeuteration, is to model the PDC with several distinct parts, including detergent/lipid molecules. Several studies have applied coarse-grain/dummy-atoms or simple geometric models of detergent/lipids (Kynde et al. 2014; Tang et al. 2010; Cardoso et al. 2009; Gohon et al. 2008) to represent the internal heterogeneity, including between different detergent/lipid moieties such as head- and tail groups. In some cases, full-atom detergent/lipid models have been produced by MD simulations (Gabel et al. 2014; Le et al. 2014). When integrating SANS datasets from different contrasts, these sophisticated approaches allow to distinguish between modestly different protein conformations as well as to fine-tune the shape and topology of the bound detergent/lipid molecules (Fig. 12.4). The combination of MD simulations with SANS (and SAXS) data from membrane protein systems is a very active field at the moment and diverse new approaches are being developed (Perez and Koutsioubas 2015; Chen and Hub 2015). Importantly, care needs to be taken that the programs that back-calculate SANS curves from membrane protein complexes take exchangeable hydrogens from detergent/lipid molecules correctly into account.

## 12.6 Comparison with Small-Angle X-Ray Scattering (SAXS)

The SANS sister technique, small-angle X-ray scattering (SAXS) has been used for several decades to characterize solubilized detergents (Bouwstra et al. 1993) and has provided models for the shapes and internal structure of isolated micelles (Oliver et al. 2013; Lipfert et al. 2007; D'Andrea et al. 2011; Gobl et al. 2010) and nanodiscs (Skar-Gislinge and Arleth 2011). SAXS has also been used as a high-throughput

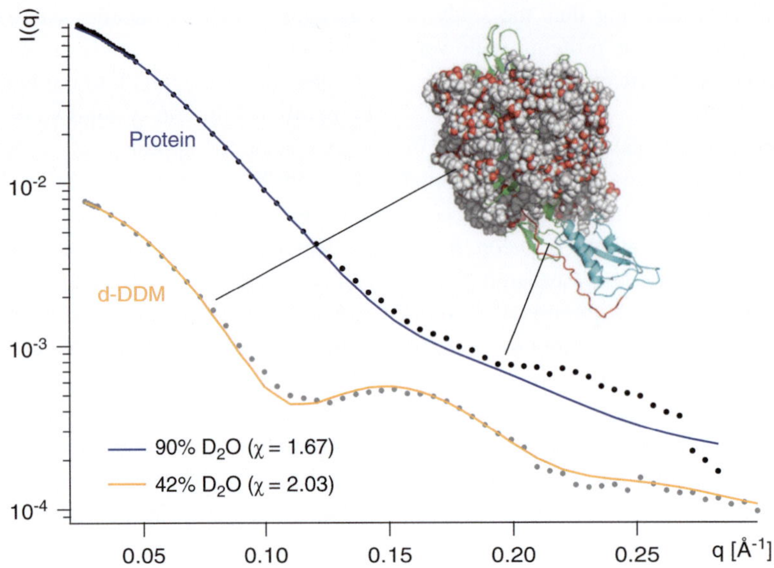


**Fig. 12.4** Full-atomic model of membrane protein FhaC, including deuterated d-DDM detergent. The structural model displayed was fitted against SANS data in solution at 42 and 90% $D_2O$ (contrast match points of protein and d-DDM, respectively) using the program CRYSON (Svergun et al. 1998). Detergent was built by MD simulations. Both SANS data sets were recorded at the instrument D22 at the Institut Laue-Langevin (ILL Grenoble, France) during 20 min at a protein concentration of 13 mg/mL. The figure was created based on data published by Gabel et al. (2014)

tool to probe crystallization phases (Joseph et al. 2011). First applications of SAXS to model solubilized membrane proteins in the presence of lipids/detergent are more recent (O'Neill et al. 2007; Columbus et al. 2006; Watanabe and Inoko 2005; Hong et al. 2004; Haas et al. 2004). While it was possible in some favorable cases to describe protein-detergent complexes by *ab initio* models using "static" SAXS experiments (Calcutta et al. 2012) the strong scattering contribution from detergent aggregates such as free micelles represents in general an important obstacle to such approaches. Due to their elevated electronic density, most detergent and lipid molecules cannot be easily contrast matched by the solvent and SAXS is therefore limited to a small number of compounds (Lipfert et al. 2007; Bu and Engelman 1999). A major breakthrough in recent years has been the advent of online size-exclusion chromatography (SEC), coupled to SAXS, which allowed to separate protein-detergent complexes from micelles and to apply sophisticated modeling approaches (Perez and Koutsioubas 2015; Dovling Kaspersen et al. 2014; Koutsioubas et al. 2013; Berthaud et al. 2012; Wright et al. 2011). Another promising recent development, so far specific to SAXS, are microfluidic platforms (Kondrashkina et al. 2013; Khvostichenko et al. 2013) that would allow the application of kinetic experiments to membrane protein systems.

## 12.7 Conclusions and Outlook

The past 10 years have witnessed an increasing number of SANS applications on solubilized membrane systems, as well as a rising level of sophisticated data interpretation (multiphase and full-atomic modelling) and biochemical sample preparation (deuteration strategies and choice of detergent/lipid/nanodisc systems). The ensemble of studies presented here illustrates that SANS is gaining momentum as a complementary structural biology technique to crystallography, cryo-EM and NMR and continues to provide valuable new insights into membrane protein function. Recently, SANS has also been used as a tool to optimize and tune detergent/lipid arrangements to solubilize and to study membrane proteins (Ashkar et al. 2015; O'Malley et al. 2011). It can be expected that over the next 5–10 years exciting new developments in instrumentation, sample environment, biochemistry and data analysis will help to improve the accuracy of SANS data from membrane proteins and the structural questions that can be addressed. Some recent developments that will gain momentum include SEC-SANS, MD simulations (Le et al. 2014; Chen and Hub 2015; Herrera et al. 2014; Perlmutter et al. 2011) and new labeling schemes for detergent/lipid compounds (Hiruma-Shimizu et al. 2016; Haertlein et al. 2016; Maric et al. 2015; Breyton et al. 2013b; Sverzhinsky et al. 2014) as well as new systems to solubilize membrane proteins (Maric et al. 2014; Midtgaard et al. 2014). Finally, newly developed or upgraded SANS instruments at various neutron sources (Dewhurst et al. 2016; Abbas et al. 2015; Jaksch et al. 2014; Heller et al. 2014; Liu et al. 2013) will improve the accessibility and impact of the technique further and open it to a broader scientific community.

## Appendix

This appendix provides detailed equations of amplitudes (i.e. weighted form factors) of the geometrical shapes presented in Figs. 12.1 and 12.3. They are modified after a book chapter by Pedersen (Pedersen 2002).

Form factor of a sphere with radius $R$ and volume $V$:

$$F(q,R) = 3\frac{\sin(qR) - qR\ \cos(qR)}{(qR)^3},\ V(R) = \tfrac{4}{3}\pi R^3$$

Scattering mass $M$ of a particle with volume $V$:

$$M = \int_V \Delta\rho dV$$

Amplitude of a micelle with radius $R$:

$$A_{micelle}(q) = \frac{1}{M_{micelle}}[\Delta\rho_{shell}V_{shell+core}F(q,R) + (\Delta\rho_{core} - \Delta\rho_{shell})V_{core}F(q,R_{core})]$$

$$M_{micelle} = \Delta\rho_{shell}V(R) + (\Delta\rho_{core} - \Delta\rho_{shell})V(R_{core})$$

Amplitude of a protein-detergent complex with radius $R$:

$$A_{complex}(q) = \frac{1}{M_{complex}}\begin{bmatrix}\Delta\rho_{head}V(R)F(R) + (\Delta\rho_{tail} - \Delta\rho_{head})V(R - R_{head})F(R - R_{head}) \\ +(\Delta\rho_{prot} - \Delta\rho_{tail})V(R - R_{head} - R_{tail})F(R - R_{head} - R_{tail})\end{bmatrix}$$

$$M_{complex} = \Delta\rho_{head}V(R) + (\Delta\rho_{tail} - \Delta\rho_{head})V(R - R_{head}) + \left(\Delta\rho_{prot} - \Delta\rho_{tail}\right)V(R - R_{head} - R_{tail})$$

## References

Abbas S et al (2015) On the design and experimental realization of a multislit-based very small angle neutron scattering instrument at the European spallation source. J Appl Crystallogr 48(4):1242–1253

Ashkar R et al (2015) Tuning membrane thickness fluctuations in model lipid bilayers. Biophys J 109 (1):106–112

Atkinson D, Small DM (1986) Recombinant lipoproteins: implications for structure and assembly of native lipoproteins. Annu Rev Biophys Biophys Chem 15:403–456

Bayburt TH, Sligar SG (2010) Membrane protein assembly into Nanodiscs. FEBS Lett 584(9):1721–1727

Berthaud A, Manzi J, Perez J, Mangenot S (2012) Modeling detergent organization around aquaporin-0 using small-angle x-ray scattering. J Am Chem Soc 134(24):10080–10088

Bouwstra JA, Gooris GS, Bras W, Talsma H (1993) Small angle X-ray scattering: possibilities and limitations in characterization of vesicles. Chem Phys Lipids 64 (1–3):83–98

Breyton C et al (2009) Micellar and biochemical properties of (hemi)fluorinated surfactants are controlled by the size of the polar head. Biophys J 97 (4):1077–1086

Breyton C et al (2013a) Small angle neutron scattering for the study of solubilised membrane proteins. Eur Phys J E Soft Matter 36(7):71

Breyton C et al (2013b) Assessing the conformational changes of pb5, the receptor-binding protein of phage T5, upon binding to its Escherichia coli receptor FhuA. J Biol Chem 288(42):30763–30772

Bu Z, Engelman DM (1999) A method for determining transmembrane helix association and orientation in detergent micelles using small angle x-ray scattering. Biophys J 77(2):1064–1073

Bu Z, Wang L, Kendall DA (2003) Nucleotide binding induces changes in the oligomeric state and conformation of Sec A in a lipid environment: a small-angle neutron-scattering study. J Mol Biol 332(1):23–30

Calcutta A et al (2012) Mapping of unfolding states of integral helical membrane proteins by GPS-NMR and scattering techniques: TFE-induced unfolding of KcsA in DDM surfactant. Biochim Biophys Acta 1818(9):2290–2301

Cardoso MB, Smolensky D, Heller WT, O'Neill H (2009) Insight into the structure of light-harvesting complex II and its stabilization in detergent solution. J Phys Chem B 113(51):16377–16383

Chacon P, Moran F, Diaz JF, Pantos E, Andreu JM (1998) Low-resolution structures of proteins in solution retrieved from X-ray scattering with a genetic algorithm. Biophys J 74(6):2760–2775

Charles M, Semeriva M, Chabre M (1980) Small-angle neutron scattering study of the association between porcine pancreatic colipase and taurodeoxycholate micelles. J Mol Biol 139(3):297–317

Chaudhuri BN (2015) Emerging applications of small angle solution scattering in structural biology. Protein Sci 24(3):267–276

Chen PC, Hub JS (2015) Structural properties of protein-detergent complexes from SAXS and MD simulations. J Phys Chem Lett 6(24):5116–5121

Clifton LA et al (2012) Low resolution structure and dynamics of a colicin-receptor complex determined by neutron scattering. J Biol Chem 287(1):337–346

Clifton LA, Neylon C, Lakey JH (2013) Examining protein-lipid complexes using neutron scattering. Methods Mol Biol 974:119–150

Columbus L et al (2006) Expression, purification, and characterization of Thermotoga maritima membrane proteins for structure determination. Protein Sci 15 (5):961–975

Compton EL, Karinou E, Naismith JH, Gabel F, Javelle A (2011) Low resolution structure of a bacterial SLC26 transporter reveals dimeric stoichiometry and mobile intracellular domains. J Biol Chem 286 (30):27058–27067

Compton EL et al (2014) Conserved structure and domain organization among bacterial Slc26 transporters. Biochem J 463(2):297–307

D'Andrea MG et al (2011) Thermodynamic and structural characterization of zwitterionic micelles of the membrane protein solubilizing amidosulfobetaine surfactants ASB-14 and ASB-16. Langmuir 27 (13):8248–8256

Dewhurst CD et al (2016) The small-angle neutron scattering instrument D33 at the Institut Laue-Langevin. J Appl Crystallogr 49(1):1–14

Dovling Kaspersen J et al (2014) Low-resolution structures of OmpADDM protein-detergent complexes. Chembiochem 15(14):2113–2124

Efremov RG, Leitner A, Aebersold R, Raunser S (2015) Architecture and conformational switch mechanism of the ryanodine receptor. Nature 517(7532):39–43

Gabel F et al (2002) Protein dynamics studied by neutron scattering. Q Rev Biophys 35(4):327–367

Gabel F et al (2014) Probing the conformation of FhaC with small-angle neutron scattering and molecular modeling. Biophys J 107(1):185–196

Gobl C et al (2010) Influence of phosphocholine alkyl chain length on peptide-micelle interactions and micellar size and shape. J Phys Chem B 114 (13):4717–4724

Gohon Y et al (2008) Bacteriorhodopsin/amphipol complexes: structural and functional properties. Biophys J 94(9):3523–3537

Haas H et al (2004) Small angle x-ray scattering from lipid-bound myelin basic protein in solution. Biophys J 86(1 Pt 1):455–460

Haertlein M et al (2016) Biomolecular Deuteration for neutron structural biology and dynamics. Methods Enzymol 566:113–157

Harroun TA et al (2005) Comprehensive examination of mesophases formed by DMPC and DHPC mixtures. Langmuir 21(12):5356–5361

Helenius A, Simons K (1975) Solubilization of membranes by detergents. Biochim Biophys Acta 415(1):29–79
Heller WT et al (2014) The bio-SANS instrument at the high flux isotope reactor of Oak Ridge National Laboratory. J Appl Crystallogr 47(4):1238–1246
Herrera FE, Garay AS, Rodrigues DE (2014) Structural properties of CHAPS micelles, studied by molecular dynamics simulations. J Phys Chem B 118 (14):3912–3921
Hiruma-Shimizu K, Shimizu H, Thompson GS, Kalverda AP, Patching SG (2016) Deuterated detergents for structural and functional studies of membrane proteins: properties, chemical synthesis and applications. Mol Membr Biol 32:1–17
Hong X, Weng YX, Li M (2004) Determination of the topological shape of integral membrane protein light-harvesting complex LH2 from photosynthetic bacteria in the detergent solution by small-angle X-ray scattering. Biophys J 86(2):1082–1088
Hunt JF, McCrea PD, Zaccai G, Engelman DM (1997) Assessment of the aggregation state of integral membrane proteins in reconstituted phospholipid vesicles using small angle neutron scattering. J Mol Biol 273 (5):1004–1019
Jacques DA, Trewhella J (2010) Small-angle scattering for structural biology – expanding the frontier while avoiding the pitfalls. Protein Sci 19(4):642–657
Jacrot B (1976) The study of biological structures by neutron scattering from solution. Rep Prog Phys 39 (10):911–953
Jaksch S et al (2014) Concept for a time-of-flight small angle neutron scattering instrument at the European spallation source. Nucl Instrum Methods Phys Res, Sect A 762:22–30
Johs A et al (2006) Modular structure of solubilized human apolipoprotein B-100. Low resolution model revealed by small angle neutron scattering. J Biol Chem 281(28):19732–19739
Jordan A et al (2016) SEC-SANS: size exclusion chromatography combined in situ with small-angle neutron scattering. J Appl Crystallogr 49(6):2015–2020
Joseph JS et al (2011) Characterization of lipid matrices for membrane protein crystallization by high-throughput small angle X-ray scattering. Methods 55 (4):342–349
Khvostichenko DS et al (2013) An x-ray transparent microfluidic platform for screening of the phase behavior of lipidic mesophases. Analyst 138 (18):5384–5395
Kondrashkina E et al (2013) Using macromolecular-crystallography beamline and microfluidic platform for small-angle diffraction studies of lipidic matrices for membrane-protein crystallization. J Phys Conf Ser 425:012013
Koutsioubas A, Berthaud A, Mangenot S, Perez J (2013) Ab initio and all-atom modeling of detergent organization around Aquaporin-0 based on SAXS data. J Phys Chem B 117(43):13588–13594
Kynde SA et al (2014) Small-angle scattering gives direct structural information about a membrane protein inside a lipid environment. Acta Crystallogr D Biol Crystallogr 70(Pt 2):371–383
Le RK et al (2014) Analysis of the solution structure of Thermosynechococcus elongatus photosystem I in n-dodecyl-beta-D-maltoside using small-angle neutron scattering and molecular dynamics simulation. Arch Biochem Biophys 550–551:50–57
Le Roy A et al (2015) AUC and small-angle scattering for membrane proteins. Methods Enzymol 562:257–286
Lipfert J, Columbus L, Chu VB, Lesley SA, Doniach S (2007) Size and shape of detergent micelles determined by small-angle X-ray scattering. J Phys Chem B 111(43):12427–12438
Liu D et al (2013) Demonstration of a novel focusing small-angle neutron scattering instrument equipped with axisymmetric mirrors. Nat Commun 4:2556
le Maire M, Champeil P, Moller JV (2000) Interaction of membrane proteins and lipids with solubilizing detergents. Biochim Biophys Acta 1508(1–2):86–111
Manet S et al (2011) Structure of micelles of a nonionic block copolymer determined by SANS and SAXS. J Phys Chem B 115(39):11318–11329
Maric S et al (2014) Stealth carriers for low-resolution structure determination of membrane proteins in solution. Acta Crystallogr D Biol Crystallogr 70 (Pt 2):317–328
Maric S et al (2015) Biosynthetic preparation of selectively deuterated phosphatidylcholine in genetically modified Escherichia coli. Appl Microbiol Biotechnol 99(1):241–254
Midtgaard SR et al (2014) Self-assembling peptides form nanodiscs that stabilize membrane proteins. Soft Matter 10(5):738–752
Nakano M et al (2009) Static and dynamic properties of phospholipid bilayer nanodiscs. J Am Chem Soc 131 (23):8308–8312
Nogales A et al (2010) Three-dimensional model of human platelet integrin alphaIIb beta3 in solution obtained by small angle neutron scattering. J Biol Chem 285(2):1023–1031
Oliver RC et al (2013) Dependence of micelle size and shape on detergent alkyl chain length and head group. PLoS One 8(5):e62488
O'Malley MA, Helgeson ME, Wagner NJ, Robinson AS (2011) Toward rational design of protein detergent complexes: determinants of mixed micelles that are critical for the in vitro stabilization of a G-protein coupled receptor. Biophys J 101(8):1938–1948
O'Neill H, Heller WT, Helton KE, Urban VS, Greenbaum E (2007) Small-angle X-ray scattering study of photosystem I-detergent complexes: implications for membrane protein crystallization. J Phys Chem B 111 (16):4211–4219
Osborne HB, Sardet C, Michel-Villaz M, Chabre M (1978) Structural study of rhodopsin in detergent micelles by small-angle neutron scattering. J Mol Biol 123(2):177–206

Pedersen JS (2002) Modelling of small-angle scattering data from colloids and polymer systems. In: Lindner P, Zemb T (eds) Neutrons, x-rays and light: scattering methods applied to soft condensed matter. Elsevier, Amsterdam, pp 391–420

Perez J, Koutsioubas A (2015) Memprot: a program to model the detergent corona around a membrane protein based on SEC-SAXS data. Acta Crystallogr D Biol Crystallogr 71(Pt 1):86–93

Perkins SJ, Weiss H (1983) Low-resolution structural studies of mitochondrial ubiquinol: cytochrome c reductase in detergent solutions by neutron scattering. J Mol Biol 168(4):847–866

Perlmutter JD et al (2011) All-atom and coarse-grained molecular dynamics simulations of a membrane protein stabilizing polymer. Langmuir 27 (17):10523–10537

Petoukhov MV, Svergun DI (2006) Joint use of small-angle X-ray and neutron scattering to study biological macromolecules in solution. Eur Biophys J 35 (7):567–576

Petoukhov MV et al (2012) New developments in the program package for small-angle scattering data analysis. J Appl Crystallogr 45(Pt 2):342–350

Qian S, Heller WT (2011) Peptide-induced asymmetric distribution of charged lipids in a vesicle bilayer revealed by small-angle neutron scattering. J Phys Chem B 115(32):9831–9837

Qian S, Heller WT (2015) Melittin-induced cholesterol reorganization in lipid bilayer membranes. Biochim Biophys Acta 1848(10 Pt A):2253–2260

Rubinson KA, Pokalsky C, Krueger S, Prochaska LJ (2013) Structure determination of functional membrane proteins using small-angle neutron scattering (sans) with small, mixed-lipid liposomes: native beef heart mitochondrial cytochrome c oxidase forms dimers. Protein J 32(1):27–38

Seddon AM, Curnow P, Booth PJ (2004) Membrane proteins, lipids and detergents: not just a soap opera. Biochim Biophys Acta 1666(1–2):105–117

Sharma KS et al (2012) Non-ionic amphiphilic homopolymers: synthesis, solution properties, and biochemical validation. Langmuir 28(10):4625–4639

Skar-Gislinge N, Arleth L (2011) Small-angle scattering from phospholipid nanodiscs: derivation and refinement of a molecular constrained analytical model form factor. Phys Chem Chem Phys 13 (8):3161–3170

Skar-Gislinge N et al (2010) Elliptical structure of phospholipid bilayer nanodiscs encapsulated by scaffold proteins: casting the roles of the lipids and the protein. J Am Chem Soc 132(39):13713–13722

Stuhrmann HB (1973) Comparison of the three basic scattering functions of myoglobin in solution with those from the known structure in crystalline state. J Mol Biol 77(3):363–369

Stuhrmann HB et al (1975) Neutron scattering study of human serum low density lipoprotein. Proc Natl Acad Sci U S A 72(6):2270–2273

Svergun DI (1999) Restoring low resolution structure of biological macromolecules from solution scattering using simulated annealing. Biophys J 76 (6):2879–2886

Svergun DI et al (1998) Protein hydration in solution: experimental observation by x-ray and neutron scattering. Proc Natl Acad Sci U S A 95(5):2267–2272

Sverzhinsky A et al (2014) Amphipol-trapped ExbB-ExbD membrane protein complex from Escherichia coli: a biochemical and structural case study. J Membr Biol 247(9–10):1005–1018

Tang KH, Urban VS, Wen J, Xin Y, Blankenship RE (2010) SANS investigation of the photosynthetic machinery of Chloroflexus aurantiacus. Biophys J 99 (8):2398–2407

Timmins PA, Zaccai G (1988) Low resolution structures of biological complexes studied by neutron scattering. Eur Biophys J 15(5):257–268

Watanabe Y, Inoko Y (2005) Physicochemical characterization of the reassembled dimer of an integral membrane protein OmpF porin. Protein J 24(3):167–174

Wise DS, Karlin A, Schoenborn BP (1979) An analysis by low-angle neutron scattering of the structure of the acetylcholine receptor from Torpedo californica in detergent solution. Biophys J 28(3):473–496

Wright GS, Hasnain SS, Grossmann JG (2011) The structural plasticity of the human copper chaperone for SOD1: insights from combined size-exclusion chromatographic and solution x-ray scattering studies. Biochem J 439(1):39–44

Yeager MJ (1976) Neutron diffraction analysis of the structure of retinal photoreceptor membranes and rhodopsin. Brookhaven Symp Biol 27:III3–III36

Zaccai NR et al (2016) Deuterium labeling together with contrast variation small-angle neutron scattering suggests how Skp captures and releases unfolded outer membrane proteins. Methods Enzymol 566:159–210

Zimmer J, Doyle DA, Grossmann JG (2006) Structural characterization and pH-induced conformational transition of full-length KcsA. Biophys J 90 (5):1752–1766

# Hybrid Applications of Solution Scattering to Aid Structural Biology

# 13

Alexander V. Grishaev

**Abstract**

Biomolecular applications of solution X-ray and neutron scattering (SAXS and SANS, respectively) started in late 1960s – early 1970s but were relatively limited in their ability to provide a detailed structural picture and lagged behind what became the two primary methods of experimental structural biology – X-ray crystallography and NMR. However, improvements in both data analysis and instrumentation led to an explosive growth in the number of studies that used small-angle scattering (SAS) for investigation of macromolecular structure, often in combination with other biophysical techniques. Such hybrid applications are nowadays quickly becoming a norm whenever scattering data are used for two reasons. First, it is generally accepted that SAS data on their own cannot lead to a uniquely defined high-resolution structural model, creating a need for supplementing them with information from complementary techniques. Second, solution scattering data are frequently applied in situations when a method such NMR or X-ray crystallography cannot provide a satisfactory structural picture, which makes these additional restraints highly desirable. Maturation of the hybrid bio-SAS approaches brings to light new questions including completeness of the conformational space sampling, model validation, and data compatibility.

**Keywords**

Solution X-ray and neutron scattering • SAXS • SANS • Hybrid modeling • Protein and RNA structure determination

## 13.1 Introduction

Scattering of X-rays or neutrons by an isotropic solution containing the macromolecule of interest produces a one-dimensional intensity signal

A.V. Grishaev (✉)
National Institute of Standards and Technology, The Institute for Bioscience and Biotechnology Research, 9600 Gudelsky Drive, Rockville, MD 20850, USA
e-mail: agrishaev@ibbr.umd.edu





B. Chaudhuri et al. (eds.), *Biological Small Angle Scattering: Techniques, Strategies and Tips*, Advances in Experimental Medicine and Biology 1009, DOI 10.1007/978-981-10-6038-0_13

dependent on the scattering angle and is commonly expressed as a function of the scattering vector $q = 4\pi \sin(\theta)/\lambda$, where $2\theta$ is the scattering angle and $\lambda$ is the wavelength of the incident radiation (X-ray photons or neutrons). A difference signal between such one-dimensional scattering intensity profiles recorded for the sample containing the bio-macromolecule of interest and the one containing an identically matching buffer can then be interpreted solely in the terms of the macromolecular structures present in solution, their surrounding surface solvent layer and, possibly, their inter-particle correlations. Solution scattering data are most often recorded at angles much smaller than the ones sampled in the macromolecular crystallography leading to a notion that these data can only provide structural information with a low nominal resolution. Considering a dramatic difference between a smooth one-dimensional solution scattering curve and an indexable three-dimensional diffraction set containing tens of thousands of reflections common in macromolecular crystallography, it seems natural to treat solution scattering as a technique with an intrinsically low information content. This notion is supported by a fairly small number of degrees of freedom associated with a solution scattering data set. With data recorded within the q-range from $q_{min}$ to $q_{max}$ for a macromolecule with the maximum dimension $d_{max}$, this number is estimated from the Shannon-Nyquist sampling theorem as $N = \pi(q_{max}-q_{min})*d_{max}$ (Svergun and Koch 2003), and in practice rarely exceeds ca. 20, indicating that only a limited number of model parameters could be obtained from such data.

The question is then, given this apparent low resolution and low information content, why are these data useful at all in modern structural biology which aims for sub-Å precision of atomic positions? The answer to this question lies in the ability of SAXS data to offset the formal resolution and information content limitations noted above with a superior signal/noise attainable when using modern synchrotron sources and single photon counting detectors. Contrary to the seemingly featureless and noisy scattering profiles that were common up till ~20 years ago unless very high protein concentrations were used (Durschlag 1975; Fedorov and Denesyuk 1978), modern synchrotron beam lines are capable of producing solution scattering data with very low visible noise and wide angular range, largely free of inter-particle correlations effects while handling relatively dilute samples. These changes occurred as SAXS, as a field, was able to capitalize on the same dramatic improvements in the photon flux and low-noise detector technologies that propelled the growth of the bio-molecular X-ray crystallography in the past two decades and are now responsible for the explosive growth in applications of cryo-electron microscopy.

These improvements in the instrumentation were occurring simultaneously with three crucial developments in data analysis capitalizing on the higher quality of SAS data, specifically: (i) generation of distance probability distribution functions *via* regularized Fourier transforms of the scattering data (Svergun 1992; Brunner-Popela and Glatter 1997), (ii) development of *ab initio* low resolution electron density reconstruction approaches (Chacon et al. 1998; Svergun 1999), and (iii) formulation of approaches for accurate calculation of the scattering data from the macromolecular atomic coordinates taking into account the surface solvent layer, groundwork for which was laid in mid-1990s (Svergun et al. 1995, 1998). Taken together, these developments now create a possibility of using solution scattering data to effectively discriminate between closely related candidate structural models with a representative example shown in Fig. 13.1. This ability to select between the candidate structural models taking advantage of accurate methodologies for connecting scattering data to the macromolecular atomic coordinates is central to the hybrid approaches that use SAXS/SANS in combination with the restraints from other experimental techniques (Putnam et al. 2007).

## 13.2 SAS Data: Advantages and Drawbacks

The main practical advantage of solution scattering data results from their ability to reflect the

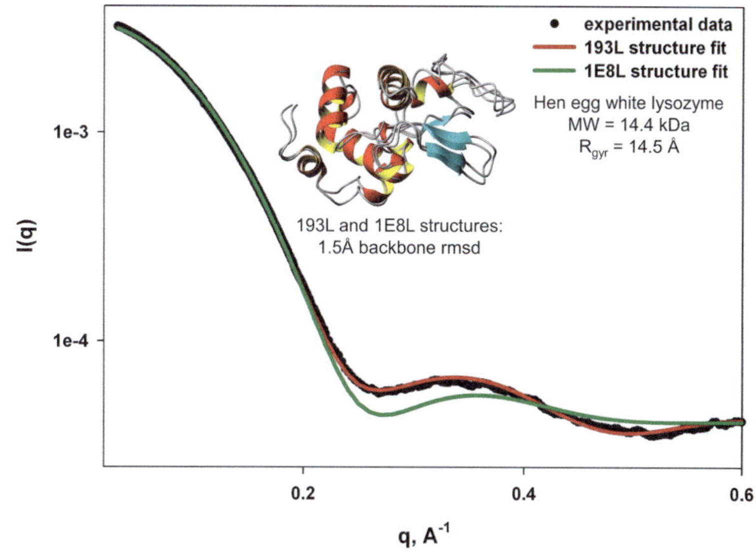


**Fig. 13.1** Solution X-ray scattering data can be very effective at distinguishing between closely related structural models describing the macromolecule

molecular geometries for the complete set of conformations present in solution, covering such important and challenging cases as intrinsically disordered proteins (IDPs) (Mittag et al. 2010), detergent micelle-solubilized membrane proteins (Lipfert et al. 2007), or macromolecules with dynamic regions, including flexibly linked multi-domain proteins (Bernado et al. 2007), or amyloid fibrils (Lu et al. 2003), as illustrated in Fig. 13.2. SAXS is applied with a similar ease to proteins, RNA, or DNA, with the latter two benefiting from an increased signal precision due to the presence of phosphorus atoms and a lower fraction of hydrogen atoms relative to the proteins (Zuo and Tiede 2004).

Solution neutron scattering brings an additional possibility of performing contrast variation measurements, particularly informative for multi-subunit system which can be prepared with selective protonation/deuteration of the individual components, or protein/oligonucleotide complexes (Whitten and Trewhella 2009). Such contrast-variation, or contrast-matched scattering data collected when varying the $H_2O$/$D_2O$ ratios in the buffer, can be distinctly different from SAXS data in $H_2O$, increasing the overall information content of the solution scattering data set.

SAS data can be collected with ease within a wide range of experimental conditions, including salt concentration up to 1–2 molar, temperature from the freezing point to ca. 90 °C, or with solute concentrations as low as 0.02–0.05 mg/mL for SAXS and 0.5–1.0 mg/mL for SANS (Grishaev 2012).

Information content of the scattering data is limited relative to the crystal diffraction data set, consistent with their one-dimensional nature. SAS intensity profile reports on the entire scattering particle, thus providing very little site-specific information unlike techniques such as macromolecular crystallography or NMR. However, this is one of the reasons for the growing popularity of the hybrid techniques that include SAXS or SANS: complementary biophysical techniques capable of providing precise site-specific restraints can be very effective at compensating the lack of such information in solution scattering.

A practical limitation of solution scattering is the requirement for a highly monodisperse and pure sample needed for the data analysis in terms of the structural model (Jacques and Trewhella 2010). From our own experience, the presence of ~5% dimer fraction relative to the monomeric species of interest produces a noticeable impact

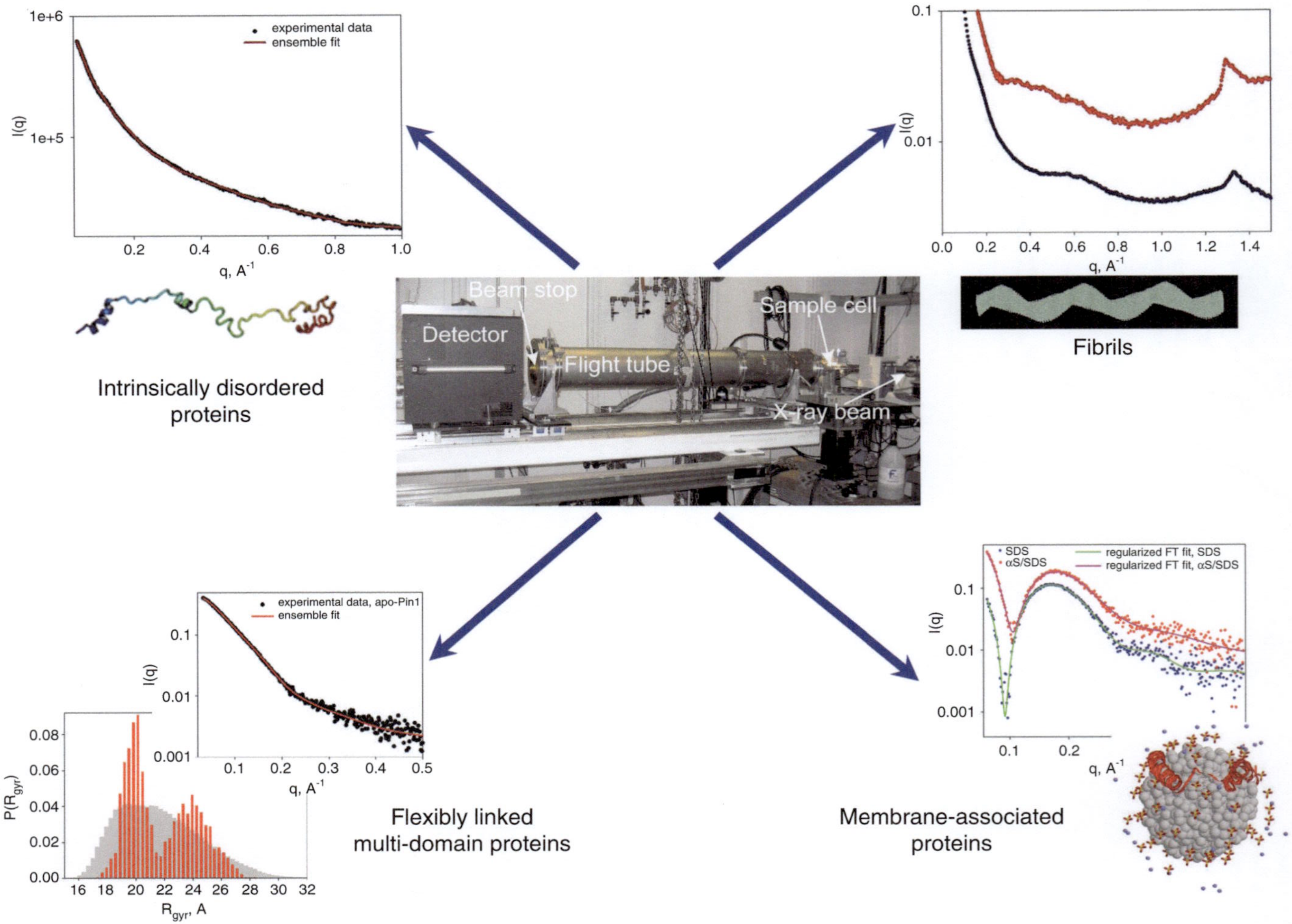


**Fig. 13.2** Solution scattering can be applied to study a diverse set of challenging systems in structural biology, frequently in combination with other biophysical techniques

on the SAS data within a wide angular range, and in the case of the aggregates ~10 fold larger relative to the mass of the molecule of interest, this threshold can become as low as ~0.1%, reminiscent of the situation encountered with light scattering data analysis. Even though the main effect of such impurities comes at low scattering angles, it cannot always be removed by simply discarding these data and becomes particularly detrimental with highly elongated macromolecular geometries such as those observed for the for the IDPs (Johansen et al. 2011).

## 13.3 SAS Data: Complementary Sources of Information

Owing to the limited information content of the solution scattering data noted above, they are now increasingly more often combined with the restraints from other techniques in structural studies. This was not the case in the early days of bio-SAS applications – 15 years ago a study involving SAXS or SANS rarely incorporated the results of complementary experimental techniques, with the end result commonly being formulated in terms of the low-resolution electron density determined solely based on the scattering intensities. This reflected both fascination with the newly found ability to derive realistic three dimensional shapes solely from 1D SAXS data – clearly, a milestone development, as well as the general lack of computational methods with which these data could be integrated with precise site-specific experimental restraints. Needless to say, it did not take long for such techniques to be formulated, with the early applications concentrating on the rigid-body refinement against SAXS data when structures of the individual subunits of the overall scattering particle were known (Petoukhov and Svergun 2005), or incorporation of the fixed relative orientational restraints from NMR (Sunnerhagen et al. 1996; Mattinen 2002).

### 13.3.1 Hybrid Structure Determination Using SAS Data and Partial Structural Information from X-Ray Crystallography

SAS data can be readily combined with the partial structural information from a complementary high-resolution technique such as X-ray crystallography, taking advantage of a number of computational methodologies developed for accurate comparison of the experimental scattering data with those predicted from the candidate all-atom models (Svergun et al. 1995; Park et al. 2009; Schneidman-Duhovny 2010; Grishaev et al. 2010; Koefinger and Hummer 2013; Chen and Hub 2014). The differences between these approaches are subtle but numerous, including: the number and nature of the fitting parameters, fit of the buffer-subtracted scattering signal or the pair of the sample and buffer scattering profiles, specifications for the exact positioning of the surface solvent layer relative to the macromolecule, and finally, the use of implicit or explicit models for the displaced and surface solvent. Implicit models are those that assume the waters displaced by the macromolecule to exactly coincide with the macromolecular atomic coordinates and use a simplified "shell" representation for the surface solvent layer. In contrast, explicit solvent models rely on the results of the molecular dynamics simulations to describe the structure of the displaced and surface solvent. It would be fair to state that all of the methods listed above were more extensively tested and optimized for proteins compared to oligonucleotides.

Several scenarios for combining SAS data with prior structural knowledge are possible. In one, the system is composed of a number of subunits with known structures whose relative arrangements are derived from fitting the candidate all-atoms models to the SAS data. The need for using SAS data in this case stems either from difficulties in finding conditions for obtaining well-diffracting crystals for the entire assembly (Comoletti et al. 2007), or from the impact of the crystal packing forces on the overall geometry for a highly non-globular particle (Heidorn and Trewhella 1988). A second scenario is when the crystal structure is either missing interpretable electron density for the fraction of its sequence due to dynamics, or when the full-length construct with flexible parts cannot be crystallized, necessitating the use of a truncated variant (Hickman et al. 2014). In these two cases either the relative arrangement of the particle subunits, or the coordinates for the missing fraction of the macromolecule have been positioned to agree with the experimental SAS data. Finally, the scattering data can become a useful constraint for determining the geometry of the assemblies such as protein/protein complexes when the structures of the individual partners are known (Schneiderman-Duhovny et al. 2010; Pons et al. 2010; Karaka and Bonvin 2013). In all of these cases, constraints based as chain connectivity, clash avoidance, and gyration radius by reference to the experimental SAS data, can all be very effective at decreasing the number of the feasible candidate models and are best applied prior to the full SAS data fit. A complication possible with all of these scenarios occurs when the experimentally determined structural models are not available for some, or all of the system subunits, leading to the use of homology-based structures with worse than ~2 Å coordinate accuracy (Comoletti et al. 2007). The consequence of relying on such lower quality subunit models is that the residual structural inaccuracies embedded in them can propagate into the structural inaccuracy of the best-fitted model for the overall assembly.

In all of the above cases, both combination of SAS information with the partial structural data from X-crystallography, and optimization of the model geometry to best-fit the scattering intensities are relatively straightforward to perform. Far more challenging is establishing that the best-fitting model obtained starting the fit from a set of initial coordinates is accurate, representing the actual set of conformations present in solution, or even unique. Possible ways to address this issue include (i) supplementing SAXS data with complementary data from other experimental techniques for model validation and refinement, and (ii) performing an exhaustive sampling of a large number of starting/candidate geometries with preset rotational and translational steps. This problem is further complicated for flexible or disordered macromolecules, for which data over-fitting becomes much more likely due to the increase in the number of model parameters. In general, proof of a non-degenerate nature of the structural model found to best-fit SAS data can be difficult to establish due to the potentially astronomical number of possible distinct candidate geometries that need to be sampled, reaching ca. $10^{11}$ even for multi-domain proteins linked by short (less than 10 amino acid) stretches of residues (Grishaev et al. 2012). This issue becomes more pronounced for protein/protein complexes where the chain connectivity constraints cannot be applied (Schneidman-Duhovny et al. 2011). Such cases will benefit from additional relative geometry constraints, which may include mutagenesis data reporting on complex breakage, or other prior information allowing approximate determination of the locations of the subunit/subunit interaction sites. It is worth keeping in mind that the structural model best-fitted to the SAS data using rigidly held high-resolution structures of the individual subunits cannot be expected to have the same structural accuracy as its constituents even though it will appear to resemble a high-resolution structure. For instance, relative domain orientations resulting from the model optimized against SAS data cannot be assumed to be uniquely and correctly identifiable even for the non-globular subunits, in the absence of validation *via* additional experimental restraints.

An additional challenge while performing SAS data driven macromolecular structure determination is that there is generally no guarantee of a perfect correlation between the structural accuracy of the candidate model and the goodness of its SAS data fit (Grishaev et al. 2011). The exact appearance of such correlation plot, which requires the knowledge of the correct structure can be impacted by a multitude of factors including the particular method used to predict SAXS/SANS data from the atomic coordinates, limited experimental signal/noise or resolution range of the scattering data, possible inaccuracies for the parts of the structural model that are held fixed, improper representation of the multiple conformations present in solution, or errors in the interpretation for SAS data due to unrecognized contributions of aggregation, sample/buffer mismatch, or sample purity or composition issues. Therefore, when performing structural model selection against solution scattering data, it is advisable to use several methodologies for linking SAS data to the atomic coordinates deriving the set of consensus models, tightly control systematic errors in the data outside of the photon counting statistics, and use a proper representation for the conformational heterogeneity when necessary.

It should be clear from the above comments that the issue of validation of the structural model restrained by SAS experimental data is of utmost importance. Such validation should rely on complementary experimental restraints weakly correlated with the fitted scattering data set. A possible way to introduce this complementarity is by recording contrast variation solution scattering data with SANS (Comoletti et al. 2007) or SAXS (Grishaev et al. 2012), allowing either a reduction in the degeneracy in the pool of the best-fitting structural models, or validation in case of a single best-fitting solution.

It is also worth pointing out that the ultimate success of using SANS data with mixed $^1H/^2H$ labeling of the complex constituents at the contrast matching conditions (~42% $D_2O$ for protonated proteins) in practice can be affected by both the dissociation constant of the complex and the relative masses of its protonated and deuterated components. In one case (Comoletti et al. 2007), accurate determination of the centers-of-mass separation between the deuterated components with protonated subunits contrast-matched was possible for a 2:2 protein/protein complex with low-nM affinity where the deuterated units comprised ca. 25% of the overall particle. In another (Schwieters et al. 2010), such determination was not successful for a nearly identical SANS data collection on a similarly sized 2:2 protein complex with ~20 μM binding affinity requiring a sixfold molar excess of the smaller deuterated component needed in order to obtain 100% binding occupancy, and deuterated components comprising only ca. 13% of the complete particle.

### 13.3.2 Hybrid NMR/SAS Macromolecular Structure Determination

Nuclear magnetic resonance (NMR) in solution had been long recognized to be a useful complementary source of structural information when combined with the SAXS data. The symbiosis of the two techniques is driven by the fact that they tend to offset each other's deficiencies such as the decrease in the density of attainable restraints with the increase in the size of the macromolecule, or difficulties in describing the molecular structure of flexible constructs, in the case of NMR; or the lack of site-specific information and model degeneracy in the case of SAS. Solution scattering data should ideally be acquired on the same sample used for NMR data collection, minimizing chances of any inconsistencies between the experimental conditions for the two measurements.

From a practical perspective, macromolecular samples used in solution NMR rarely require significant modifications for application of SAXS (Grishaev 2012). $^2H/^{13}C/^{15}N$ isotopic labeling has no effect on the X-ray scattering data while the presence of $^2H$ does contribute to the signal measured by SANS. Therefore, the least expensive (in practice, $^{15}N$–labeleld) material frequently ends up being used for solution scattering measurements. Slight changes in the SAXS buffer composition relative to typical NMR conditions may include increase in the salt concentration to suppress inter-particle repulsion, replacement of the commonly used phosphate in the NMR buffers by agents containing lighter elements (TRIS, HEPES, etc), and addition of the free radical scavengers such as DTT or TCEP to the dialysis buffers in order to suppress protein radiation damage. 5–10% $D_2O$ typically included in the NMR samples for frequency lock does not need to be present for SAXS data collections, most commonly carried out in pure $H_2O$. X-ray scattering measurements are typically done at the same (or lower) concentrations as those used in solution NMR and should be performed at the temperature exactly matching NMR data collection for seamless combination of the restraints from the two techniques. Monodispersity requirements for SAXS are more stringent compared to NMR where the aggregated populations simply becomes invisible, and preliminary characterization by techniques such as analytical ultracentrifugation or light scattering is a must. A growing number of bio-SAXS beam lines at synchrotrons now offer in-line size-exclusion chromatography setups immediately preceding SAXS measurement which greatly enhance the quality of the collected scattering data at the expense of the decrease in sample concentration and, therefore, the signal/noise of the recorded data. While planning SAXS measurement in the context of an NMR study it is always a good idea to predict theoretical scattering curves beforehand when structural models are available. This step will help to both select the appropriate experimental angular range, and estimate the resolving power of the SAXS by reference to the uncertainty of the structural models resulting from NMR-only structure determination.

It is now generally accepted that out of all types of currently accessible NMR restraints, residual dipolar couplings (RDCs) observed *via* weak alignment of the macromolecules induced by the strongly aligned liquid crystalline media (Bax 2003), are by far the most useful type of data when combined with the solution scattering intensity profiles. This complementarity is easily rationalized keeping in mind both relative insensitivity of the solution scattering data to domain rotations around their centers of mass, and independence of the orientational restraints from NMR from the translations of the particle subunits. In our own early work on combining SAXS and NMR data in a hybrid structure determination we observed no improvement of the structural accuracy when the NMR restraints were composed entirely from the short-range inter-proton distances and torsion angles; such improvement occurred only when RDCs were included in the NMR data set (Grishaev et al. 2005).

Hybrid structure determination using RDC and SAXS data can proceed under two scenarios – rigid body structure optimization involving solely translational and orientational degrees of freedom that specify relative domain positioning, or a fully flexible refinement in which all of the internal degrees of freedom are active. The choice between using the two depends on the density of the available site-specific NMR restraints and the coordinate accuracy for the individual domains, with lower restraint density and higher structural accuracy generally favoring rigid body techniques. As a rule of thumb, fully flexible refinement is warranted when the structures of the individual domains can be determined solely from NMR data with the coordinate accuracy better than ca. 1.5 Å, or backbone N-H RDC cross-validation Q-factors (Bax 2003) better than ca. 0.4. On the other hand, rigid-body refinement techniques are preferable if the available structures of the individual subunits can be fitted to the experimental backbone N-H RDC data with Q-factors better than ca. 0.3, likely associated with higher resolution (better than ca. 2.4 Å) crystal structures.

A joint rigid-body fit of RDC and SAXS data relies on accurate determination of the molecular alignment tensors for the individual domains, which in practice requires measurement of at least ca. 35 backbone RDCs for each domain. In cases when the molecular alignment tensors determined by a singular value decomposition (SVD) fits (Losonczi et al. 1999) for the individual domains are strongly correlated (generalized scalar product between the corresponding alignment tensors larger than ca. 0.9, corresponding to the tensor orientation difference not exceeding ca. 20°), their relative positions can be assumed to be rigid, allowing single-model representation during the structure refinement. In cases of small numbers or large certainties of the experimentally attainable RDCs an approximate criterion for a single conformation representation would be for the Q-factor of the joint SVD fit of the RDCs for the two domains allowed to reorient not exceeding those fitted separately to the individual domains by more than ~0.05. Otherwise, an ensemble representation for the complete macromolecule would be required, possibly coupled with a scheme for prediction of the conformation-dependent molecular alignment tensor (Zwezkstetter et al. 2004; Marsh et al. 2008; Venditti et al. 2015). When performing both single-model and ensemble-averaged rigid body refinement against RDCs it should be kept in mind that even though the relative orientation of the two domains can be fitted from the experimental NMR data with a precision of ca 3–5°, it comes with a fourfold degeneracy corresponding to 180° rotations around the three axes of the alignment tensor. Even though in theory this degeneracy could be resolved by collecting similarly precise RDCs from a different alignment medium with a weakly correlated alignment tensor, in practice this situation is relatively rare for proteins, and even less common for oligonucleotides. Therefore, all four distinct orientations often need to be sampled in a joint RDC/SAXS data fit. When present, domain connectivity constraints with short linkers not exceeding ca. 10 residues often reduce this fourfold degeneracy by a factor of 2, with the burden of distinguishing between the remaining ones placed solely on the SAXS data. In cases of the protein/protein complexes, unless the locations of the interacting sites on each of the subunits are determined independently, all four possible conformations have to be distinguished based on the SAXS data fits alone, with several successful examples reported in the literature (Parsons et al. 2008; Zuo et al. 2008). It can shown that use of the RDCs as orientational restraints corresponds to an approximately 100-fold reduction in the number the possible candidate structural models when rigid-body refinement is employed (Grishaev et al. 2012).

With a sufficient density of site-specific NMR restraints, fully flexible model refinement against the combined NMR and SAXS data sets becomes possible, often performed either when the complete structure cannot be separated into the individual domains, or when the structural accuracy of the individual domains is insufficient for the application of the rigid body methods according to the criteria listed above. In practice sufficiently high density can correspond to as few as ca. 1 backbone torsion angle restraint and 1–2 distance restraints per residue, supplemented by backbone RDCs in two independent alignment media, or RDCs combined with the backbone anisotropic chemical shifts in a single alignment medium. A number of such joint structure determinations were performed starting from ca. 2005, with many of the early applications centered on validation – establishing proof that a joint fully flexible NMR/SAXS structure refinement leads to a clear improvement in the structural accuracy, whether by reference to a previously determined crystal structure (Grishaev et al. 2005, 2007), or *via* cross-validation with respect to the experimental RDC data (Grishaev et al. 2008). One of the conclusions that emerged from this work is the need for a set of locally rigid restraints from both NMR experimental data and database or homology constraints that also allow global flexibility of the macromolecule when performing refinement against the solution scattering data. The number of such applications increases at a steady pace, now including flexibly linked proteins and oligonucleotides, as well as IDPs.

## 13.4 Hybrid Applications of Solution Scattering: Computational Modeling Tools

The software that is capable of using solution scattering data for determination of macromolecular structure can be grouped into three broadly defined classes. The first of these contains standalone packages dealing primarily with SAXS/SANS data that allow very limited input from complementary techniques and rely on simple molecular simulation engines favoring sampling efficiency over force field accuracy, exemplified by EOM (Bernado et al. 2007) or SASREF (Petoukhov and Svergun 2005). The second class of programs adds capability of fitting against SAXS/SANS data to a previously

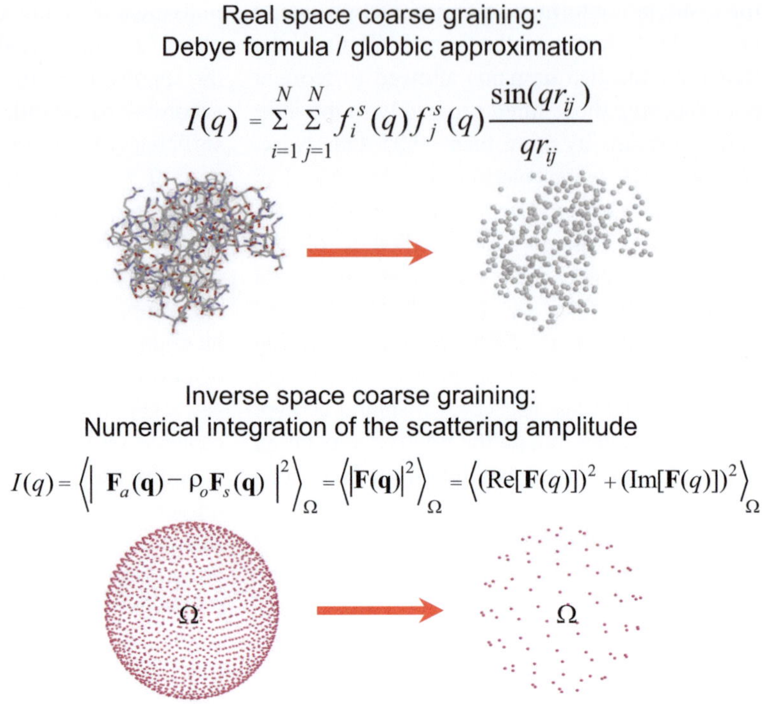


**Fig. 13.3** Refinement against solution scattering data in molecular simulation requires application of coarse graining, either in the direct or inverse space, in order to speed up calculation of these data from the macromolecular coordinates

developed package including either a sophisticated molecular simulation core with advanced force fields, or a structure optimization engine with integrative modeling, with examples such as SASSIE (Curtis et al. 2012), IMP (Russel et al. 2012), flexible-Meccano (Ozenne et al. 2012), and RASREC-Rosetta (Rossi et al. 2015). The software belonging to this class would have a somewhat limited ability of handling diverse types of experimental data from complementarity techniques but will be typically capable of including subsets of such data across different techniques in an integrative fashion. Finally, the third class of software adds an ability to refine against SAXS or SANS data to a package whose primary task is macromolecular structure determination against experimental crystal diffraction or NMR data driven by a somewhat simplified molecular simulation engine, with examples including CNS (Brunger et al. 1998) and Xplor-NIH (Schwieters et al. 2003). Therefore, the three classes correspond to emphases on either the scattering data, or the advanced simulation engines and force fields, or effective handling of the complementary experimental restraints from crystallography or NMR. The choice of the class of software would likely depend on the amount of complementary site-specific experimental restraints, with higher restraint density best coupled to the third class of programs. The decision between using the first two classes of software in practice depends on the user's familiarity with the more advanced simulation setup of the second class, or a preference for the greater simplicity of operation generally exhibited by the first class.

When using any software for refining against SAS data, it is important to keep in mind the mode of scattering data calculation, and to be aware of its limitations as the codes coupled to the molecular simulation engines often emphasize the speed of calculation over accuracy, and will likely fall behind dedicated scattering data simulations tools for a single-model analysis listed in Sect. 13.2.2, in terms of the fidelity of data prediction and fit quality. This is an unavoidable consequence of the generally high computational costs associated with the calculation of the predicted scattering data, and the associated molecular forces along the molecular dynamics trajectory. For example, all of the structure refinement packages that include SAS data use the faster but not necessarily more accurate implicit model for the displaced and surface solvent. Calculation speedup is achieved either by coarse-grading the representation of the macromolecule in the real space *via* globbic approximation to the Debye formula (Grishaev et al. 2005), or by coarse-graining in the inverse space *via* approximations to the spherical averaging of the complex scattering amplitude (Grishaev et al. 2008), illustrated in Fig. 13.3. Recently formulated algorithm for a fast accurate hierarchical approximation to the spherical harmonic expansion of the Debye formula (Berlin et al. 2014) is of particular interest, yet to be implemented in a structure refinement package. In many cases the approximation noted above are a reasonable price to pay for the ability to optimize the structure against SAS data within a molecular dynamics simulation, and to perform joint refinement with restraints from complementary techniques.

## 13.5 Summary and Conclusions

The past decade has brought a rapid expansion in the number of studies that use solution X-ray and neutron scattering data for derivation of macromolecular structures. Even though the initial methodology developments in bio-SAS focused on the benefits of applying SAXS or SANS data in isolation, an increasing fraction of structural studies involving SAS nowadays use these data in combination with experimental restraints from a growing set of complementary techniques that include crystallography, NMR, FRET or electron microscopy. This fact reflects both universal appreciation of the value of combining data across multiple techniques, and a rapidly expanding repertoire of computational methodologies that allow such combination. Hybrid applications involving SAS are expected to become a norm as structural biology shifts towards studies of challenging architectures that cannot be a analyzed with a single technique, including flexible or dynamic macromolecules, weak macromolecular interactions, or membrane-associated assemblies.

## References


Bax A (2003) Weak alignment offers new NMR opportunities to study protein structure and dynamics. Protein Sci 12:1–16

Berlin K, Gumerov NA, Fushman D, Duraiswami R (2014) Hierarchical O(N) computation of small-angle scattering profiles and their associated derivatives. J Appl Crystallogr 47:755–761

Bernado P, Mylonas E, Petoukhov MV, Blackledge M, Svergun DI (2007) Structural characterization of flexible proteins using small-angle scattering. J Amer Chem Soc 129:5656–5664

Brunger AT, Adams PD, Clore GM, Delano WL, Gros P, Grosse-Kunstleve RW, Jiang JS, Kuszewski J, Nilges M, Pannu NS, Read RJ, Rice LM, Simonson T, Warren GL (1998) Crystallography & NMR system (CNS), a new software suite for macromolecular structure determination. Acta Crystallogr D54:905–921

Brunner-Popela J, Glatter O (1997) Small-angle scattering of interacting particles. I. Basic principles of a global evaluation method. J Appl Crystallogr 30:431–442

Chacon P, Moran F, Diaz JF, Pantos E, Andreu JM (1998) Low-resolution structures of proteins in solution retrieved from X-ray scattering with a genetic algorithm. Biophys J 74:2760–2775

Chen PC, Hub JS (2014) Validating solution ensembles from molecular dynamics simulation by wide-angle X-ray scattering data. Biophys J 107:435–447

Comoletti D, Grishaev A, Whitten A, Tsigelny I, Taylor P, Trewhella J (2007) Synaptic arrangement of the neuroligin/beta-neurexin complex revealed by X-ray and neutron scattering. Structure 15:693–705

Curtis JE, Raghunandan S, Nanda H, Krueger S (2012) SASSIE: a program to study disordered biological molecules and macromolecular ensembles using experimental scattering restraints. Comput Phys Commun 183:382–389

Durchschlag H (1975) X-ray small-angle studies of pyruvate-dehydrogenase core complex from Escherichia-coli K-12.2. Biophys Struct Mechanism 1:153–188

Fedorov BA, Denesyuk AI (1978) Sperm whale myoglobin structure in solution differs from its structure in crystal by a shift of hairpin GH. FEBS Lett 88:114–117

Grishaev A (2012) Sample preparation, data collection and preliminary data analysis in biomolecular solution X-ray scattering. Curr Protocols Protein Sci 17 (14):1–18

Grishaev A, Wu J, Trewhella Bax A (2005) Refinement of multi-domain protein structures by combination of solution small angle X-ray scattering and NMR data. J Amer Chem Soc 127:16621–16628

Grishaev A, Ying J, Canny M, Pardi A, Bax A (2008) Solution structure of tRNA-Val from joint refinement against dual alignment residual dipolar couplings and SAXS data. J Biomol NMR 42:99–109

Grishaev A, Guo L, Irving T, Bax A (2010) Improved fitting of solution X-ray scattering data to macromolecular structures and structural ensembles by explicit water modeling. J Amer Chem Soc 132:15484–15486

Grishaev A, Anthis N, Clore GM (2012) Contrast-matched small angle X-ray scattering from a heavy atom-labeled protein in structure determination: application to a lead-substituted calmodulin-peptide complex. J Amer Chem Soc 134:14686–14689

Heidorn DB, Trewhella J (1988) Comparison of the crystal and solution structures of calmodulin and troponin C. Biochemistry 27:909–905

Hickman A, Ewis H, Li X, Knapp J, Laver T, Doss AL, Tolun G, Steven A, Grishaev A, Bax A, Atkinson P, Craig N, Dyda F (2014) Structural basis of hAT transposon end recognition by Hermes, an octameric transposase from Musca domestica. Cell 158:353–367

Jacques DA, Trewhella J (2010) Small-angle scattering for structural biology – expanding the frontier while avoiding the pitfalls. Protein Sci 19:642–657

Johansen D, Trewhella J, Goldenberg DP (2011) Fractal dimensions of an intrinsically disordered protein: small-angle X-ray scattering and computational study of the bacteriophage lambda N protein. Protein Sci 20:1955–1970

Karaka E, Bonvin AMJJ (2013) On usefulness of ion-mobility mass spectrometry and SAXS data in scoring docking decoys. Acta Cryst D69:683–694

Koefinger J, Hummer G (2013) Atomic-resolution structural information from scattering experiments on macromolecules in solution. Phys Rev E 87:052712

Lipfert J, Columbus L, Chu VB, Doniach S (2007) Analysis of small-angle X-ray scattering data of protein–detergent complexes by singular value decomposition. J Appl Crystallogr 40:s235–s239

Losonczi JA, Andrec M, Fischer MWF, Prestegard JH (1999) Order matrix analysis of residual dipolar couplings using singular value decomposition. J Magn Reson 138:334–342

Lu K, Jacob J, Thiyagarajan P, Conticello VP, Lynn DG (2003) Exploiting amyloid fibril lamination for nanotube self-assembly. J Amer Chem Soc 125:6391–6393

Marsh JA, Baker JMR, Tollinger M, Forman-Kay JD (2008) Calculation of residual dipolar couplings from disordered state ensembles using local alignment. J Amer Chem Soc 130:7804–7805

Mattinen ML, Paakkonen K, Ikonen T, Craven J, Drakenberg T, Serimaa R, Waltho J, Annila A (2002) Quaternary structure built from subunits combining NMR and small-angle X-ray scattering data. Biophys J 83:1177–1183

Mittag T, Marsh JA, Grishaev A, Orlicky S, Lin H, Sicheri F, Kay LE, Tyers M, Forman-Kay JD (2010) Structure/function implications in an intrinsically disordered protein and its dynamic complex. Structure 18:494–506

Ozenne V, Bauer F, Salmon L, Huang JR, Jensen MR, Segard S, Celine BP, Charavay C, Blackledge M (2012) Flexible-meccano: a tool for the generation of explicit ensemble descriptions of intrinsically disordered proteins and their associated experimental observables. Bioinformatics 28:1463–1470

Park S, Bardhan JP, Roux B, Makowski L (2009) Simulated X-ray scattering of protein solution using explicit solvent models. J Chem Phys 130:134114

Parsons L, Grishaev A, Bax A (2008) The periplasmic domain of TolR forms a large hydrophobic groove and may contact TolQ through its C-terminus: NMR solution structure and its comparison to SAXS data. Biochemistry 47:3131–3142

Petoukhov MV, Svergun DI (2005) Global rigid body modeling of macromolecular complexes against small-angle scattering data. Biophys J 89:1237–1250

Pons C, D'Abramo M, Svergun DI, Orozco M, Bernado P, Fernandez-Recio J (2010) Structural characterization of protein-protein complexes by integrating computational docking with small-angle scattering data. J Mol Biol 403:217–230

Putnam CD, Hammel M, Hura GM, Tainer JA (2007) X-ray solution scattering (SAXS) combined with crystallography and computation: defining accurate macromolecular structures, conformations and assemblies in solution. Q Rev Biophys 40:191–285

Rossi P, Shi L, Liu GH, Barbieri CM, Lee HW, Grant TD, Luft JR, Xiao R, Actorn TB, Snell EH, Montelione GT, Baker D, Lange OF, Sgourakis NG (2015) A hybrid NMR/SAXS-based approach for discriminating oligomeric protein interfaces using Rosetta. Protein Struct Func Bioinform 83:309–317

Russel D, Lasker K, Webb B, Velazques-Muriel J, Tjioe E, Schneidman-Duhovny D, Peterson B, Sali A (2012) Putting the pieces together: integrative modeling platform software for structure determination of macromolecular assemblies. PLoS Biol 10: e1001244

Schneidman-Duhovny D, Hammel M, Sali A (2010) FoXS: a web server for rapid computation and fitting of SAXS profiles. Nucl Acids Res 38:W540–W544

Schneidman-Duhovny D, Hammel M, Sali A (2011) Macromolecular docking restrained by a small-angle X-ray scattering profile. J Struct Biol 173:461–471

Schwieters CD, Kuszewski J, Tjandra N, Clore GM (2003) The Xplor-NIH NMR molecular structure determination package. J Magn Reson 160:66–74

Schwieters C, Suh JY, Grishaev A, Takayama Y, Clore GM (2010) Solution conformation of the 128 kDa enzyme I dimer from Escherichia Coli and its 146kDa complex with HPr using residual dipolar couplings and small angle x-ray scattering. J Amer Chem Soc 132:13026–13045

Sunnerhagen M, Olah GA, Stenflo J, Forsen S, Drakenberg T, Trewhella J (1996) The relative orientation of Gla and EGF domains in coagulation factor X is altered by Ca2+ binding to the first EGF domain. A combined NMR small angle X-ray scattering study. Biochemistry 35:11547–11559

Svergun DI (1992) Determination of the regularization parameter in indirect-transform methods using perceptual criteria. J Appl Crystallogr 25:495–503

Svergun DI (1999) Restoring low resolution structure of biological macromolecules in solution using simulated annealing. Biophys J 76:2879–2886

Svergun DI, Koch MHJ (2003) Small-angle studies of biological macromolecules in solution. Rep Progr Phys 66:1735–1782

Svergun DI, Barberato C, Koch MHJ (1995) CRYSOL – a program to evaluate X-ray solution scattering of biological macromolecules from atomic coordinates. J Appl Crystallogr 28:768–773

Svergun DI, Richard S, Koch MHJ, Sayers Z, Kuprin S, Zaccai G (1998) Protein hydration in solution: experimental observation by x-ray and neutron scattering. Proc Natl Acad Sci U S A 95:2267–2272

Vendittii V, Schwieters CD, Grishaev A, Clore GM (2015) Dynamic equilibrium between closed and partially closed states of the bacterial enzyme I unveiled by solution NMR and X-ray scattering. Proc Natl Acad Sci U S A 37:11565–11570

Whitten AE, Trewhella J (2009) Small-angle scattering and neutron contrast variation for studying bio-molecular complexes. Micro and nano technologies in bioanalysis: methods and protocols book series. Methods Mol Biol 544:307–323

Zuo X, Tiede DM (2004) Resolving conflicting crystallographic and NMR models for DNA with solution X-ray diffraction. J Amer Chem Soc 127:16–17

Zuo X, Wang J, Foster TR, Schwieters CD, Tiede DM, Butcher SE, Wang YX (2008) Global molecular structure and interfaces: refining an RNA:RNA complex structure using solution X-ray scattering data. J Amer Chem Soc 130:3292–3293

Zweckstetter M, Hummer G, Bax A (2004) Prediction of charge-induced molecular alignment of biomolecules dissolved in dilute liquid-crystalline phases. Biophys J 86:3444–3460

# A Practical Guide to iSPOT Modeling: An Integrative Structural Biology Platform

14

An Hsieh, Lanyuan Lu, Mark R. Chance, and Sichun Yang

**Abstract**

Integrative structure modeling is an emerging method for structural determination of protein-protein complexes that are challenging for conventional structural techniques. Here, we provide a practical protocol for implementing our integrated iSPOT platform by integrating three different biophysical techniques: small-angle X-ray scattering (SAXS), hydroxyl radical footprinting, and computational docking simulations. Specifically, individual techniques are described from experimental and/or computational perspectives, and complementary structural information from these different techniques are integrated for accurate characterization of the structures of large protein-protein complexes.

**Keywords**

iSPOT • Integrative structural biology • SAXS • Hydroxyl radical footprinting • Computational docking simulation • Structural mass spectrometry • Protein-protein interaction

## 14.1 Introduction

Macromolecular interactions provide the molecular underpinning for virtually every biological process. Despite decades of effort, however, structure determination of protein-protein complexes is still a daunting task for conventional techniques due to size, stability, and/or complexity of protein complexes of interest. To advance the ability to characterize these complexes, we have recently established a multi-technique iSPOT platform by integrating small-angle X-ray scattering (SAXS), hydroxyl radical footprinting and computational docking

A. Hsieh • M.R. Chance • S. Yang (✉)
Center for Proteomics and Bioinformatics and Department of Nutrition, Case Western Reserve University, 10900 Euclid Ave, Cleveland, OH 44106-4988, USA
e-mail: sichun.yang@case.edu

L. Lu
School of Biological Sciences, Nanyang Technological University, 60 Nanyang Drive, Singapore 637551, Singapore





B. Chaudhuri et al. (eds.), *Biological Small Angle Scattering: Techniques, Strategies and Tips*, Advances in Experimental Medicine and Biology 1009, DOI 10.1007/978-981-10-6038-0_14

simulations (Huang et al. 2016). iSPOT leverages the widespread availability of individual protein or domain structures and in particular enables the structure determination of complexes in the range of 50–200 kDa that are often challenging for nuclear magnetic resonance (too big) or electron microscopy (too small). Driven by its potential as an emerging technique towards large-scale applications, this iSPOT platform is described here to facilitate broad adoption.

The iSPOT platform overcomes the limitation of individual techniques and succeeds in combining multiple sources of structural information from different techniques that are complementary to each other. For example, computational docking benefits from its combination with experimental scattering/footprinting data, while molecular shape information from SAXS is complemented by solvent accessibility of specific protein sites probed by hydroxyl radical footprinting.

## 14.2 Implementation of the Integrated iSPOT Platform

The entire iSPOT platform has three major sources of structural information for each protein-protein complex of interest: (1) molecular shape and structural arrangement from small-angle X-ray scattering (SAXS), (2) solvent accessibility of specific sites probed by hydroxyl radical footprinting, and (3) model prediction by computational protein-protein docking. Figure 14.1 outlines a schematic demonstrating the integration of three different, complementary biophysical techniques in the iSPOT platform.

It is worth noting that while the integration of all three techniques is emphasized here, a combination of any two approaches can be utilized to generate structure ensembles for a specific question of interest, while the remaining data are used for a validation purpose if available. For this consideration, we describe each component of this iSPOT platform, followed by the integration of all three.

Figure 14.2 provides an overview of the iSPOT workflow. It is arbitrarily divided into four components: (a) computational protein-protein docking for generating structural candidates (or "poses"), (b) parallel SAXS and footprinting data acquisition, (c) candidate scoring against experimental data, and (d) selection and optimization of ensemble structures. A proof-of-principle demonstration of this iSPOT platform has been shown in an earlier publication on several protein-protein complexes with their crystal structures known (Huang et al. 2016). By using the atomic structures of individual proteins (not the complex), iSPOT is able to accurately predict the structures of a large protein-protein complex (TGFβ-FKBP12) and a multidomain nuclear receptor homodimer (HNF-4α), by using simulated SAXS and footprinting data of each complex.

### 14.2.1 Computational Protein-Protein Docking

Computational studies of protein-protein interaction have been a long-term focus of research (Janin et al. 2003). Quite a few algorithms are now available for docking two proteins into a bound complex. As such, computationally docked conformations or "poses" can be evaluated and compared against experimental data (discussed later). Specifically, rigid-body and flexible docking are described below, as well as post-docking clustering analysis.

#### 14.2.1.1 Rigid-Body Docking

Rigid-docking techniques have been successfully developed over the years (Chen et al. 2003; Dominguez et al. 2003; Gabb et al. 1997; Tovchigrechko and Vakser 2006). These docking algorithms, such as ClusPro (Comeau et al. 2004) and ZDock (Pierce et al. 2014), are computationally robust and efficient. For this reason, it is a good idea to try rigid-body docking as a first diagnostic step, or even use docking results for evaluating with experimental data if the proteins are relatively non-flexible upon binding. Notably, ZDock is particularly easy to use and

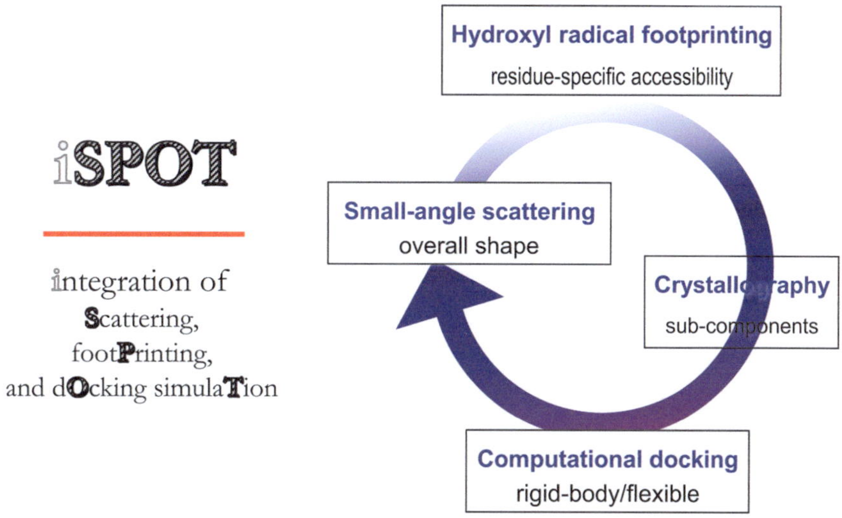


**Fig. 14.1** Multi-technique iSPOT platform for integrated structural modeling of protein-protein complexes. iSPOT represents the integration of small-angle X-ray scattering (SAXS), hydroxyl radical footprinting, and computational docking simulation (Huang et al. 2016). iSPOT also takes advantage of available cystal structures of individual protein components within the complex

provides a simple web interface (http://zdock.umassmed.edu), as well as executable files available for download.

#### 14.2.1.2 Flexible Docking by RotPPR-CGMD Molecule Dynamics Simulation

To account for structural flexibility in protein-protein interaction, we have developed a molecular dynamics (MD) based docking method, termed RotPPR-CGMD (described below), which combines an exhaustive generation of initial poses and subsequent coarse-grained molecular dynamics simulations. This RotPPR-CGMD is composed of (a) conformational sampling by RotPPR and (b) coarse-grained (CG) simulation. The former is to make sure that the conformational space is properly and exhaustively searched; the latter is to use a one-bead-per-residue Cα model to simplify the protein representation as we have shown previously (Ravikumar et al. 2012; Yang et al. 2010a). A suite of source codes and executable files for the setup and configurations of RotPPR-CGMD simulations will be made available for this type of RotPPR-CGMD docking simulations.

Specially, the RotPPR sampling, a combination of a pull-push-release (PPR) strategy along the inter-protein translational axis and a rotational pose generator, collectively enables an extensive conformational sampling in the docking space (Huang et al. 2016). The translation-centric PPR sampling is achieved *via* a harmonic spring between the centers-of-mass of two proteins to facilitate the docking (Ravikumar et al. 2012), while the pose generator provides a set of different initial docking poses to account for all five rotational degrees-of-freedom (as illustrated in Fig. 14.3).

The energy function used in RotPPR-CGMD simulations is a predictive coarse-grained Cα model, where interaction between two proteins is defined by residue-residue interactions whose parameters are tabulated in a previous publication (Huang et al. 2014). It is worth noting that although the structure of each protein is used for the modeling, it does not require structural knowledge of the entire complex (Ravikumar et al. 2012). Because of its coarse-grained nature, this CGMD is expected to significantly enhance the protein-protein docking, compared to atom-level simulations.

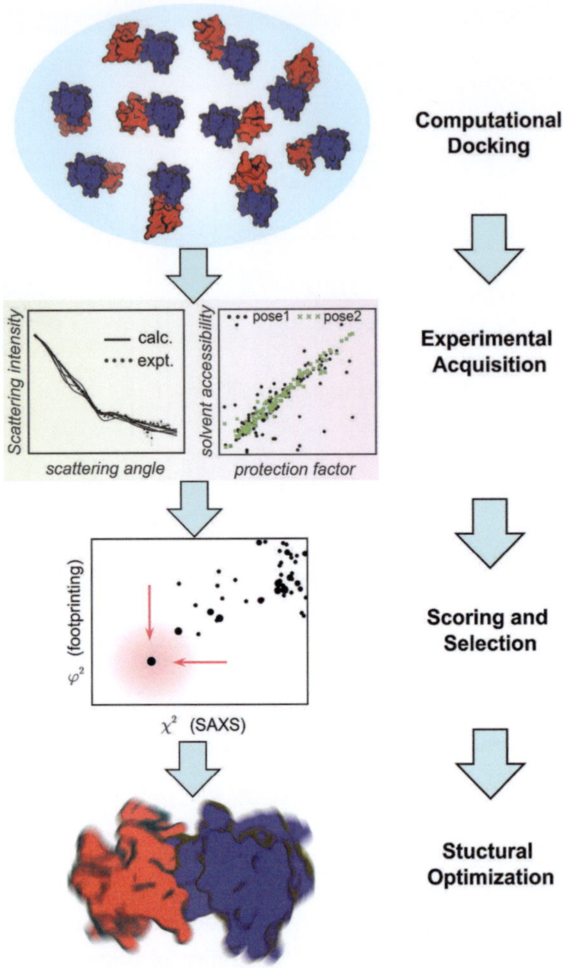


**Fig. 14.2** The iSPOT workfolow. It consists of four compoments: (**a**) computational protein-protein docking, (**b**) experimental SAXS and footprinting data acquisition, (**c**) scoring and selection, and (**d**) structural model optimization

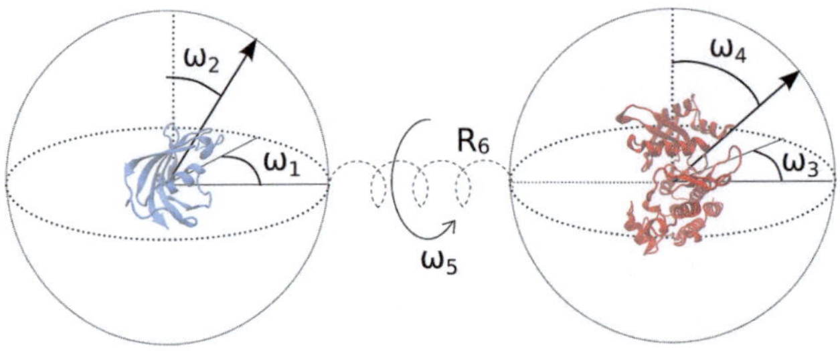


**Fig. 14.3** Computational protein-protein flexible docking. Shown are the six degrees of feedom (five rotational and one translational) involved in two-body protein docking that are extensively sampled by RotPPR-CGMD simulations (Modified with permission Huang et al. 2016)

#### 14.2.1.3 Structure Clustering

For post-docking data analysis, structure clustering of RotPPR-CGMD simulation data can be achieved on the basis of structural similarity *via* two specific metrics: fRMSD and oRMSD. The former is a regular RMSD measure of Cα atoms from the entire complex and the latter is an extension of fRMSD by accounting for the difference in relative orientation between two proteins (Huang et al. 2016). The resulting oRMSD clustering improves the structural ambiguity observed in traditional fRMSD clustering since the measure of oRMSD is more sensitive to protein-protein orientations. As a result, oRMSD clustering is able to group similar simulation-generated structures into one cluster or conformation that appear more homogenous than what was based on fRMSD clustering.

Another notable difference is the input parameter needed for clustering. Traditionally, the number of clusters is used as an input, while a RMSD cutoff value is used in the oRMSD clustering here. Overall, the oRMSD clustering is able to outline top structural candidates to explicitly account for the relative orientations between two proteins.

We have recently illustrated that RotPPR-CGMD is capable of searching various docking conformations (Huang et al. 2016), where the docking conformational space has been visited extensively. Thus, the RotPPR-CGMD provides an MD-based docking strategy to account for the structural flexibility for protein-protein docked conformations, ranging from compacted to extended shapes and from assembled to fully disassembled.

### 14.2.2 Small-Angle X-Ray Scattering (SAXS)

For characterizing protein-protein complexes, small-angle X-ray scattering (SAXS) data are particularly informative with regard to molecule shape of the entire complex and specifically, subcomponent arrangements. Quite a few excellent reviews have already discussed the basic principles and applications of SAXS (Bernado and Blackledge 2010; Blanchet and Svergun 2013; Kikhney and Svergun 2015; Putnam et al. 2007), and hence we describe the current state-of-the-art SAXS data acquisition and SAXS computing methods below.

#### 14.2.2.1 Experimental SAXS Data Collection

While acquisition of reliable SAXS data is non-trivial, experimental procedures have been recently described in detail (Jeffries et al. 2016; Skou et al. 2014), in addition to what has been covered in this book. Here, we point out that it is becoming a standard option for SAXS data acquisition to use an online chromatography-coupled setup, as illustrated in Fig. 14.4. This chromatography-coupled setup is particularly useful for aggregation-prone samples to allow the separation of a target complex from larger aggregates and/or smaller, excess substrates and thus improve sample homogeneity.

#### 14.2.2.2 SAXS Computing Methods

For the interpretation of experimental SAXS data, how to compute the SAXS profile from a given protein conformation, e.g. those generated from above RotPPR-CGMD simulations, is of particular importance because it is essentially the theoretical foundation of most SAXS data analyses.

*CRYSOL* and *Fast-SAXS-pro* are representative among currently available SAXS computing methods. Specially, *CRYSOL* requires the atomic coordinates (Svergun et al. 1995), while *Fast-SAXS-pro* takes the coordinates of either all atoms or just Cα atoms alone (Ravikumar et al. 2013). Additional differences include the treatment of excess electron density in a hydration layer by explicitly placing dummy water molecules surrounding the biomolecule. Comparison between these two methods is listed in Table 14.1. It should be noted that *CRYSOL* can be used for next-step optimization for iSPOT-derived atomic-structure ensembles since it provides an additional capability of best-fitting theoretical and experimental SAXS profiles.

Given its ability of handling the coordinates generated from RotPPR-CGMD docking simulations, *Fast-SAXS-pro* is thus used for SAXS computing to calculate theoretical scattering profiles, resulting from a collection of efforts (Ravikumar et al. 2013; Tong et al. 2016; Yang et al. 2009, 2010b). A web interface for *Fast-*

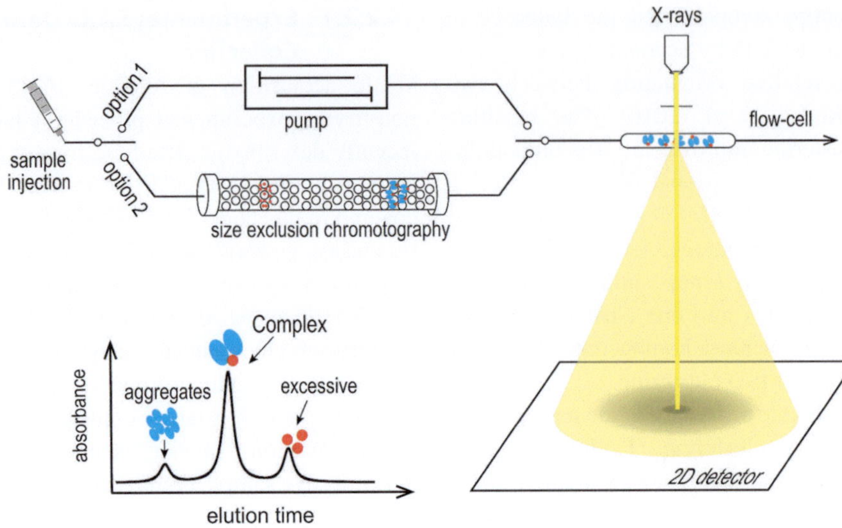


**Fig. 14.4** Experimental SAXS data acquisition. Two setups are routinely used: one with a simple flow-cell (*top*) and the other coupled with online chromatography (*bottom*) (Modified with with permission Yang 2014)

**Table 14.1** Comparison between *CRYSOL* and *Fast-SAXS-pro*

| | Structural coordinates | | Surrounding hydration layer (water) | | |
|---|---|---|---|---|---|
| | Atomic | Cα | Implicit solvent | Explicit solvent | Scattering difference between protein and RNA/DNA |
| *CRYSOL* | √ | | √ | | |
| *Fast-SAXS-pro* | √ | √ | | √ | √ |

*SAXS-pro* computing is available from the website at http://www.theyanglab.org/saxs.html, as well as executable files will be made available upon request.

### 14.2.3 Hydroxyl Radical Footprinting

Complementary to shape information obtained from SAXS is the solvent accessibility of specific sites probed by hydroxyl radical footprinting (Huang et al. 2015; Kaur et al. 2015; Xu and Chance 2007). The sites probed can be at the peptide level or at the single-residue level. As described below, specific rate constant measurements from footprinting are correlated to the solvent accessibility of probed amino acids, thereby providing structural information at a rather local residue-specific level.

#### 14.2.3.1 Experimental Footprinting Rate Measurement

The rate constant measurements of probed sites each from a different protein region are illustrated in Fig. 14.5. Typically, irradiation of water by X-rays generates hydroxyl radicals (OH•) that react protein residues *via* covalent modification. These OH•-modified samples are analyzed *via* proteolysis and the level of modification or "footprinting" is quantified *via* mass spectrometry (MS). This MS quantification is normally conducted at a single time point of X-ray exposure or repeated at various time points. In the latter, a dose-response curve of footprinting can be determined for each probed site, thereby establishing a footprinting rate $k_{fp}$ to characterize the overall footprinting effect on each individual site.

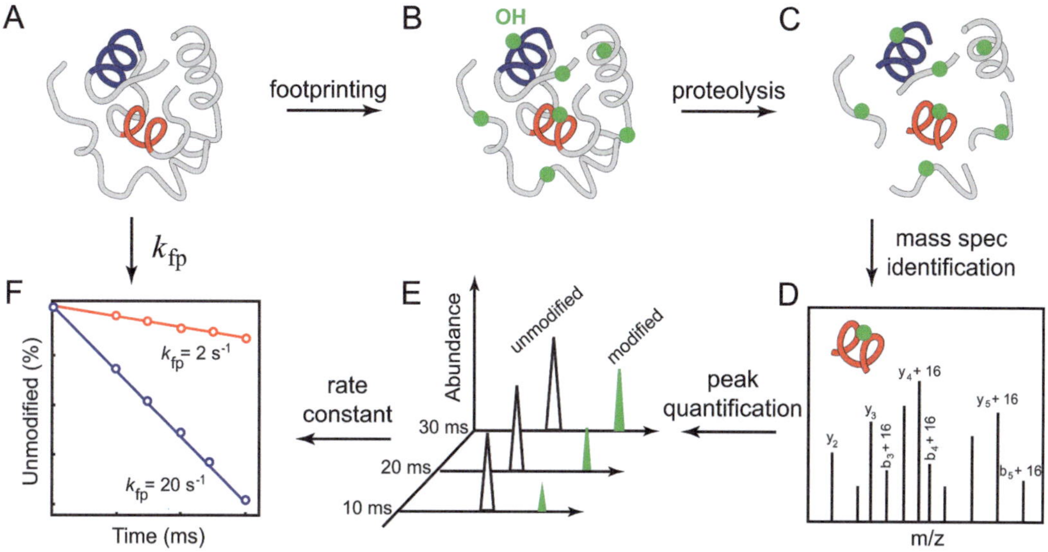


**Fig. 14.5** Site-specific rate measurement from hydroxyl radical (OH•) footprinting. Following **a–f**, different regions of a protein are covalently modified by OH• generated from X-ray irradiation of water, which is subsequently quantified by mass spectrometry. A dose-response measurement yields a kinetic rate constant for each site probed (Modified with permission Huang et al. 2015)

#### 14.2.3.2 Protection Factor Analysis and Structural Parameters

To use the footprinting rates $k_{fp}$ for structural characterization, we have established a protection factor (PF) analysis method (Huang et al. 2015; Kaur et al. 2015). This PF analysis can be applied at a single-residue or a peptide level. For example, PFs for single residues (or multiple residues within a peptide) are calculated by dividing the intrinsic reactivity $k_{\text{intrinsic}}$ of the residue (or the sum of the intrinsic reactivity for all of the residues within the peptide) by the observed rate $k_{fp}$,

$$\text{PF} = \frac{k_{\text{intrinsic}}}{k_{\text{fp}}}. \tag{14.1}$$

This simple conversion to PF values provides structural interpretation of footprinting measurements, enabling for the first time a structural comparison between different amino acid types that were previously impossible because footprinting rates alone are not correlated to any known structural properties. A key advantage of this PF analysis is absolute comparison between different sites that are probed simultaneously within an intact protein, as opposed to the previously limited comparison of a singular site crossing different conformational states. Specially, high-PF regions are structurally buried, while low-PF regions are solvent-exposed.

The PF data are correlated with structural features/parameters of protein sites probed. This is typically examined on a case-by-case basis partially due to the extent of footprinting being dependent on the protein sequence composition and its 3D structure. A list of structural parameters that reflect the related solvent accessibility are solvent accessible surface area (SASA), number of structural contacts (NC), and even the simple binary measure of being exposed or buried. These structural parameters are compared with experimental PF values to quantitatively evaluate the agreement between a protein structure candidate and its corresponding experimental footprinting data.

The intrinsic reactivity data can be from the website at http://www.theyanglab.org/protection.html. This weblet also provides the rate-PF conversion for single-residue footprinting data.

### 14.2.4 Data Integration by iSPOT

The multi-technique iSPOT platform is a result of these developments made in computational docking, SAXS and footprinting (illustrated in Fig. 14.1). These techniques are different but complementary, so the integration enabled by iSPOT provides a novel approach for structure determination of previously uncharacterized protein-protein complexes. Following the iSPOT workshop described in Fig. 14.2, we here show that each docking pose is used for evaluation against experimental SAXS and footprinting data *via* two specific scoring functions $\chi^2$ and $\varphi^2$ as detailed below.

#### 14.2.4.1 The Goodness of Fit to SAXS Data $\chi^2$

For each docked pose (or conformational cluster), the goodness of fit between the theoretical ($I_{\mathrm{cal}}$) and experimental ($I_{\exp}$) SAXS profiles is scored by a unitless $\chi^2$ (Yang et al. 2010a),

$$\chi^2 = \frac{1}{N}\sum_q \frac{\left(\log I_{cal}(q) - \log I_{\exp}(q)\right)^2}{\sigma^2(q)}, \quad (14.2)$$

where $\sigma(q)$ is the uncertainty of $\log I_{\exp}(q)$ and $N$ is the number of data points in $I_{\exp}(q)$. Theoretical SAXS profiles $I_{cal}(q)$ can be calculated from the docking configuration by either Fast-SAXS-pro or CRYSOL as described earlier. Specifically, a lower $\chi^2$ value represents a better fit between theoretical and experimental SAXS data. For example, $\chi^2$ often approaches 1–3 when experimental and theoretical SAXS profiles start to agree well.

#### 14.2.4.2 The Goodness of Fit to Footprinting Data $\varphi^2$

For the same docked pose, the goodness of fit between experimental footprinting PFs and structural parameters is scored by another unitless $\varphi^2$ (Huang et al. 2016),

$$\varphi^2 = \frac{1}{N_{fp}}\sum_i \frac{\left(\log PF_i - c \cdot SA_i\right)^2}{\delta_i^2}, \quad (14.3)$$

where $\log(PF_i)$ is the protection factor of each site $i$ probed by footprinting (either at a single-residue or peptide level) (Huang et al. 2015; Kaur et al. 2015), $\delta_i$ is the uncertainty of $\log PF_i$, and $N_{fp}$ is the total number of probed sites. As aforementioned, a list of structural parameters of solvent accessibility $SA_i$ include solvent accessible surface area (SASA) and number of neighboring contacts (NC). The scaling constant of $c$ is to offset the linear fitting between SA and *log*PF. Similar to $\chi^2$, here $\varphi^2$ is the difference between experimental footprinting PFs and theoretical solvent accessibility of each docked conformation. For example, a lower $\varphi^2$ value indicates a better fit of the candidate toward the target structure.

#### 14.2.4.3 iSPOT Model Selection and Refinement

The best-fit structural models that are selected by iSPOT are among the lowest $\chi^2$ and $\varphi^2$ values. This selection is illustrated in Fig. 14.2, where the orthogonal information provided by SAXS (about overall shape) and footprinting (about local solvent accessibility) is able to accurately select the crystal-like ensemble structures of a large complex. By testing on several protein-protein complexes with known structures, we have showed that the iSPOT is able to narrow down the correct target structure of bound complexes such as TGFβ-FKBP12 (Huang et al. 2016).

Refinement of the iSPOT-derived structure models of a protein-protein complex can be achieved by force-field based molecular dynamics (MD) simulations. Based on the atomic coordinates of individual protein components of the complex, a realistic structure of the complex can be constructed for all-atom, explicit-solvent MD simulations, as illustrated in the bottom of Fig. 14.2. As such, iSPOT is able to generate atomic structure ensembles of protein-protein complexes that can be further tested for model validation.

## 14.3 Summary

Structure determination of protein-protein complexes has been a challenging task. The multi-technique iSPOT platform is therefore a niche method available to structurally characterize such biomolecular complexes that are in the range of 50–200 kDa, although the method will work well for complexes of any size. We should stress that compared to other structural techniques that are quite matured or currently in their prime time, the development and application of iSPOT is still at its infancy. This early-stage technology development thus provides a critical step for future iSPOT applications to many biologically and biomedically important protein complexes.

**Acknowledgements** This work was supported by the NIH (R01GM114056 and P30EB009998), the DoD (W81XWH-11-1033), and by the Ministry of Education of Singapore (2014-T2-1-065). Beamtime access was supported *via* the BioCAT at the Advanced Photo Source and the LiX beamline at the NSLS-II by the DoE (DE-AC02-06CH11357 and KP1605010) and by the NIH (9P41GM103622 and P41GM111244). Additional support was provided by the Ohio Supercomputer Center and the Clinical and Translational Science Collaborative of Cleveland (4UL1TR000439). The content is solely the responsibility of the authors and does not necessarily reflect the official views of the National Institute of General Medical Sciences or the National Institutes of Health.

## References

Bernado P, Blackledge M (2010) Structural biology: proteins in dynamic equilibrium. Nature 468:1046–1048

Blanchet CE, Svergun DI (2013) Small-angle X-ray scattering on biological macromolecules and nanocomposites in solution. Annu Rev Phys Chem 64:37–54

Chen R, Li L, Weng Z (2003) ZDOCK: an initial-stage protein-docking algorithm. Proteins 52:80–87

Comeau SR, Gatchell DW, Vajda S, Camacho CJ (2004) ClusPro: a fully automated algorithm for protein-protein docking. Nucleic Acids Res 32:W96–W99

Dominguez C, Boelens R, Bonvin AM (2003) HADDOCK: a protein-protein docking approach based on biochemical or biophysical information. J Am Chem Soc 125:1731–1737

Gabb HA, Jackson RM, Sternberg MJE (1997) Modelling protein docking using shape complementarity, electrostatics and biochemical information. J Mol Biol 272:106–120

Huang W, Ravikumar KM, Yang S (2014) A newfound cancer-activating mutation reshapes the energy landscape of estrogen-binding domain. J Chem Theory Comput 10:2897–2900

Huang W, Ravikumar KM, Chance MR, Yang S (2015) Quantitative mapping of protein structure by hydroxyl radical footprinting mediated structural mass spectrometry: a protection factor analysis. Biophys J 108:1–9

Huang W, Ravikumar KM, Parisien M, Yang S (2016) Theoretical modeling of multiprotein complexes by iSPOT: integration of small-angle X-ray scattering, hydroxyl radical footprinting, and computational docking. J Struct Biol 196:340

Janin J, Henrick K, Moult J, Eyck LT, Sternberg MJ, Vajda S, Vakser I, Wodak SJ, Critical Assessment of, P.I (2003) CAPRI: a critical assessment of predicted interactions. Proteins 52:2–9

Jeffries CM, Graewert MA, Blanchet CE, Langley DB, Whitten AE, Svergun DI (2016) Preparing monodisperse macromolecular samples for successful biological small-angle X-ray and neutron-scattering experiments. Nat Protoc 11:2122–2153

Kaur P, Kiselar J, Yang S, Chance MR (2015) Quantitative protein topography analysis and high-resolution structure prediction using hydroxyl radical labeling and tandem-ion mass spectrometry. Mol Cell Proteomics (in press)

Kikhney AG, Svergun DI (2015) A practical guide to small angle X-ray scattering (SAXS) of flexible and intrinsically disordered proteins. FEBS Lett 589:2570–2577

Pierce BG, Wiehe K, Hwang H, Kim BH, Vreven T, Weng Z (2014) ZDOCK server: interactive docking prediction of protein-protein complexes and symmetric multimers. Bioinformatics 30:1771–1773

Putnam CD, Hammel M, Hura GL, Tainer JA (2007) X-ray solution scattering (SAXS) combined with crystallography and computation: defining accurate macromolecular structures, conformations and assemblies in solution. Q Rev Biophys 40:191–285

Ravikumar KM, Huang W, Yang S (2012) Coarse-grained simulations of protein-protein association: an energy landscape perspective. Biophys J 103:837–845

Ravikumar KM, Huang W, Yang S (2013) Fast-SAXS-pro: a unified approach to computing SAXS profiles of DNA, RNA, protein, and their complexes. J Chem Phys 138:024112

Skou S, Gillilan RE, Ando N (2014) Synchrotron-based small-angle X-ray scattering of proteins in solution. Nat Protoc 9:1727–1739

Svergun D, Barberato C, Koch MHJ (1995) CRYSOL – a program to evaluate x-ray solution scattering of biological macromolecules from atomic coordinates. J Appl Crystallogr 28:768–773

Tong D, Yang S, Lu L (2016) Accurate optimization of amino acid form factors for computing small-angle

X-ray scattering intensity of atomistic protein structures. J Appl Crystallogr 49:1148–1161
Tovchigrechko A, Vakser IA (2006) GRAMM-X public web server for protein-protein docking. Nucleic Acids Res 34:W310–W314
Xu G, Chance MR (2007) Hydroxyl radical-mediated modification of proteins as probes for structural proteomics. Chem Rev 107:3514–3543
Yang S (2014) Methods for SAXS-based structure determination of biomolecular complexes. Adv Mater 26:7902–7910
Yang S, Park S, Makowski L, Roux B (2009) A rapid coarse residue-based computational method for x-ray solution scattering characterization of protein folds and multiple conformational states of large protein complexes. Biophys J 96:4449–4463
Yang S, Blachowicz L, Makowski L, Roux B (2010a) Multidomain assembled states of Hck tyrosine kinase in solution. Proc Natl Acad Sci U S A 107:15757–15762
Yang S, Parisien M, Major F, Roux B (2010b) RNA structure determination using SAXS data. J Phys Chem B 114:10039–10048

# Small Angle Scattering for Pharmaceutical Applications: From Drugs to Drug Delivery Systems

# 15

Aaron Alford, Veronika Kozlovskaya, and Eugenia Kharlampieva

**Abstract**

The sub-nanometer scale provided by small angle neutron and X-ray scattering is of special importance to pharmaceutical and biomedical investigators. As drug delivery devices become more functionalized and continue decreasing in size, the ability to elucidate details on size scales smaller than those available from optical techniques becomes extremely pertinent. Information gathered from small angle scattering therefore aids the endeavor of optimizing pharmaceutical efficacy at its most fundamental level. This chapter will provide some relevant examples of drug carrier technology and how small angle scattering (SAS) can be used to solve their mysteries. An emphasis on common first-step data treatments is provided which should help clarify the contents of scattering data to new researchers. Specific examples of pharmaceutically relevant research on novel systems and the role SAS plays in these studies will be discussed. This chapter provides an overview of the current applications of SAS in drug research and some practical considerations for selecting scattering techniques.

**Keywords**

Small angle scattering • Pharmaceutical • Structural characterization • Delivery vehicles • Nanoparticles • Micelles • Vesicles • Multilayer capsules

## 15.1 SAS Data Analysis

There are many methods of modeling and interpreting SAS data. Figure 15.1 shows an overview of the most common techniques used *ab initio* to interpret the data gathered in structural studies. Starting from the common Porod

A. Alford • V. Kozlovskaya • E. Kharlampieva (✉)
Department of Chemistry, University of Alabama at Birmingham, 901 14th Street South, CHEM 272, Birmingham, AL 35294, USA
e-mail: ekharlam@uab.edu





B. Chaudhuri et al. (eds.), *Biological Small Angle Scattering: Techniques, Strategies and Tips*, Advances in Experimental Medicine and Biology 1009, DOI 10.1007/978-981-10-6038-0_15

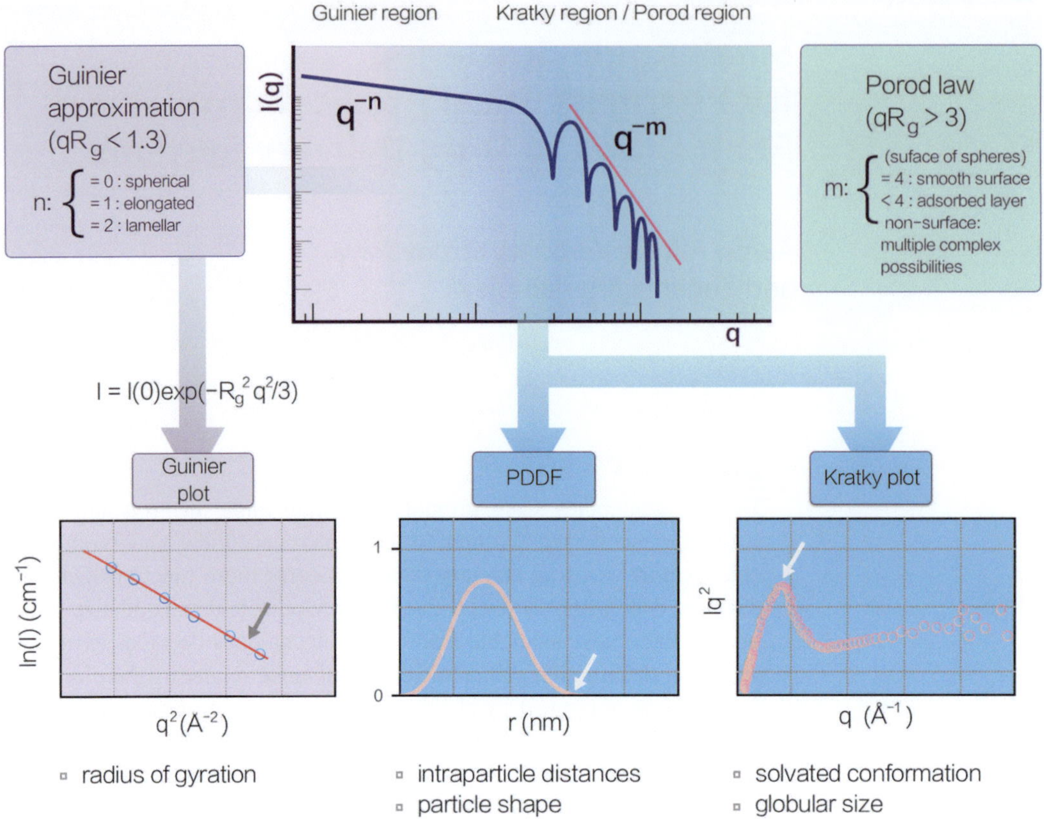


**Fig. 15.1** Different regimes in scattering curves with the relevant structural information defined for each method of analysis

(log-log) plot most often encountered from reduced data, the first of the model-free analyses is the Guinier approximation used to estimate the radius of gyration, $R_g$, and generalized particle form. This is a common first step for data on drug delivery carriers in the 50–200 nm size range. The Guinier range is the region of $q$ such that $qR_g$<1.3 and the $q^n$ dependency of the form factor in this range of the log-log plot describes in general the morphology of the scattering body with n = 0, 1, and 2 representing globular (spherical), rod-like, and lamellar (flat) morphologies, respectively (Schnablegger and Singh 2013). It is important to note that several assumptions of ideality must be taken in this analysis. First, since the intensity of scattering $I$ as a function of the scattering vector $q$ is a product of two factors, following Eq. 15.1,

$$I(q) = P(q)\,S(q) \tag{15.1}$$

where $P(q)$ and $S(q)$ are the form and structure factors, respectively, sufficient dilution must be achieved such that the form factor can be analyzed independently of the contribution from the structure factor. The reason for this is that the scattering contribution from the structure factor due to interparticle interactions and arrangement in diluted systems is minimal such that it tends to approach unity and the intensity therefore becomes equal to the form factor (Guinier and Fournet 1955). Guinier's approximation is made in reference to the form factor with sufficient dilution being assumed. This is a particularly important consideration for drug delivery carriers that exist as insoluble suspensions. Secondly, monodispersity of the particles is a requirement for the most accurate approximation, as small polydispersities affect the slope of the scattering curve significantly at very small $q$. With these assumptions fulfilled, an expansion of Guinier's original theory (Glatter and Kratky 1982) takes the form of

$$I = I(0)\exp\left(\frac{-R_g^2 q^2}{3}\right) \tag{15.2}$$

where $I$ is the scattering intensity. Plotting $\ln(I)$ vs $q^2$ therefore gives the $R_g$ from the slope of the fit line once it is extrapolated to the intercept $I(0)$.

The next region in the log-log plot encompasses the intermediate values of $q$ between the Guinier range and Porod range ($qR_g > 3$). Indirect Fourier Transform (IFT) and Kratky analyses both produce useful plots from this region with the IFT providing a Pair Distance Distribution Function (PDDF). The PDDF represents a distribution of the intraparticle distances elucidated by the scattered radiation. The shapes produced by this plot are characteristic of specific form factors and again present model-free approximations as well as another representation of the particle size with the largest intraparticle distance(s) being the distance from one terminal edge of the particle to another (Schnablegger and Singh 2013). This distance is found at the end of the rightmost shoulder for spherical particles and other shapes have multiple points at which critical dimensions are represented. Figure 15.2 shows a representation of various PDDF plot shapes with their corresponding morphologies and size indications.

Flowing from the same region of $q$ is the Kratky plot of $Iq^2$ vs $q$ which gives insight into the conformation of polymers in solution from their mid to high $q$ scattering. At the higher part of this $q$ range, Gaussian coiled polymers result in scattering that follows $1/q^2$ behavior and so tend to produce curves that level out in the Kratky plot. Deviation from this line-shape indicates non-Gaussian conformations, the source of which is usually interpreted by the researcher based on the sample composition. The shape of the curve in a Kratky plot can therefore be used to distinguish between stretched Gaussian coil polymers or non-ideal conformations due to attractive/repulsive forces

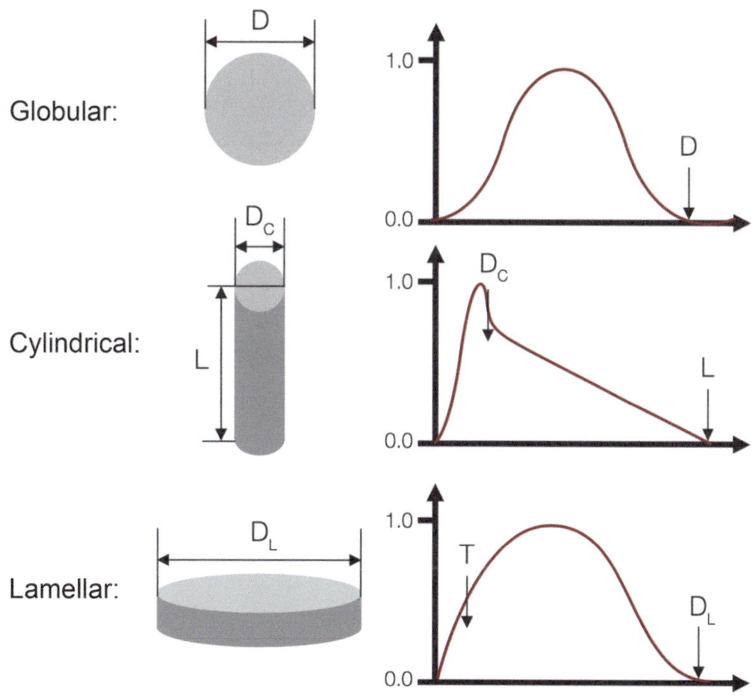


**Fig. 15.2** Determination of critical dimension and shape from PDDF plots (Schnablegger and Singh 2013) (Reprinted with permission from Anton Paar GmbH)

that result in excluded volume (Hammouda 2016). For drug delivery applications involving self-assembled systems, polymer conformation can be an important piece of information.

Deeper application of this analysis in drug delivery vehicle structural research will be clear when the following is considered: if Gaussian chains result in a curve with a slow rise that levels into a plateau, self-associative forces that lead to precipitation and the formation of nano-objects are characterized by a peak in the initial region of the Kratky plot. The position of this peak is inversely proportional to the $R_g$ (Filippov et al. 2012). Especially in systems of complex multi-component structure, such as drug-conjugated copolymers, this analysis provides detail of the solvated or structural morphology that can be focused using contrast variation techniques.

The Kratky and PDDF analysis techniques were used to probe constructs of 2-hydroxypropylmethacrylamide (HPMA) copolymers containing cholesterol moieties and doxorubicin in elegant characterizations that revealed the conformation of polymers within the nanoparticle structures as rounded cylinders (Filippov et al. 2012, 2013). Kratky analysis and cross-sectional PDDF in this case showed that the nanoparticle structure changed from Gaussian coiled coronas to more compact objects with higher excluded volume as a function of the concentration of cholesterol moieties (0–3.0 mol%) in the chain (Filippov et al. 2012). The power of modeling software and the true richness contained within scattering data can be seen in Fig. 15.3 (Kratky and PDDF) from the same study in which computational analysis

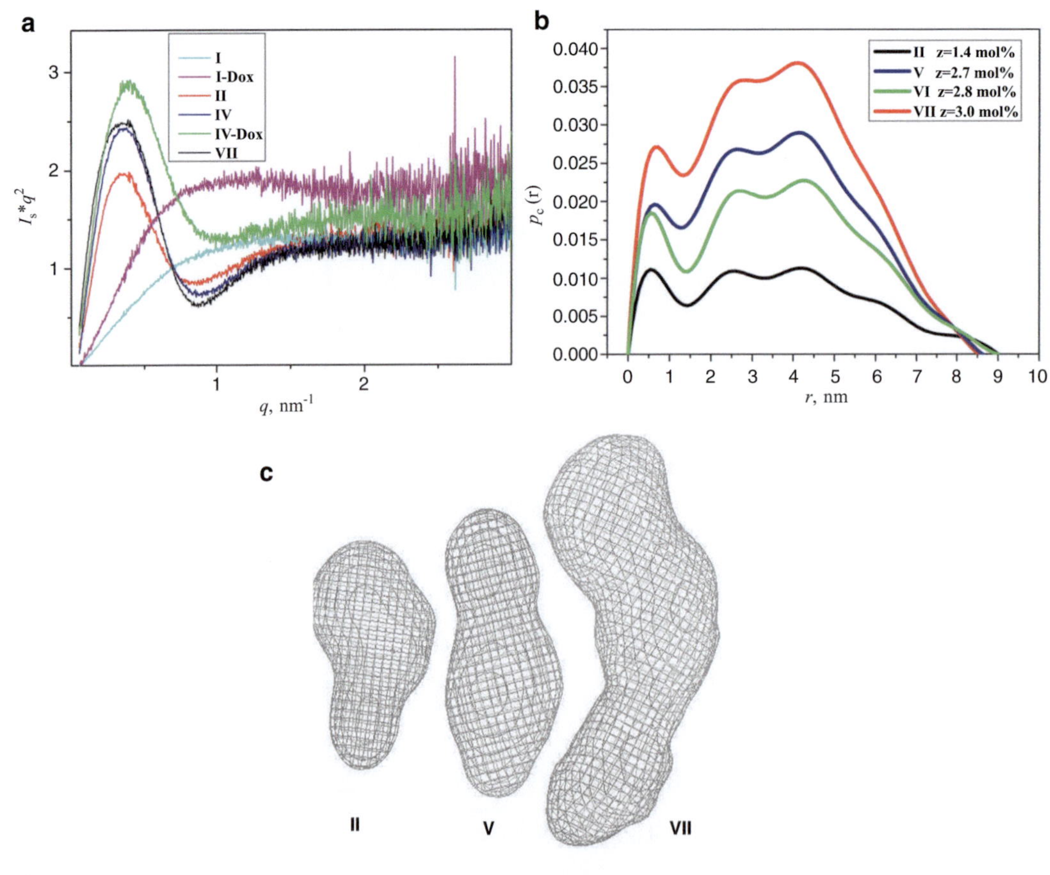


**Fig. 15.3** Kratky (**a**) and cross-sectional PDDF (**b**) plots showing morphology and cross-sectional distances of conjugates used to compute the hypothetical structure (**c**) (Reprinted with permission from Filippov et al. 2012, copyright American Chemical Society)

yielded hypothetical shapes of the nanoparticles from cross-sectional PDDF as ellipsoids with lengths of 12.1 nm and widths of 4.6 nm.

After the Guinier and Kratky analyses, the final type of model-free data to be obtained comes from the region of the standard scattering plot known as the Porod range where $qR_g > 3$. The slope of the $q^{-m}$ dependency in this region is dependent on the surface per volume according to Porod's law and the value of $m$ can speak about the physical properties of the surface scattered at large angles. An $m$ value of 4 at high $q$ is indicative of a smooth surface for spherical objects while $2 < m < 4$ represents scattering due to a rougher adsorbed layer (Estrela-Lopis et al. 2009). Extensive multi-slope plots, their $q^{-m}$ laws, and their shape implications can be found in the works of Boualem Hammouda (2010a, b, 2016). Plots built from combined concepts such as Kratky-Porod reveal even greater detail from advanced methods and will be discussed later in the chapter.

Outside of these model-free techniques there are a multitude of model-intensive approaches often used within analysis environments such as SASview, PRIMUS, and others that can fit scattering curves to relevant physical parameters. However, since scattering curves are not fingerprints, the models must be informed to avoid unreasonable parameter definitions. This is an important point for characterization of drug delivery vehicles and is one of the predominate reasons that supplementary characterizations are often performed alongside SAS. The fewer parameters that are left to float during model fitting routines, the more accurate the remaining parameters can describe the scattering curve and the better the picture that can be obtained with SAS. Modeled approaches are often used if the shape or form of the particle is known *a priori*; therefore data fitting approaches should start from model-free analysis before moving into precise characterization with fitting models. The workings of model fitting are not discussed in detail due to their highly specific and customizable nature. However, model analysis is still present in nearly every SAS study, as the parameters are highly relevant especially in terms of the properties of drug delivery devices. This will be seen in the following sections.

## 15.2 Structural Characteristics of Drugs and Drug Delivery Vehicles

### 15.2.1 Delivery Vehicles

#### 15.2.1.1 Nanoparticles

Nanoparticles can be defined as nanometer-sized clusters of atoms or molecules. Nanoparticles come in many shapes and have been gaining popularity as drug delivery vehicles (Urbina et al. 2008; Mattingly et al. 2015). Owing to their small size (1–100 nm), tunable composition, and potential inclusion of metals to imbue them with fluorescent/IR/magnetic sensitivities they often have unique properties in terms of cellular uptake, bio-distribution, and reactivity to stimuli (Remant Bahadur et al. 2012). These properties also make them common candidates for SAS studies because, while other techniques can give size information, they fall short in characterizing the internal structures and inter-particle interactions that make nanoparticles unique and useful for drug delivery.

#### 15.2.1.2 Micelles

Micelles are single or dual-component nanometer-sized formations of amphiphilic molecules dispersed in a solvent, often water, that generally have spherical morphology but are not strictly limited to spheres (Biggs et al. 2007). Micelles have a core-corona architecture featuring hydrophobic cores and hydrophilic coronas when used in aqueous environments. Micelles often self-assemble in water to stabilize the interactions of the two portions of the molecule with the solvent, with reverse micelles possible in the case of hydrophobic solvents (Nunes and Car 2013; Correa et al. 2012). Block copolymers are widely utilized as micelle forming delivery devices and often include PEG paired with a less hydrophilic polymer such as poly(lactic acid) or polystyrene (Cho et al. 2006; Bae et al. 2009). SAS studies with micelles often utilize contrast variation techniques and/or selective deuteration of components to isolate them from the overall scattering in efforts to characterize their structural components (Almgren et al. 2007; Naruse et al. 2009; Smith et al. 2014).

#### 15.2.1.3 Vesicles

Polymeric vesicles are multi-component nanometer- or micrometer-sized assemblies of amphiphilic molecules dispersed in a solvent. Vesicles are similar to micelles in terms of their components, but in contrast, form hollow spherical or worm-shaped structures with solvent trapped in the inner cavity and a hydrophobic bilayer between the core cavity and hydrophilic corona (Srinivas et al. 2013; Sauer and Meier 2004). Vesicle applications within drug delivery are often similar to those of micelles, with some systems able to switch from one morphology to the other in response to stimuli (Lagzi et al. 2010).

#### 15.2.1.4 Multilayer Capsules

Multilayer capsules are nanometer to micron-sized polymer constructs made of ionic, hydrogen-bonded, or covalently bonded layers of polymers or other constituents (Kharlampieva et al. 2009; Kozlovskaya et al. 2009, 2012a, b; Erel-Unal and Sukhishvili 2008). Hollow hydrogel microcapsules are formed by layer-by-layer (LbL) assembly of polyelectrolytes on sacrificial particulate templates, followed by chemical cross-linking of the assembled polyelectrolyte multilayer and template dissolution (Sukhorukov and Möhwald 2007; Wang et al. 2008; Skirtach et al. 2011). LbL technology offers unique opportunities to fabricate capsules of any size, geometry, composition and thickness controlled at the nanoscale (Lvov et al. 1993; Decher 2002; Kharlampieva and Sukhishvili 2006; Lutkenhaus and Hammond 2007; Hammond 2011; Mjahed et al. 2008; Tang et al. 2006). The technique can synthetically recreate shapes and easily impart a desired elasticity and responsiveness to the thin capsule wall (Elsner et al. 2006; Lisunova et al. 2011). The capsule's wall can be easily functionalized, with an internal cavity which can be used for loading functional compounds. Unlike other competing delivery systems such as nanoparticles, liposomes, dendrimers, and block copolymer vesicles and micelles, nanothin capsules have a high loading capacity and can deliver cargo on demand in response to a stimulus (Donath et al. 1998; Caruso et al. 1998; Becker et al. 2010; Matsusaki and Akashi 2009; Tong and Gao 2008). Because of their templated construction and multilayer or hydrogel architecture, capsules have greater structural stability compared to vesicles and micelles (Liu et al. 2014; Kozlovskaya et al. 2012a, b). Their core-shell structure and potential for selective deuteration both make polymer capsules interesting candidates for SAS studies where characterization of the shell thickness, capsule morphology, or cargo in the cavity can only be determined through precise methods.

While efforts are being directed toward making drugs more selective by their structural design, at least as many efforts are aimed at developing smart carriers that can chaperone drugs that have high efficacy but limited selectivity to their intended target in the body, thus minimizing collateral damage from rogue drug molecules by releasing them in a controlled manner *via* biologically relevant triggers such as pH, temperature, and presence of cellular peptides (Xue et al. 2015; Liu et al. 2015; Kozlovskaya et al. 2014).

The structures of these drug carriers range from micelles and vesicles between 5 and 200 nm in diameter (Fig. 15.4) to larger constructs including capsules and gel particles in the 1–10 μm range and, although they function according to their intended purpose, researchers are often left hypothesizing about *exactly* what happens during response to the stimulus, release of the drugs, or sometimes even exactly what size or shape the carriers are. Due to their extremely diminutive nature, characterizing the shape and structure of a drug carrier such as a nanoparticle can be challenging without advanced techniques (Mariani et al. 2016). SANS and SAXS are among these techniques: they can probe the structures of these innovative materials to evaluate carrier's shape and size, interior and exterior

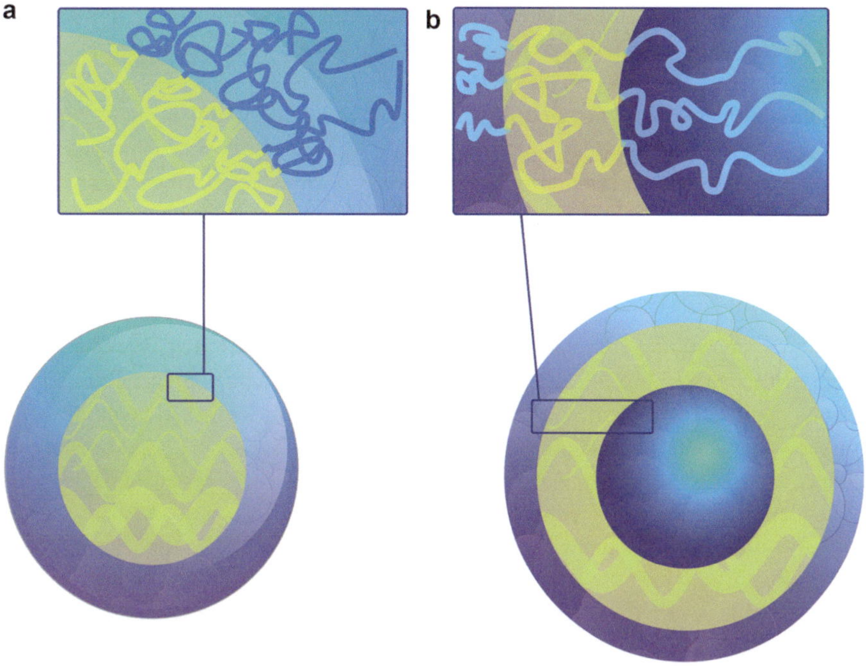


**Fig. 15.4** Schematic of the internal structure of micelles (**a**) and vesicles (**b**) where hydrophobic portions of the molecule are shown as *yellow* and the hydrophilic portions are *colored blue*

dimensions, layer spacing and stratification, interparticle spacing and interaction behaviors, and dynamic assembly with the proper experimental design and material construction (Lu et al. 2016; Fong et al. 2010; Yin et al. 2016). For instance, consider an amphiphilic molecule or polymer in $D_2O$ (Fig. 15.4). Since the scattering length density (SLD) values for the protium and deuterium isotopes of hydrogen lie on opposite sides of a spectrum of values, hydrogen-containing molecules in $D_2O$ will have an SLD that lies somewhere within this spectrum depending on the extent of solvation. A simple experimental setup such as this allows for core-corona dimensions to be modeled since the hydrophobic core will have a different extent of solvation compared to the hydrophilic corona. Selective deuteration opens many possibilities for structural characterization and methods of determining dynamic solution behaviors, fine examples of which can be seen in the works of the Lodge group (Lu et al. 2012, 2015, 2016).

### 15.2.2 Pharmaceutically Relevant Structural Considerations of Self-Assembled Carriers

Besides effective delivery of drugs, there are aspects of drug behavior on the nanoscale that are important to the pharmaceutical researcher. For example, while it can be easily observed that a chemical entity once PEGylated gains water solubility (Harris et al. 2012; Veronese and Mero 2012), the conformation that clusters of the amphiphilic molecule adopt in solution has implications on how the drug will exist and act in the body.

For example, it has been shown in computational analyses that each PEG repeat unit can bind at least two water molecules which end up forming a hydrated corona around the molecule (Rissanen et al. 2014). Increased hydration due to PEG chains, however, leads to an increased entropic barrier for entering a lipid bilayer. Furthermore, this hydration increases association of $Na^+$ ions which compounds repulsive forces between drug conjugates and cellular membranes as demonstrated by modeled computations of distance between PEGylated (~2000 Da PEG) biochanin and tetrahydroxyphenylporphyrin (THPP) drug conjugates and a modeled 1-palmitoyl-2-oleoylphosphatidylcholine (POPC) lipid bilayer (Rissanen et al. 2014). The results from these studies demonstrated that exposed hydrophobic portions of the PEG-THPP conjugates had solvent accessible surface areas (SASA) of 54–57% depending on the presence or absence of 140 mM NaCl while the PEG-biochanin conjugates had SASAs of only 41–45% in the same conditions. Higher SASA helped to locate the PEG-THPP conjugates on or near the modeled membrane, with typical distances of 0.8–1 nm from the membrane center, while PEG-biochanin conjugates more extensively wrapped with a PEG corona (lower SASA) associated more $Na^+$ ions which prevented any association with the POPC membrane.

Stopped-flow SANS was used in a related kinetic study of the interaction of amphotericin B-sodium cholesteryl sulfate (AmB-SCS) ellipsoidal micelles (45 × 2.5 nm) with 100 nm POPC-cholesterol vesicles as model 3D membranes (Foglia et al. 2015). Incorporation of SCS into the micelle decreased the time to swell the POPC-cholesterol vesicles by 0.5 nm (indication of the interaction) to 5 s vs 25 min for SCS-free micelles. It should be clear from these illustrations how information on the sizes of the core and corona in drug conjugates and the creative use of powerful SAS techniques can significantly enhance the quality of results obtained from development research.

Size characterizations such as dynamic light scattering (DLS) can give information on the size of assemblies, but with proper sample preparation, SAS studies enable "visualization" of the core and corona with nanometer accuracy and therefore give much deeper insight into the adopted conformation of the vehicle. In fact, the interior structure of an intricate self-assembled poly-ion complex system was precisely characterized using contrast variation (CV) SAXS in a study that depicted the particle morphology as a function of the ratio of cationic to neutral repeat units in a copolymer (Sakamoto et al. 2014). This investigation elegantly used form factor analysis of SAXS curves of the copolymer dispersion in water to elucidate the morphology as micelle- or vesicle-like. In addition, CV SAXS performed in a series of 5–20% PEG/$H_2O$ solvent systems yielded electron density profiles of the vesicular structures from simultaneous model fitting of the scattering curves at different solvent contrasts. The model fitting allowed extremely precise analysis of the vesicle structure: a compact 18 nm diameter core with a 4 nm thick coronal layer and 100 nm thick hydrated shell. One of the interesting features of SAS studies such as this is the exposition of radial electron density that shows gradients of hydration. Micelles, for example, are often depicted with a strictly hydrophobic and hydrophilic region, and techniques such as DLS that reveal only size distribution cannot resolve the picture any further. However, CV SAXS and CV SANS reveal a clearer picture of the non-uniform nature of each structural component, with micelles that have interpenetrating hydrophilic polymers in the core and a shell that slowly becomes more hydrated as the polymers stretch out into the solvent.

While DLS and SAS are often complementary techniques, going from DLS to SANS characterization of the same drug-loaded micelle means going from knowing the size distribution and polydispersity of the particle to knowing such details as the intermixing of hydrophobic and hydrophilic sections when drugs are present and absent. For example, two SAXS studies on the partially benzyl esterified poly(aspartic acid)-PEG block copolymer PEG-P(Asp(Bzl)) found differing behaviors of the hydrophobic core upon loading of drug mimics of various hydrophobicities (Akiba et al. 2010; Sanada et al. 2013). Model fitting of scattering curves showed that the more hydrophobic retinoid LE540 was shown to increase the radius of the core from 5.9 to 6.9 nm when the weight % of the drug mimic in the micelle was increased from 0 to 8.3%. In analysis of a linear-linear scattering curve at high $q$ (Fig. 15.5), the decrease of a characteristic ordering peak prompted the

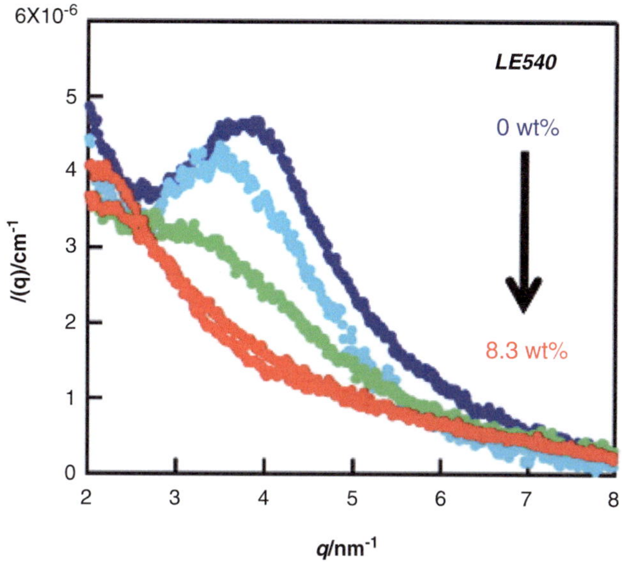


**Fig. 15.5** Linear-linear plot of decreasing ordering from LE540 displacing hexagonal chain arrangement (Reprinted with permission from Akiba et al. 2010, copyright American Chemical Society)

discovery that a decrease in ordering of the core occurred as the LE540 concentration increased (Sanada et al. 2013). This result was taken to be indicative that the drug mimic was uniformly dispersed throughout the core which interrupted the interchain ordering from the aspartate groups.

In contrast, using the less hydrophobic cargo molecule tetrabromocatechol (TBC) caused no appreciable growth of the core, but anomalous SAXS showed that the TBC concentration sphere extended slightly into the hydrophilic PEG region (Akiba et al. 2010). The highly precise characterizations from both of these studies, which are uniquely available through SAS, led to the significant conclusion that the relative hydrophobicity of the cargo plays a critical role in influencing the micelle morphology, even revealing that cargo molecules may exist in both micellar regions to a small extent based on the relative hydrophobicity.

The architecture of micelles in different backbone-selective solvents is also an interesting topic that has been addressed with SANS (Alexander et al. 2014). In the investigation of grafted polyisoprene(PIP)-*g*-Pluronic polymers (where *Pluronic* is a trade name for PEO-poly(propylene oxide)-PEO triblock copolymers), the micelle architecture in solvent systems of mixed polymer selectivity (ethanol: Pluronic-selective, hexane: backbone-selective, THF: good solvent for both) was characterized to gain understanding on the solution behavior of the micelles. The contrast between deuterated solvents and hydrogenated polymers was used in model fitting to elucidate the structures. For PIP-*g*-P123 in 100% EtOH, fuzzy micelles with 16.5 nm PIP cores and 7.3 nm thick P123 chain coronas were formed. One hundred percent hexane did not produce stable particles, and after iterative solvent trials it was found that 40% THF/hexane was able to produce stable flower-like micelles with 6.4 nm P123 cores and 15.5 nm thick PIP loops in the corona. Although CV techniques are powerful as addressed above, intensive model analysis can sometimes provide similar results. In the case of the PIP-*g*-Pluronic micelles, the core and coronas were separated by their solvation densities from within model fitting analysis to provide the thicknesses.

### 15.2.3 Drugs

The solubility of drugs plays a key role in their behavior in and out of the body as well as in their bioavailability. For example, the loading of hydrophobic drugs into drug delivery nanocarriers is a common way to increase their water solubility but the mechanism of drug entrapment is often assumed because it is not easily characterized without precise information on the drug's localization in the nanocarrier (Yokoyama et al. 1996).

It is clear that SAS can help illuminate the answers to these questions, but it is also used for more complex characterizations. SAS studies have been used to find the kinetics of drug release from conjugates, drug/polymer co-crystallization and association characteristics, and the relationship between conformation of formulations and their respective efficacy (Zhu et al. 2011; Paul et al. 2007; Vicent et al. 2005). For instance, time-resolved SAXS showed that the lamellar long period of 3300 Da PEG increased by 5 nm when co-crystallized with 20% benzocaine, visualized by a low $q$ peak around 0.04 $nm^{-1}$ that increased in size over 30 min (Zhu et al. 2011). After 24 h, this peak reduced to nothing as benzocaine was expelled from the PEG and the long period returned to 11.6 nm, close to the value of pure PEG of 11.4 nm. Ibuprofen at the same concentration was found to increase the PEG long period by 4.9 nm but as the drug crystallized separately, it was rapidly excluded from the PEG lamellae after 100 min.

In a separate study, the inclusion of a targeting ligand, galactosamine, on a polymer-drug conjugate was shown to drastically change its maximum tolerated dose (MTD). To evaluate the effect of the ligand on the in-solution conformation, solutions of 30 kDa 2-hydroxypropylmethacrylamide (HPMA)-doxorubicin (DOX) and HPMA-DOX-galactosamine in $D_2O$ were analyzed with USANS (Vicent et al. 2005). Fitting of Gaussian coil models to the scattering data showed that the formulation containing galactosamine had a larger $R_g$ of 9.0 nm compared to that of the standard HPMA-DOX conjugate of 7.8 nm. While both conjugates showed the same overall antitumor activity and total release of DOX, the larger $R_g$ of the targeted conjugate was hypothesized to be responsible for the lower MTD as it indicated a more open solution geometry which allowed greater exposure of the drug to the biological locale.

## 15.3 SAS Studies of Drugs and Drug Delivery Vehicles

### 15.3.1 Nanoparticles and Micelles

One of the caveats of inorganic and drug nanoparticle preparation is the need to stabilize the particles by attaching organic molecules or polymers to their surfaces. The choice of stabilizer obviously has significant effects on the morphology and efficacy of the drug nanoparticle agent and several studies have therefore identified relevant drug-stabilizer interactions to develop rational design approaches (Choi et al. 2013; Ghosh Chaudhuri and Paria 2012).

Detailed structural analysis of the adsorption of polymers on nanoparticles is therefore an important endeavor but is only feasible through SAS. To address this, hydroxypropylcellulose (HPC)-coated nabumetone and halofantrine particles 300 nm in size were prepared and analyzed in a CV SANS experiment to gain further understanding of polymer-drug adsorption parameters (Goodwin et al. 2013). Tabulated nabumetone and halofantrine SLDs were used to initiate CV calibration and the respective $D_2O/H_2O$ ratios of 31.3 and 33.8% were found to match the drug suspensions and agree with the tabulated values. CV SANS studies in these solvent ratios therefore eliminated scattering due to the drug core from the form factor $P(q)$ and enabled strict characterization of the adsorbed HPC corona. First, a relationship between the known $R_g$ of free HPC of molecular weights ranging from 47 to 110 kDa and the thickness of the respective coronas was established which showed that adsorbed layers may approach thicknesses nearing the $R_g$ (Goodwin et al. 2013). However, the adsorbed layers were found to have thicknesses below the $R_g$ which implied that the polymers were attaching to the surface in a compacted state. For example, the $R_g$ of 110 kDa HPC was measured to be 23.9 nm while the thickness of its nanoparticle corona layer was only 15.4 nm.

Interestingly, the molecular weight was not found to proportionally affect the layer thickness, as polymer $R_g$ values between 16 and 24 nm ($M_W$ 47–110 kDa) resulted in layer thicknesses that varied randomly between 11.2 and 15.4 nm. This indicated that some of the lower molecular weight HPC adopted a less condensed conformation at the nanoparticle surface. This example shows how model fitting analysis can be effectively used to characterize the polymer-stabilized surface of micellar structures, but model-free analysis is a more concrete tool for analyzing scattering curves. Its use in conjunction with model fitting will be discussed in the next example.

In an investigation of DNA-coated gold nanoparticles (NPs), the effect of ionic strength on the architecture of 15-mer polythymine coated NPs (named T15) and 15-mer mixed sequence of 7 polythymine-CTCATGAG coated NPs (named T7–8) was characterized with SAXS/SANS (Yang et al. 2015b). This approach demonstrates intelligent solutions to some of the considerations of using both X-rays and neutrons. First, the X-ray scattering of coated metal NPs is dominated by the scattering due to the metal because of its much higher electron density in comparison to the organic molecules, which makes elucidation of coronal parameters such as thickness and chain conformation difficult (Von White et al. 2011). Therefore, SAXS was used to evaluate core parameters such as size and polydispersity independently; finding these values allowed fixing of those particular parameters during model fitting of SANS data, where, since neutron SLD values are not dependent on the size of the nucleus, core and corona scattering would not be highly differentiated between gold and DNA.

Guinier analysis of SANS curves at minimum $q$ was used to find the $R_g$ of the particles with T15 and T7–8 equaling 9.06 and 9.03 nm, respectively. These values were slightly higher than those found with DLS; this is a good place to note that the Guinier approximation, being made in the assumption of sufficient dilution ($S(q) \sim 1$ so $I(q)=P(q)$), is sensitive to interparticle interactions that give rise to structure factor $S(q)$ scattering. In light of this, indirect Fourier transform (IFT) analysis was used to populate a pair distance distribution function (PDDF), from which the $R_g$ values of 8.80 and 8.79 nm for the T15 and T7–8 NPs, respectively, agreed with DLS measurements.

After the addition of 0.5 M NaCl, the PDDF of T15 retained a bell shape indicative of spherical geometry but shifted to reveal an $R_g$ value of 8.34 nm. The T7–8 sample was found to increase in size to 13.21 nm at the same salt concentration. Interestingly, at 1 M NaCl, the $R_g$ of T15 decreased to 8.21 nm but T7–8 aggregated which resulted in separate peaks in the PDDF. DLS data confirmed the IFT indication of aggregates.

Finally, model fitting revealed a decrease in coronal thickness for T15 from 6.58 to 5.91 nm after addition of 1 M NaCl. The most interesting analysis came from this result, where the difference between the SLD of the solvated corona and solvent was found to be $3.4 \times 10^{-7}$ Å$^{-2}$ at both 0.5 and 1 M salt concentrations. Since the shell decreased in size, the contrast difference would be expected to increase as solvent was excluded from the DNA chain. However, a constant SLD indicated that the material density of the corona increased; in other words, the solvent density was higher in the corona than in the bulk solvent. This study is an excellent example of the extremely interesting nanoscale results that can be illuminated by well-designed SAXS/SANS experiments and model-free/modeled scattering analyses.

A final example of model-free nanoparticle analysis is the SANS study of single chain nanoparticles (SCNP) in which the gap between current efforts and the goal of synthetic, globule-state, protein-like nanomachines is demonstrated (Pomposo et al. 2014). In general, the goal of SCNP research is to produce a synthetic polymer that, by carrying out the crosslinking in such a way as to eliminate interchain crosslinks in favor of purely intrachain linkages, can adopt a compact globular state. Computer simulations and SAXS/SANS data of poly(methyl methacrylate) (PMMA) SCNPs with 20 mol% reactive groups pointed to particles in an open state. The clearest presentation of the difference between SCNPs and true compact globules is presented in a Kratky plot from this study (Fig. 15.6) which shows the PMMA SCNP curve resembling

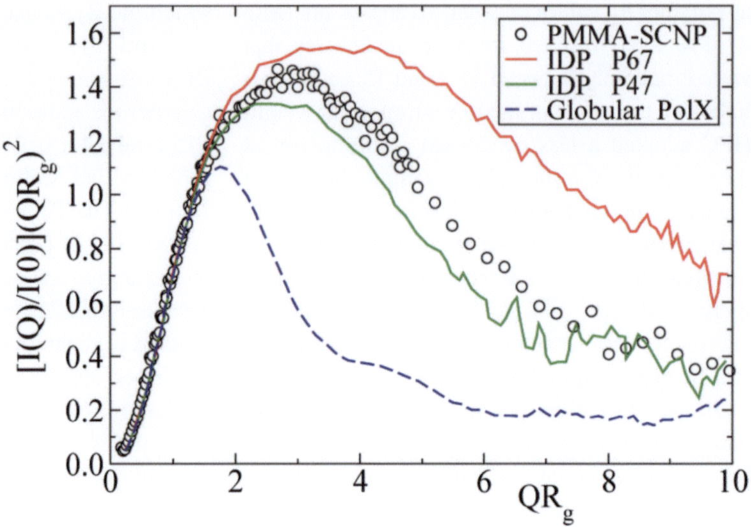


**Fig. 15.6** Kratky plot of the SCNPs compared to a true globular protein and various intrinsically disordered proteins (IDPs) with open solution geometries (Reprinted with permission from Pomposo et al. 2014 copyright (2014) the American Chemical Society)

those of intrinsically disordered proteins and lacking the characteristic sharp curve of globular scattering objects. This should serve as a good example for why Kratky plots are so useful in SAS research: the Kratky plot provides quick, model-free, clear presentation of chain/particle conformation.

Although micelles are well established delivery candidates, structural studies are being carried out on even classic examples like poly(lactic acid)-poly(ethylene glycol) (PLA-PEG) micelles (Riley et al. 2003) and the poly(lactic-co-glycolide)-poly(ethylene glycol) (PLGA-PEG) system employed in one study to create micelle-like nanoparticles (Yang et al. 2015a). While the diameter of the particles was in the range of 186 nm from nanoprecipitation in THF/$H_2O$ as observed using DLS, information on the internal particle structure was only obtainable through CV SANS which was collected using 70 and 100% $D_2O$ in simultaneous fits (Yang et al. 2015a).

The characterization of the hydrophobic core was separated from the overall particle dimension *via* CV SANS which elucidated the core as a 7–9 nm PLGA sphere. The DLS and CV SANS $R_g$ values were in good agreement averaging ~98 nm for particles prepared in THF/$H_2O$. The size results provided validation for the use of SANS to characterize these particles. A total particle $R_g$ of ~98 nm with a core sized at only 7 nm in radius would seem to suggest a micelle with a giant corona in comparison to the core. A detailed and advanced modeling of the CV SANS data was able to reveal the large 180 nm diameter particle as a hydrophilic PEG/water environment with a fractal arrangement of multiple PLGA/PEG micelles (Fig. 15.7). It should be noted that this eventual conclusion was drawn through multiple-trial fitting routines of different models when describing the assemblies as simple spherical core-shells or solid multicomponent particles did not give reasonable conclusions.

As is often the case in SAS studies, detailed modeling is complex and iterative. For example, in one study of hexaethylene glycol monododecyl ether ($C_{12}E_6$) micelles prepared by simply mixing the surfactant in $D_2O$, four separate models including rounded elliptical rod, 3-axial ellipsoid, polydisperse sphere, and ellipsoid of revolution were fit to the scattering data and all showed reasonable agreement (Gapiński et al. 2010). For each model, the calculated diffusion coefficients from the dimensions were compared to DLS and fluorescence correlation spectroscopy (FCS) results before finally agreeing on the morphology of a rounded rod with elliptical

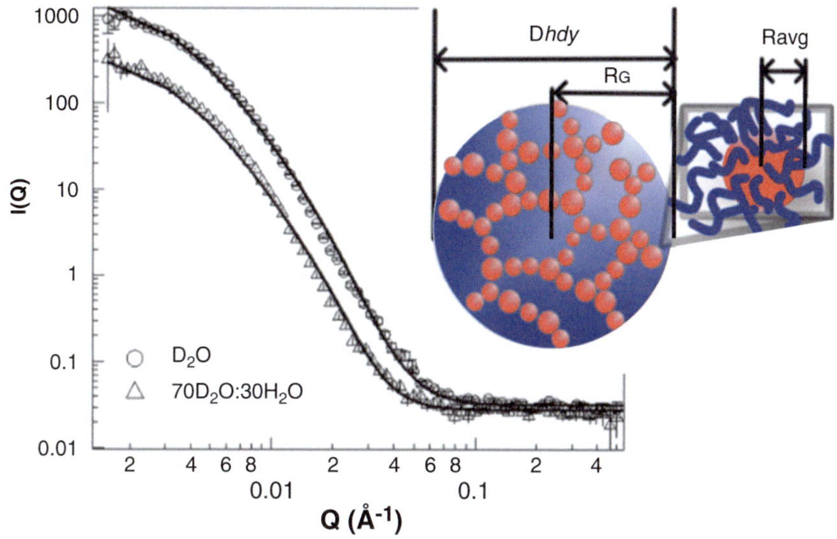


**Fig. 15.7** CV SANS on PLGA-PEG nanoparticles used to elucidate the structure as fractal micelles forming a nanoparticle (Reprinted with permission from Yang et al. 2015a, copyright American Chemical Society)

cross-section resulting in the dimensions of L = 6.8 nm, a = 2.55 nm, b = 1.95 nm.

While other characterization techniques are often used to support SAS studies, as demonstrated above, SAS is equally used to validate the results from physical characterizations such as transmission electron microscopy (TEM). For example, pixel analysis of cryo-TEM images was used to determine the coronal dimension of micelles formed from pincer-ligand end-functionalized poly(N-isopropylacrylamide) (PNIPAM) of 20 kDa in water (Patterson et al. 2013). However, to validate this method and extend the characterization, SANS curves from mixing the polymer in $D_2O$ were fit with core-shell models including a corona of radially decreasing hydration. The cryo-TEM image analysis identified the micelles as ~20 nm spheres, which agreed with DLS results. The SANS results extended this characterization to a 1–2 nm core with 18–20 nm corona. Interestingly, the core size from SANS was indicative of a fully extended $C_{12}$ chain, which was consistent with the size of the end-functional pincer ligand. It should be noted that SANS is an ensemble technique; scattering from a sample represents an average of all the constituents of that sample. It therefore gives a representative picture of the size and shape of all the particles in solution, in contrast to techniques like cryo-TEM which offer direct imaging of a very localized scope. Using both techniques in parallel offers the best range of size scales and the most complete level of precision.

In a characterization of simpler PLA/PEG micelles, the PLA core could be modeled as homogenous with a constant SLD (Riley et al. 2003). However, the PEG shell was better modeled by an SLD that decreased radially from the core as the polymer became hydrated to the point of equaling the solvent in SLD. This accurate model based on contrast variation was used to show that the micelle radius increased from 16 to 54.5 nm as the ratio of PLA:PEG repeat units increased from 3:5 to 45:5 and that the thickness, structure, and density of the PEG layer depended on the same parameter. CV SANS was used to isolate the core parameters from the overall scattering by subtracting the shell from the overall size. In the case of the PLA/PEG micelles, different solvent ratios and pure polymers were scanned as building blocks for the CV analysis (Fig. 15.8).

It should be noted, however, especially considering these successful characterizations through CV SANS, that the actual experimental

**Fig. 15.8** CV SANS on mixed solvents showing the range of SLDs obtainable and the scattering due to PLA(D)-PEG 15:5 micelles at (*a*) 100% $D_2O$, (*b*) 80% $D_2O$, (*c*) 65% $D_2O$ (Reprinted with permission from Riley et al. 2003, copyright American Chemical Society)

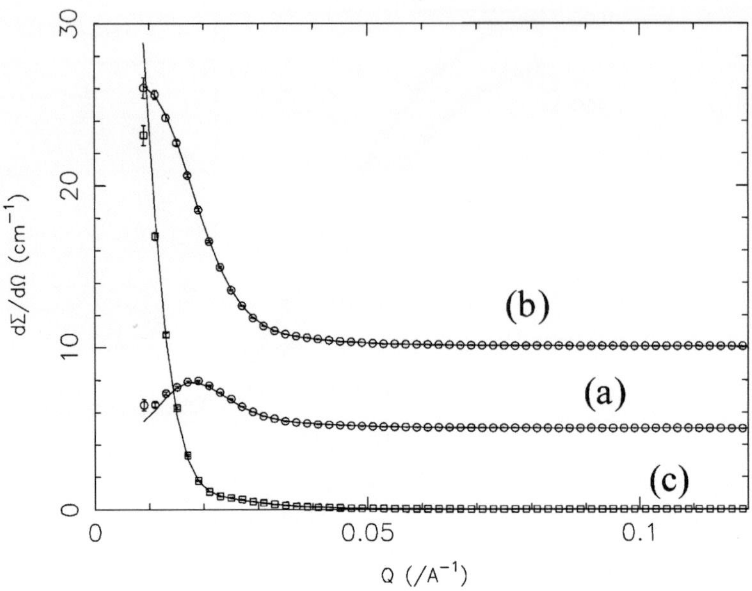


case of contrast management is slightly more complicated than the usual diagram depictions and results sections may appear. What would be expected to be acutely disparate contrast differences (i.e. hydrophobic core/hydrophilic shell) are in reality blurred by factors like chain exchange, core interpenetration, and partial hydration (Patterson et al. 2013; Lu et al. 2016). Thus, in an effort to clarify the properties affecting chain exchange, polystyrene (PS)- poly(ethylene-*alt*-propylene) (PEP) block copolymers were assembled into micelles in the PEP-selective solvent squalane in thoughtfully designed CV SANS studies which monitored dynamic contrast over time to evaluate the kinetics of chain exchange (Lu et al. 2015, 2016). The experimental design in these studies is a wonderful example of inventive CV SANS use. Firstly, PS-PEP-PS and PEP-PS-PEP triblock copolymers with comparable solvophilic PEP blocks form micelles in squalane, with the former resulting in flower-like micelles with PEP loops extending from the core and the latter forming the classic fuzzy micelles with PEP chains extending from the core. The molecular weight of the copolymers was ~230 kDa, with PEP blocks of ~140 kDa. Standard copolymers and identical versions with perdeuterated PS blocks were prepared as micelles in the solvent and dried before being mixed in equal ratios. The inventive concept was as follows: when the dried, pure-core micelles, half containing hydrogenated PS and half containing deuterated PS, were suspended in a 58% D-squalane/H-squalane solvent mixture which has an SLD matching that of a perfectly mixed PS core, the initial contrast from the pure micelle cores, which is given by the contrast factor $I \sim (SLD_{d,core} - SLD_{h,core})^2$ would be at a maximum value and would decay to zero as the chains exchanged and eventually cause evenly mixed cores (Lu et al. 2012). From time resolved (TR) SANS studies on these micelles it was found that the exchange rate of PS-PEP-PS micelles was nine orders of magnitude greater than the PEP-PS-PEP micelles which resulted in the conclusion that the rate limiting mechanism was the extraction of PS block cores into the solvent.

### 15.3.2 Vesicles

The ability of delivery vehicles to entrap and release drugs of various hydrophobicities is correlated to the carrier structure (Chemelli et al. 2012; Nguyen et al. 2011). In this regard, stimuli responsive shape change in a drug carrier offers a viable mechanism for controlled delivery. However, detailed characterization of these shape changes is crucial for understanding and optimizing the mechanism of delivery.

Using pH change as a stimulus, nicergoline/glycerol monooleate (GMO) colloids prepared by sonication of both solids (5 w/w% total, nicergoline varied between 10 and 30 wt%) in 1% Pluronic F108/water were found to have pH-dependent morphology, switching from vesicles to inverse micelles during a change from below pH = 5.1 to above pH = 6.6 (Salentinig et al. 2014). GMO bearing a relatively small $C_{17}$ chain resulted in micelles 4 nm in diameter when the nicergoline concentration was >20% at pH >6.6, with the nicergoline entrapped in the center.

However, nicergoline has multiple lone-pair nitrogens that can be protonated at different pH values and is responsible for the pH-driven shape change. Using crystallographic parameters outside the scope of this description allowed the determination of the apparent $pK_a$ of the nicergoline-GMO system. Generalized IFT from $q$ >0.09 $Å^{-1}$ was used to provide a PDDF that characterized the size and shape of the nano-objects and revealed that the monooleate bilayer of the vesicle resulting from shifting the pH <5.1 was 5 nm in thickness.

This type of analysis for vesicle bilayers, or generically, locally flat sheets with respect to the scattering range of $q$, from thickness PDDF plots is an analysis shown in early SAXS works by Glatter (Glatter 1980; Glatter and Hainisch 1984). In these crystallographic studies, Glatter worked with simulated scattering data to develop mathematical analysis techniques for interpreting physical parameters of solid objects. It was shown that deconvolution using a function he derived allows radial electron density profiles to be calculated. Electron densities within the particle radius can thusly reveal the thickness of interlayers and their molecular ordering (crystallographic unit cell packing parameters) in addition to providing information on the location of drug molecules as shown in this study. For example, the packing of GMO was found from Bragg reflections in the scattering data to be indicative of Pn3m cubosomes and hexosomes, which shifted to an *Im3m* bicontinuous cubic phase as pH went from >8.0 to <7.0.

Further, radially dependent electron density moving outward from the micellar monooleate core was also used to evaluate the water hydration sphere in the glycerol corona. From the examples in this study, along with the chain exchange, solvent selective morphology, and drug effect on micelle conformation studies discussed in the previous section it can be seen that questions concerning self-assembly are being addressed very quantitatively with SAS techniques.

Vesicles, owing to their unique architecture and larger size compared to micelles, are often the focus of studies aiming to uncover details about the hydrophobic layer or bilayer. The first important consideration for vesicles is that their size can easily exceed 1 μm and, therefore, the range of $q$ in scattering studies is not low enough to encompass the Guinier region (Borchert et al. 2006). Instead, the scattering intensity in the reasonable $q$ range of the instrument is best approximated by that due to a flat sheet. For vesicles this enables detailed analysis of the hydrophobic interlayer when CV conditions or modeling analyses are used.

For example, the bilayer thickness of poly (2-vinylpyridine-*b*-ethylene oxide) (PVP-PEO) block copolymer vesicles 1–10 μm in diameter formed from nanoprecipitation in $CHCl_3/H_2O$ was evaluated with SANS in $D_2O$ (Rubner and Cohen n.d.). The PVP polymer contains ionizable nitrogens that render the polymer insoluble above the $pK_a$ and result in the formation of vesicles. By fitting scattering data in the $q$ range of 0.005–0.05 Å$^{-1}$, the PVP bilayer thickness of vesicles formed from PVP-PEO copolymers with ~30, 40, and 70 PVP units at pH >5 was found to be 10.7, 13.6, and 13.5 nm, respectively.

Similarly, in an investigation of polydimethylsiloxane-*g*-poly(ethylene oxide) (PDMS-*g*-PEO) graft polymers and PEO-PDMS-PEO triblock copolymers, the hydrophobic bilayer was characterized using SANS in $D_2O$ (Salva et al. 2013). The vesicles were prepared by mixing the polymers in water and extruding 19 times through 0.1 μm filters. However, cryo-TEM still revealed large, polydisperse vesicles which caused oscillations at low $q$ such that quantification in that range was not possible. However, modeling of the scattering curves revealed an average of 33 nm PEO cores with 5.75 nm PDMS bilayer membranes and 15 nm outer shells. An interesting use of a log-log scaled Kratky plot was also utilized for intermediate $q$ in which $\pi/qmax$ (the local maximum of intermediate $q$) provided the characteristic length of the membrane, 6.0 nm, which agreed well with the model fitting analysis.

For the PEO-PDMS-PEO triblock copolymers, high polydispersity disrupted higher $q$ such that the log scaled Kratky plot was not useful. However, the slope of the fit line from a Kratky-Porod plot of $\ln(Iq^2)$ vs $q^2$ provided the membrane thickness as 11 nm which agreed with cryo-TEM images. Model fitting of the entire curve gave a core of 14.9 nm with a PDMS interlayer membrane of 11.4 nm and a 26.5 nm hydrated PEO outer corona, giving a final agreement for multiple analyses.

While the two studies discussed above focused on the characteristics of the vesicle membrane as a homogenous entity, there are a number of works focused on investigating small nano-domains within vesicle walls. The nanodomains have completely different form factors compared to the assumed flat sheet of the vesicle surfaces which make them difficult to analyze without SAS. Similar principles to those in the studies above apply: that in a polydisperse system of large spheres, surface curvature can be neglected and the scattering is treated as coming from randomly dispersed flat surfaces (Vogtt et al. 2010).

In a study of vesicles formed from multiple extrusions of a mixed lipid system (molar ratio of 45:27:28 DOPC:DPPC-d62:cholesterol) through 30-nm filters in water, CV SANS in $D_2O/H_2O$ was used to characterize the vesicles as 14.5 nm spheres with 2.5 nm bilayers. Selection of a single deuterated component and a contrast matched $D_2O/H_2O$ solvent system also allowed characterization of small nanodomains in the disordered liquid phase as flat cylinders 2.5 nm long with a radius of 14.5 nm. This general flattened shape was confirmed from the low $q$ behavior of the log-log scattering curve in the region of $0.1 < q < 1.5$ nm$^{-1}$ where the scattering curve displayed $q^{-2}$ dependence indicative of flat objects.

The formation of nanodomains is not limited to vesicles composed purely of lipids, as demonstrated by the study of grafted PDMS-g-PEG mixed with DPPC/POPC (50:50) ratio to form 80:20 polymer:lipid vesicles (Dao et al. 2015). In this work, thin film rehydration/extrusion through 100-nm pores was used to create vesicles 52 nm in diameter. CV SANS in 10% $D_2O/H_2O$ provided PDMS-matched conditions in which independent observation of the lipids was possible, while 80% $D_2O/H_2O$ matched the SLD of the lipids and allowed polymer "visualization" in the SANS curve. From Kratky-Porod plots the vesicle membrane thickness was found to be 5.8 nm, but the vesicle model was found to be insufficient for the lipid-matched polymer scattering curves. Instead, the polydisperse flat cylinder model was found to describe the polymer nanodomains as 30 nm radius, 5.1 nm length cylinders. It should be noted that the apparent size of the flat cylinders from the model does not seem reasonable considering the diameter of the overall vesicles (~52 nm). A major consideration in this case was the polydispersity and variable content of the nanodomains and the ever-present fact that SAS curves can be acceptably fit with multiple models.

Without information from other characterization techniques it can be difficult to pinpoint some structural features in SAS modeling due to the number of uninformed floating parameters. While characterization of the vesicle's overall structure was confirmable with DLS and shown to be reliable from several modeled approaches, characterization of the nanodomains suffered from being necessarily independent from other techniques. Both the immense strength and characteristic weaknesses of SAS can be seen in these studies.

### 15.3.3 Capsules

Hollow polymer capsules present one of the most interesting subjects of SAS drug delivery subjects. Firstly, they have an extremely wide range of compositional aspects that can be varied quite easily, including polymer composition, molecular weight, interlayer interaction, and size (Li et al. 2011; Shutava and Lvov 2006; Shutova et al. 2007; Zavgorodnya et al. 2015). Small nm-range sizes of capsules can be studied encompassingly with SANS which provides very interesting details on their morphology before and after template dissolution (Estrela-Lopis et al. 2002). Since polymer capsules can also exist as large particles >1 μm, very low $q$ ranges can be needed to fully characterize their size and shape in SAS. However, when this information is gathered in other more facile ways these parameters can be included in fitting models. Following this, the ratio of shell thickness to overall particle diameter is very small, which makes characterizing the thickness accurately impossible without SAS techniques. Additionally, specific modeling and analyses such as those stemming from Kratky-Porod plots are able to shed light on this topic (Balgavy et al. 2001; Richter et al. 2011). In this regard the smallest of small angles can be fully utilized and scattering instruments can be pushed to their physical limits. For perspective, the D11 SANS instrument at Institut Laue-Langevin uses a 39 m sample-to-detector distance to separate low $q$ range scattering measurements from the main beam.

Capsules also have the potential to be extremely monodisperse due to their templated assembly which provides highly uniform scattering in contrast to self-assembled structures that can be quite polydisperse. Common goals for SAS research on hollow capsules are finding the capsule wall thickness in solution and characterizing the interior environment (Estrela-Lopis et al. 2007). The hollow morphology of these particles means that the solvent-filled core can be modeled as equal to the bulk solvent in SLD and scattering due to the polymeric shell can be analyzed separately.

It is important to remember that scattering data may look somewhat similar and that the important conclusions come from careful modeling. For acquisitions leading to information on capsule shell thickness, fewer parameters left floating during data fitting such as particle radius, SLD of the core/shell, interparticle factors, and polydispersity lead to more accurate results when the models are being pushed. This is born from the fact that the $q$ range of most instruments falls short of fully wrapping the scale of such objects. Polydispersity and size can be measured with DLS (Rube et al. 2005), or confocal laser scanning microscopy (Kozlovskaya et al. 2014) while the SLD value of polymeric material can be closely approximated to known values. Since polydispersity smoothens scattering profiles quite effectively, having guidelines for these parameters can enhance model accuracy for other parameters such as core size which can be independently measured by scanning electron microscopy (Chen et al. 2013) and shell thickness which can be determined by atomic force microscopy (AFM) (Liu et al. 2014; Espinosa-Dzib et al. 2015). Outside of some of the more advanced modeling routines, a good example of model-free analysis is the modified Guinier plot which can give bilayer thickness of vesicles (Kim et al. 2014) from the slope of $\ln(Iq^2)$ vs $q^2$ in the Guinier thickness range (Fig. 15.9).

Analyses such as these have also been conducted on polyelectrolyte multilayer capsules made *via* ionic pairing of poly(styrene sulfonate) with poly(allylamine hydrochloride) (PSS/PAH) and ranging from 130 nm to 3 μm in size for separate samples with successful modeling of

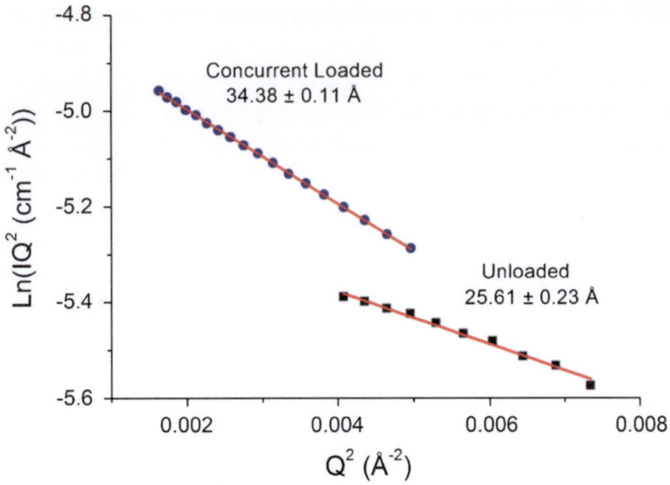


**Fig. 15.9** Modified Guinier plot with linear fits giving $R_g$ which can be used to calculate shell thickness (Reprinted with permission from Kim et al. 2014, copyright American Chemical Society)

capsule wall thicknesses (Estrela-Lopis et al. 2009). The capsules in this study were modeled as core-shell particles, and removal of interparticle structure factors (by sufficient dilution) allowed isolation of the form factors representing scattering from the core, core-shell interface, and shell polymers.

As mentioned, for hollow capsules a core-shell model where the core matches the solvent in SLD can be used to isolate scattering from the polymer shell. However, analysis of scattering from separate samples with reserved core templates enabled the conclusion that supported shell layers, i.e., those still standing on the templates, are 25% thinner than free capsule shells, i.e., those with templates having been dissolved (Balgavý et al. 2001). While this result may seem intuitive, it is quite significant in that SANS is one of the only ways to characterize detailed shell swelling and prove this reasonable hypothesis.

Moreover, the hydration of the $(PSS/PAH)_{8,12,16}$ shells where the subscript denotes the number of PSS/PAH polymer bilayers within the capsule shell was determined to be higher in freestanding layers (52%) compared to template-bound layers (42%) from the difference in SLD of hydrated polymers compared to pure solvent (Balgavý et al. 2001). The thickness of PSS/PAH capsule shells in solution, found from SANS analysis to be ~2 nm per PSS/PAH bilayer, was found to be in agreement with the thickness of collapsed capsules in the dry state found using AFM.

However, AFM thickness data cannot always accurately approximate thicknesses of hydrated layers. For example, in a study of PSS/PAH multilayer capsules formed on red blood cells, AFM measurements revealed capsule wall thickness that were artificially higher than the thickness measured with SANS (Estrela-Lopis et al. 2007). In this work, six bilayers of PSS/PAH were found from modeled SANS data to have a total thickness of only 4.5 nm, while AFM results were much higher. The difference stems from the removal of up to 90% of the PSS during core removal, which was shown to be trapped within the core with SANS. However, when drying the capsule suspension for AFM it was hypothesized that the PSS trapped in the core contributed to the height found from the AFM probe. These results are a significant example of how SANS can validate results brought to hypothesis from other techniques, while also shedding light on the intrinsic weakness of other techniques that can lead to inaccurate conclusions.

Investigating the core environment of capsules is another useful application for SAS techniques where in-solution techniques like DLS are not sensitive to scattering from the object's interior (Lof et al. 2007). For encapsulated biological macromolecules that contain metals, SAXS is particularly helpful due to the disparate contrast from the electron density of metals compared to organic molecules as shown in the characterization of hemoglobin inside (chitosan/alginate)$_{10}$ capsules (Mandal et al. 2012). In this study, the aggregation of hemoglobin inside the capsules compared to free hemoglobin in solution was compared to evaluate the effect of encapsulation on redox response found from cyclic voltammetry of the hemoglobin. Porod analysis was used heavily with a focus being the Porod exponent aggregation behavior.

First, a theoretical $R_g$ for hemoglobin was established by a simulation and found to be 22.37 Å. This value was compared to the value determined from SAXS of the polymer-encapsulated hemoglobin (26.6 Å) and found to be in agreement. The more interesting result came from the experimentally determined polydispersity figure for encapsulated hemoglobin of 0.11, which was much lower than the free hemoglobin value of 0.71. This implied that the hemoglobin was highly monodisperse when encapsulated which is indicative of non-aggregation. This study is a good example of how SAXS can be used to take advantage of the increased scattering from biological macromolecules that contain or incorporate metals to evaluate their properties independently of the system in which they are being studied.

### 15.3.4 Drug Modeling, Behaviors, Interactions

The constant development of new drugs that are very often hydrophobic comes with the parallel development of methods to increase the solubility of these drugs. Characterizing the solubility of drugs is an effort being pursued by medicinal laboratories but often the acquisition timescale is slower than the drug dissolution species change. Powder X-ray diffraction is often used to characterize solids but available instruments typically have measurement timescales around 10 min. Efforts to elucidate more detail in shorter time intervals for rapidly solubilizing or crystallizing drugs are being supported by the speed and resolution of SAXS (Boetker et al. 2012), with intermediate species and kinetics being discovered from this technique. Both anhydrous and monohydrate forms of theophylline were characterized kinetically during dissolution and recrystallization using SAXS at complete experimental timescales under 100 s.

At the other end of the spectrum, encapsulating protein drugs is an application that has potential for drug delivery (Besanger et al. 2004; Ciriminna et al. 2013; Zeno et al. 2014). However, characterizing the encapsulated proteins is not simple using optical techniques. Instead, CV SANS and deuterated proteins were used in a study of green fluorescent protein *in silica* sol-gels to evaluate the conformation of the protein and its effects on the structure of the gel (Luo et al. 2009). The mixed solvent system of 60% $D_2O/H_2O$ was found to match silica which allowed direct access to scattering from the protein, while 42% $D_2O$ was used to match hydrogenated protein and evaluate scattering from the silica matrix. From simultaneous fits of SANS data at 0, 30, and 100% $D_2O$ the typical particle size in the silica sol-gel was found to have a radius of 13.73 Å with pores having a radius of 33.69 Å. Entrapment of the protein did not significantly change the particle size (12.24 Å) but was shown to increase the pore size significantly to 66.01 Å. A Fourier transform approach from the silica-matched SANS curve with deuterated protein to increase the contrast without having to significantly increase protein concentration allowed the elucidation of protein quaternary structure to be an end-to-end dimer.

## 15.4 Conclusions and Practical Tips

SAS techniques are powerful tools to the researchers looking to discover the most fundamental details of drug delivery systems. The elucidation of these details has rapidly increased the knowledge available to polymer, material, and pharmaceutical researchers over the last decade. While other techniques have allowed discovery of information about the size and dispersity of nanometer sized particles, SAS has helped answer questions that will likely increase the efficiency of drug delivery systems moving forward. However, as the subjects of SAS studies continue to increase in complexity, the instrumental techniques are also increasing in power, resolution, etc. The proliferation of USANS and USAXS instruments is heralding the characterization of larger particles while increased flux in both types of scattering sources is increasing throughput of samples and the speed of information being gained on both novel and fundamental systems.

*Practical tips:*

- When using model-free data analysis of the Guinier region, keep in mind that particle aggregation has a significant effect on the line slope and apparent $R_g$: this may provide estimates that are different than would be expected. Aggregation of particles also effects parameters in model fitting in ways that are not always apparent. For instance, aggregation increases polydispersity which intuitively affects the apparent particle size, but can decrease the apparent shell thickness for core-shell models.
- SAS characterizations of drug delivery particles and systems provide the most reliable data only when supplemented with as many other techniques as possible. Model parameters that can be fixed due to prior knowledge make the model fit more accurate. Model-free analyses give the best results when they are informed by relevant physical characterizations. The same scattering data can be fit equally well by many completely unrelated models.
- Always keep the size frame of reference for SAS scattering in mind when looking at data: Porod slopes describe surfaces or fractals; Guinier region slopes describe particle shape if the particle size is measurable at that q; remember also that non-dilute samples show interparticle peaks and/or structure factor scattering.
- Conclusions drawn from SAS techniques are often the result of an iterative approach with a sizable amount of trial and error where good intuition can help clarify the results; particle scattering could be fit equally well by a fractal micelle model or one that indicates a micelle with unreasonable dimensions, such as a 2 nm core and 100 nm corona. It is left to the user to distinguish whether the conclusion is reasonable or not and to interpret the model from an informed standpoint to give reasonable results.
- Polymers, drug molecules, and polymeric materials do not always have the same behavior in $D_2O$ and $H_2O$. It is best to check simple things like the solubility, $R_g$, and dispersity before beginning scattering experiments.

**Acknowledgement** This work was supported by NSF Career Award 1350370.

## References

Akiba I, Terada N, Hashida S, Sakurai K, Sato T, Shiraishi K, Yokoyama M, Masunaga H, Ogawa H, Ito K, Yagi N (2010) Encapsulation of a hydrophobic drug into a polymer-micelle core explored with synchrotron SAXS. Langmuir 26(10):7544–7551

Alexander S, Cosgrove T, de Vos WM, Castle TC, Prescott SW (2014) Aggregation behavior of polyisoprene–pluronic graft copolymers in selective solvents. Langmuir 30(20):5747–5754

Almgren M, Garamus VM, Asakawa T, Jiang N (2007) Contrast variation SANS investigation of composition distributions in mixed surfactant micelles. J Phys Chem B 111(25):7133–7141

Bae JW, Lee E, Park KM, Park KD (2009) Vinyl sulfone-terminated PEG−PLLA diblock copolymer for thiol-reactive polymeric micelle. Macromolecules 42 (10):3437–3442

Balgavý P, Dubničková M, Kučerka N, Kiselev MA, Yaradaikin SP, Uhríková D (2001) Bilayer thickness and lipid interface area in unilamellar extruded 1,2-diacylphosphatidylcholine liposomes: a small-angle neutron scattering study. Biochim Biophys Acta Biomembr 1512(1):40–52

Becker AL, Johnston APR, Caruso F (2010) Layer-by-layer-assembled capsules and films for therapeutic delivery. Small 6(17):1836–1852

Besanger TR, Easwaramoorthy B, Brennan JD (2004) Entrapment of highly active membrane-bound receptors in macroporous sol–gel derived silica. Anal Chem 76(21):6470–6475

Biggs S, Labarre M, Hodges C, Walker LM, Webber GB (2007) Polymerized rodlike micelle adsorption at the solid–liquid interface. Langmuir 23 (15):8094–8102

Boetker J, Rades T, Rantanen J, Hawley A, Boyd BJ (2012) Structural elucidation of rapid solution-mediated phase transitions in pharmaceutical solids using in situ synchrotron SAXS/WAXS. Mol Pharm 9(9):2787–2791

Borchert U, Lipprandt U, Bilang M, Kimpfler A, Rank A, Peschka-Süss R, Schubert R, Lindner P, Förster S (2006) pH-induced release from P2VP–PEO block copolymer vesicles. Langmuir 22(13):5843–5847

Caruso F, Caruso RA, Möhwald H (1998) Nanoengineering of inorganic and hybrid hollow spheres by colloidal templating. Science 282 (5391):1111

Chemelli A, Maurer M, Geier R, Glatter O (2012) Optimized loading and sustained release of hydrophilic proteins from internally nanostructured particles. Langmuir 28(49):16788–16797

Chen J, Kozlovskaya V, Goins A, Campos-Gomez J, Saeed M, Kharlampieva E (2013) Biocompatible shaped particles from dried multilayer polymer capsules. Biomacromolecules 14(11):3830–3841

Cho J, Hong J, Char K, Caruso F (2006) Nanoporous block copolymer micelle/micelle multilayer films with dual optical properties. J Am Chem Soc 128 (30):9935–9942

Choi I, Malak ST, Xu W, Heller WT, Tsitsilianis C, Tsukruk VV (2013) Multicompartmental microcapsules from star copolymer micelles. Macromolecules 46(4):1425–1436

Ciriminna R, Fidalgo A, Pandarus V, Béland F, Ilharco LM, Pagliaro M (2013) The sol–gel route to advanced silica-based materials and recent applications. Chem Rev 113(8):6592–6620

Correa NM, Silber JJ, Riter RE, Levinger NE (2012) Nonaqueous polar solvents in reverse micelle systems. Chem Rev 112(8):4569–4602

Dao TPT, Fernandes F, Er-Rafik M, Salva R, Schmutz M, Brûlet A, Prieto M, Sandre O, Le Meins JF (2015) Phase separation and nanodomain formation in hybrid polymer/lipid vesicles. ACS Macro Lett 4(2):182–186

Decher G (2002) Polyelectrolyte multilayers, an overview. In: Multilayer thin films. Wiley-VCH Verlag GmbH & Co. KGaA, Weinheim, pp 1–46

Donath E, Sukhorukov GB, Caruso F, Davis SA, Möhwald H (1998) Novel hollow polymer shells by colloid-templated assembly of polyelectrolytes. Angew Chem Int Ed 37(16):2201–2205

Elsner N, Kozlovskaya V, Sukhishvili SA, Fery A (2006) pH-triggered softening of crosslinked hydrogen-bonded capsules. Soft Matter 2(11):966–972

Erel-Unal I, Sukhishvili SA (2008) Hydrogen-bonded multilayers of a neutral polymer and a polyphenol. Macromolecules 41(11):3962–3970

Espinosa-Dzib A, Chen J, Zavgorodnya O, Kozlovskaya V, Liang X, Kharlampieva E (2015) Tuning assembly and enzymatic degradation of silk/poly (N-vinylcaprolactam) multilayers *via* molecular weight and hydrophobicity. Soft Matter 11(25):5133–5145

Estrela-Lopis I, Leporatti S, Moya S, Brandt A, Donath E, Möhwald H (2002) SANS studies of polyelectrolyte multilayers on colloidal templates. Langmuir 18 (21):7861–7866

Estrela-Lopis I, Leporatti S, Typlt E, Clemens D, Donath E (2007) Small angle neutron scattering investigations (SANS) of polyelectrolyte multilayer capsules templated on human red blood cells. Langmuir 23 (13):7209–7215

Estrela-Lopis I, Leporatti S, Clemens D, Donath E (2009) Polyelectrolyte multilayer hollow capsules studied by small-angle neutron scattering (SANS). Soft Matter 5 (1):214–219

Filippov SK, Chytil P, Konarev PV, Dyakonova M, Papadakis C, Zhigunov A, Plestil J, Stepanek P, Etrych T, Ulbrich K, Svergun DI (2012) Macromolecular HPMA-based nanoparticles with cholesterol for solid-tumor targeting: detailed study of the inner structure of a highly efficient drug delivery system. Biomacromolecules 13(8):2594–2604

Filippov SK, Franklin JM, Konarev PV, Chytil P, Etrych T, Bogomolova A, Dyakonova M, Papadakis CM, Radulescu A, Ulbrich K, Stepanek P, Svergun DI (2013) Hydrolytically degradable polymer micelles for drug delivery: a SAXS/SANS kinetic study. Biomacromolecules 14(11):4061–4070

Foglia F, Rogers SE, Webster JRP, Akeroyd FA, Gascoyne KF, Lawrence MJ, Barlow DJ (2015) Neutron scattering studies of the effects of formulating amphotericin B with cholesteryl sulfate on the drug's interactions with phospholipid and phospholipid–sterol membranes. Langmuir 31(29):8042–8051

Fong WK, Hanley TL, Thierry B, Kirby N, Boyd BJ (2010) Plasmonic nanorods provide reversible control over nanostructure of self-assembled drug delivery materials. Langmuir 26(9):6136–6139

Gapiński J, Szymański J, Wilk A, Kohlbrecher J, Patkowski A, Hołyst R (2010) Size and shape of micelles studied by means of SANS, PCS, and FCS. Langmuir 26(12):9304–9314

Ghosh Chaudhuri R, Paria S (2012) Core/shell nanoparticles: classes, properties, synthesis mechanisms, characterization, and applications. Chem Rev 112(4):2373–2433

Glatter O (1980) Evaluation of small-angle scattering data from lamellar and cylindrical particles by the indirect transformation method. J Appl Crystallogr 13:577–584

Glatter O, Hainisch B (1984) Improvements in real-space deconvolution of small-angle scattering data. J Appl Crystallogr 17:435–441

Glatter O, Kratky O (1982) Small angle X-ray scattering. Academic, New York

Goodwin DJ, Sepassi S, King SM, Holland SJ, Martini LG, Lawrence MJ (2013) Characterization of polymer adsorption onto drug nanoparticles using depletion measurements and small-angle neutron scattering. Mol Pharm 10(11):4146–4158

Guinier A, Fournet G (1955) Small-angle scattering of X-rays. Wiley, New York

Hammond PT (2011) Engineering materials layer-by-layer: challenges and opportunities in multilayer assembly. AICHE J 57(11):2928–2940

Hammouda B (2010a) A new guinier porod model. J Appl Crystallogr 43:716–719

Hammouda B (2010b) Analysis of the beaucage model. J Appl Crystallogr 43:1474–1478

Hammouda B (2016) Probing nanoscale structures using neutron scattering. NIST NCNR distance learning course

Harris JM, Martin NE, Modi M (2012) Pegylation. Clin Pharmacokinet 40(7):539–551

Kharlampieva E, Sukhishvili SA (2006) Hydrogen-bonded layer-by-layer polymer films. J Macromol Sci Part C 46(4):377–395

Kharlampieva E, Kozlovskaya V, Sukhishvili SA (2009) Layer-by-layer hydrogen-bonded polymer films: from fundamentals to applications. Adv Mater 21 (30):3053–3065

Kim MD, Dergunov SA, Richter AG, Durbin J, Shmakov SN, Jia Y, Kenbeilova S, Orazbekuly Y, Kengpeiil A, Lindner E, Pingali SV, Urban VS, Weigand S, Pinkhassik E (2014) Facile directed assembly of hollow polymer nanocapsules within spontaneously formed catanionic surfactant vesicles. Langmuir 30 (24):7061–7069

Kozlovskaya V, Kharlampieva E, Erel I, Sukhishvili SA (2009) Multilayer-derived, ultrathin, stimuli-responsive hydrogels. Soft Matter 5(21):4077–4087

Kozlovskaya V, Baggett J, Godin B, Liu X, Kharlampieva E (2012a) Hydrogen-bonded multilayers of silk fibroin: from coatings to cell-mimicking shaped microcontainers. ACS Macro Lett 2012(1):384–387

Kozlovskaya V, Zavgorodnya O, Chen Y, Ellis K, Tse HM, Cui W, Thompson JA, Kharlampieva E (2012b) Ultrathin polymeric coatings based on hydrogen-bonded polyphenol for protection of pancreatic islet cells. Adv Funct Mater 22(16):3389–3398

Kozlovskaya V, Alexander JF, Wang Y, Kuncewicz T, Liu X, Godin B, Kharlampieva E (2014) Internalization of red blood cell-mimicking hydrogel capsules with pH-triggered shape responses. ACS Nano 8 (6):5725–5737

Lagzi I, Wang D, Kowalczyk B, Grzybowski BA (2010) Vesicle-to-micelle oscillations and spatial patterns. Langmuir 26(17):13770–13772

Li H, Pang S, Wu S, Feng X, Mullen K, Bubeck C (2011) Layer-by-layer assembly and UV photoreduction of graphene-polyoxometalate composite films for electronics. J Am Chem Soc 133(24):9423–9429

Lisunova MO, Drachuk I, Shchepelina OA, Anderson KD, Tsukruk VV (2011) Direct probing of micromechanical properties of hydrogen-bonded layer-by-layer microcapsule shells with different chemical compositions. Langmuir 27 (17):11157–11165

Liu F, Kozlovskaya V, Zavgorodnya O, Martinez-Lopez-C, Catledge S, Kharlampieva E (2014) Encapsulation of anticancer drug by hydrogen-bonded multilayers of tannic acid. Soft Matter 10(46):9237–9247

Liu F, Kozlovskaya V, Medipelli S, Xue B, Ahmad F, Saeed M, Cropek D, Kharlampieva E (2015) Temperature-sensitive polymersomes for controlled delivery of anticancer drugs. Chem Mater 27 (23):7945–7956

Löf D, Schillén K, Loh W, Olofsson G (2007) A calorimetry and light scattering study of the formation and shape transition of mixed micelles of EO20PO68EO20 triblock copolymer (P123) and nonionic surfactant (C12EO6). J Phys Chem B 111 (21):5911–5920

Lu J, Choi S, Bates FS, Lodge TP (2012) Molecular exchange in diblock copolymer micelles: bimodal distribution in core-block molecular weights. ACS Macro Lett 1(8):982–985

Lu J, Bates FS, Lodge TP (2015) Remarkable effect of molecular architecture on chain exchange in triblock copolymer micelles. Macromolecules 48 (8):2667–2676

Lu J, Bates FS, Lodge TP (2016) Addition of corona block homopolymer retards chain exchange in solutions of block copolymer micelles. Macromolecules 49 (4):1405–1413

Luo G, Zhang Q, Castillo ARD, Urban V, O'Neill H (2009) Characterization of sol−gel-encapsulated proteins using small-angle neutron scattering. ACS Appl Mater Interfaces 1(10):2262–2268

Lutkenhaus JL, Hammond PT (2007) Electrochemically enabled polyelectrolyte multilayer devices: from fuel cells to sensors. Soft Matter 3(7):804–816

Lvov Y, Decher G, Moehwald H (1993) Assembly, structural characterization, and thermal behavior of layer-by-layer deposited ultrathin films of poly(vinyl sulfate) and poly(allylamine). Langmuir 9 (2):481–486

Mandal SS, Bhaduri S, Amenitsch H, Bhattacharyya AJ (2012) Synchrotron small-angle x-ray scattering studies of hemoglobin nonaggregation confined inside polymer capsules. J Phys Chem B 116(32):9604–9610

Mariani G, Moldenhauer D, Schweins R, Grohn F (2016) Elucidating electrostatic self-assembly: molecular parameters as key to thermodynamics and nanoparticle shape. J Am Chem Soc 138(4):1280–1293

Matsusaki M, Akashi M (2009) Functional multilayered capsules for targeting and local drug delivery. Expert Opin Drug Deliv 6(11):1207–1217

Mattingly SJ, O'Toole MG, James KT, Clark GJ, Nantz MH (2015) Magnetic nanoparticle-supported lipid bilayers for drug delivery. Langmuir 31 (11):3326–3332

Mjahed H, Porcel C, Senger B, Chassepot A, Netter P, Gillet P, Decher G, Voegel J-C, Schaaf P, Benkirane-Jessel N, Boulmedais F (2008) Micro-stratified architectures based on successive stacking of alginate gel layers and poly(l-lysine)-hyaluronic acid multilayer films aimed at tissue engineering. Soft Matter 4 (7):1422–1429

Naruse K, Eguchi K, Akiba I, Sakurai K, Masunaga H, Ogawa H, Fossey JS (2009) Flexibility and cross-sectional structure of an anionic dual-surfactant wormlike micelle explored with small-angle x-ray scattering coupled with contrast variation technique. J Phys Chem B 113(30):10222–10229

Nguyen T-H, Hanley T, Porter CJH, Boyd BJ (2011) Nanostructured liquid crystalline particles provide long duration sustained-release effect for a poorly water soluble drug after oral administration. J Control Release 153(2):180–186

Nunes SP, Car A (2013) From charge-mosaic to micelle self-assembly: block copolymer membranes in the last 40 years. Ind Eng Chem Res 52(3):993–1003

Patterson JP, Kelley EG, Murphy RP, Moughton AO, Robin M, Lu A, Colombani O, Chassenieux C, Cheung D, Sullivan MO, Epps TH 3rd, O'Reilly RK (2013) Structural characterization of amphiphilic homopolymer micelles using light scattering, SANS, and cryo-TEM. Macromolecules 46(15):6319–6325

Paul A, Vicent MJ, Duncan R (2007) Using small-angle neutron scattering to study the solution conformation of N-(2-hydroxypropyl)methacrylamide copolymer–doxorubicin conjugates. Biomacromolecules 8(5):1573–1579

Pomposo JA, Perez-Baena I, Lo Verso F, Moreno AJ, Arbe A, Colmenero J (2014) How far are single-chain polymer nanoparticles in solution from the globular state? ACS Macro Lett 3(8):767–772

Remant Bahadur KC, Thapa B, Xu P (2012) pH and redox dual responsive nanoparticle for nuclear targeted drug delivery. Mol Pharm 9(9):2719–2729

Richter AG, Dergunov SA, Ganus B, Thomas Z, Pingali SV, Urban V, Liu Y, Porcar L, Pinkhassik E (2011) Scattering studies of hydrophobic monomers in liposomal bilayers: an expanding shell model of monomer distribution. Langmuir 27(7):3792–3797

Riley T, Heald CR, Stolnik S, Garnett MC, Illum L, Davis SS, King SM, Heenan RK, Purkiss SC, Barlow RJ, Gellert PR, Washington C (2003) Core–shell structure of PLA–PEG nanoparticles used for drug delivery. Langmuir 19(20):8428–8435

Rissanen S, Kumorek M, Martinez-Seara H, Li Y-C, Jamróz D, Bunker A, Nowakowska M, Vattulainen I, Kepczynski M, Róg T (2014) Effect of PEGylation on drug entry into lipid bilayer. J Phys Chem B 118 (1):144–151

Rube A, Hause G, Mader K, Kohlbrecher J (2005) Core-shell structure of miglyol/poly(D,L-lactide)/poloxamer nanocapsules studied by small-angle neutron scattering. J Control Release 107(2):244–252

Rubner MF, Cohen RE. Layer-by-layer processed multilayers: challenges and opportunities. In: Decher G, Schlenoff JB (eds) Multilayer thin films: sequential assembly of nanocomposite materials. 2nd edn. Wiley-VCH Verlag GmbH & Co. KGaA, Weinheim

Sakamoto S, Sanada Y, Sakashita M, Nishina K, Nakai K, Yusa S-I, Sakurai K (2014) Chain-length dependence of polyion complex architecture bearing phosphobetaine block explored using SAXS and FFF-MALS. Polym J 46(9):617–622

Salentinig S, Tangso KJ, Hawley A, Boyd BJ (2014) pH-driven colloidal transformations based on the vasoactive drug nicergoline. Langmuir 30(49):14776–14781

Salva R, Le Meins J-F, Sandre O, Brûlet A, Schmutz M, Guenoun P, Lecommandoux S (2013) Polymersome shape transformation at the nanoscale. ACS Nano 7 (10):9298–9311

Sanada Y, Akiba I, Sakurai K, Shiraishi K, Yokoyama M, Mylonas E, Ohta N, Yagi N, Shinohara Y, Amemiya Y (2013) Hydrophobic molecules infiltrating into the poly(ethylene glycol) domain of the core/shell interface of a polymeric micelle: evidence obtained with anomalous small-angle x-ray scattering. J Am Chem Soc 135(7):2574–2582

Sauer M, Meier W (2004) Polymer nanocontainers for drug delivery. In: Carrier-based drug delivery, vol 879. American Chemical Society, pp 224–237

Schnablegger H, Singh Y (2013) The SAXS guide, 3rd edn. Anton Paar GmbH, Austria

Shutava TG, Lvov YM (2006) Nano-engineered microcapsules of tannic acid and chitosan for protein encapsulation. J Nanosci Nanotechnol 6 (6):1655–1661

Shutova TG, Agabekov VE, Lvov YM (2007) Reaction of radical cations with multilayers of tannic acid and polyelectrolytes. Russ J Gen Chem 77(9):1494–1501

Skirtach AG, Yashchenok AM, Mohwald H (2011) Encapsulation, release and applications of LbL polyelectrolyte multilayer capsules. Chem Commun 47 (48):12736–12746

Smith GN, Alexander S, Brown P, Gillespie DAJ, Grillo I, Heenan RK, James C, Kemp R, Rogers SE, Eastoe J (2014) Interaction between surfactants and colloidal latexes in nonpolar solvents studied using contrast-variation small-angle neutron scattering. Langmuir 30(12):3422–3431

Srinivas G, Mohan RV, Kelkar AD (2013) Polymer micelle assisted transport and delivery of model hydrophilic components inside a biological lipid vesicle: a coarse-grain simulation study. J Phys Chem B 117(40):12095–12104

Sukhorukov GB, Möhwald H (2007) Multifunctional cargo systems for biotechnology. Trends Biotechnol 25(3):93–98

Tang Z, Wang Y, Podsiadlo P, Kotov NA (2006) Biomedical applications of layer-by-layer assembly: from biomimetics to tissue engineering. Adv Mater 18 (24):3203–3224

Tong W, Gao C (2008) Multilayer microcapsules with tailored structures for bio-related applications. J Mater Chem 18(32):3799–3812

Urbina MC, Zinoveva S, Miller T, Sabliov CM, Monroe WT, Kumar CSSR (2008) Investigation of magnetic nanoparticle−polymer composites for multiple-controlled drug delivery. J Phys Chem C 112 (30):11102–11108

Veronese FM, Mero A (2012) The impact of PEGylation on biological therapies. BioDrugs 22(5):315–329

Vicent MJ, Greco F, Nicholson RI, Paul A, Griffiths PC, Duncan R (2005) Polymer therapeutics designed for a combination therapy of hormone-dependent cancer. Angew Chem Int Ed 44(26):4061–4066

Vogtt K, Jeworrek C, Garamus VM, Winter R (2010) Microdomains in lipid vesicles: structure and distribution assessed by small-angle neutron scattering. J Phys Chem B 114(16):5643–5648

Von White G, Mohammed FS, Kitchens CL (2011) Small-angle neutron scattering investigation of gold nanoparticle clustering and ligand structure under antisolvent conditions. J Phys Chem C 115 (38):18397–18405

Wang Y, Angelatos AS, Caruso F (2008) Template synthesis of nanostructured materials *via* layer-by-layer assembly. Chem Mater 20(3):848–858

Xue B, Kozlovskaya V, Liu F, Chen J, Williams JF, Campos-Gomez J, Saeed M, Kharlampieva E (2015) Intracellular degradable hydrogel cubes and spheres for anti-cancer drug delivery. ACS Appl Mater Interfaces 7(24):13633–13644

Yang B, Lowe JP, Schweins R, Edler KJ (2015a) Small angle neutron scattering studies on the internal structure of poly(lactide-co-glycolide)-block-poly(ethylene glycol) nanoparticles as drug delivery vehicles. Biomacromolecules 16(2):457–464

Yang W, Lu J, Gilbert EP, Knott R, He L, Cheng W (2015b) Probing soft corona structures of DNA-capped nanoparticles by small angle neutron scattering. J Phys Chem C 119(32):18773–18778

Yin P, Wu B, Mamontov E, Daemen LL, Cheng Y, Li T, Seifert S, Hong K, Bonnesen PV, Keum JK, Ramirez-Cuesta AJ (2016) X-ray and neutron scattering study of the formation of core-shell-type polyoxometalates. J Am Chem Soc 138:2638–2643

Yokoyama M, Okano T, Sakurai Y, Suwa S, Kataoka K (1996) Introduction of cisplatin into polymeric micelle. J Control Release 39(2):351–356

Zavgorodnya O, Kozlovskaya V, Kharlampieva E (2015) Nanostructured highly-swollen hydrogels: complexation with amino acids through copper (II) ions. Polymer 74:94–107

Zeno WF, Hilt S, Aravagiri KK, Risbud SH, Voss JC, Parikh AN, Longo ML (2014) Analysis of lipid phase behavior and protein conformational changes in nanolipoprotein particles upon entrapment in sol–gel-derived silica. Langmuir 30(32):9780–9788

Zhu Q, Harris MT, Taylor LS (2011) Time-resolved SAXS/WAXS study of the phase behavior and microstructural evolution of drug/PEG solid dispersions. Mol Pharm 8(3):932–939

# Index





B. Chaudhuri et al. (eds.), *Biological Small Angle Scattering: Techniques, Strategies and Tips*, Advances in Experimental Medicine and Biology 1009, DOI 10.1007/978-981-10-6038-0

**E**



**F**



**G**



**H**



**I**

MIX
Papier aus verantwortungsvollen Quellen
Paper from responsible sources
FSC® C105338

If you have any concerns about our products,
you can contact us on
**ProductSafety@springernature.com**

In case Publisher is established outside the EU,
the EU authorized representative is:
**Springer Nature Customer Service Center GmbH**
**Europaplatz 3, 69115 Heidelberg, Germany**

Printed by Libri Plureos GmbH
in Hamburg, Germany